AS & A2 Chemistry

Exam Board: OCR A

Complete Revision and Practice

AS-Level Contents

How Science Works

Unit 1: Module 1 — Atoms and Reactions

Unit 1: Module 2 — Electrons, Bonding and Structure

Unit 1: Module 3 — The Periodic Table

Unit 2: Module 1 — Basic Concepts and Hydrocarbons

Unit 2: Module 2 — Alcohols, Halogenoalkanes and Analysis

Unit 2: Module 3 — Energy

Unit 2: Module 4 — Resources

A2-Level Contents

Unit 4: Module 1 — Rings, Acids and Amines

Unit 4: Module 2 — Polymers and Synthesis

Unit 4: Module 3 — Analysis

Unit 5: Module 1 — Rates, Equilibrium and pH

Unit 5: Module 2 — Energy

Unit 5: Module 3 — Transition Elements

Published by CGP

Editors:
Amy Boutal, Mary Falkner, David Hickinson, Sarah Hilton, Paul Jordin, Sharon Keeley, Simon Little, Andy Park, Michael Southorn, Hayley Thompson.

Contributors:
Antonio Angelosanto, Mike Bossart, Rob Clarke, Vikki Cunningham, Ian H. Davis, John Duffy, Max Fishel, Emma Grimwood, Richard Harwood, Lucy Muncaster, Glenn Rogers, Jane Simoni, Derek Swain, Paul Warren, Chris Workman.

Proofreaders:
Barrie Crowther, Julie Wakeling.

ISBN: 978 1 84762 421 5

With thanks to Jan Greenway and Laura Jakubowski for the copyright research.

With thanks to Science Photo Library for permission to reproduce the photographs used on pages 93 and 94.

Graph to show trend in atmospheric CO_2 concentration and global temperature on page 92 based on data by EPICA Community Members 2004 and Siegenthaler et al 2005.

Groovy website: www.cgpbooks.co.uk
Jolly bits of clipart from CorelDRAW®
Printed by Elanders Ltd, Newcastle upon Tyne.

Based on the classic CGP style created by Richard Parsons.

AS-Level Chemistry

Exam Board: OCR A

The Scientific Process

'How Science Works' is all about the scientific process — how we develop and test scientific ideas. It's what scientists do all day, every day (well except at coffee time — never come between scientists and their coffee).

Scientists Come Up with **Theories** — Then **Test Them**...

Science tries to explain **how** and **why** things happen. It's all about seeking and gaining **knowledge** about the world around us. Scientists do this by **asking** questions and **suggesting** answers and then **testing** them, to see if they're correct — this is the **scientific process**.

1) **Ask** a question — make an **observation** and ask **why or how** whatever you've observed happens.
   E.g. Why does sodium chloride dissolve in water?
2) **Suggest** an answer, or part of an answer, by forming a **theory** or a **model** (a possible **explanation** of the observations or a description of what you think is happening actually happening).
   E.g. Sodium chloride is made up of charged particles which are pulled apart by the polar water molecules.
3) Make a **prediction** or **hypothesis** — a **specific testable statement**, based on the theory, about what will happen in a test situation.
   E.g. A solution of sodium chloride will conduct electricity much better than water does.
4) Carry out **tests** — to provide **evidence** that will support the prediction or refute it.
   E.g. Measure the conductivity of water and of sodium chloride solution.

The evidence supported Quentin's Theory of Flammable Burps.

A theory is only scientific if it can be tested.

...Then They **Tell** Everyone About Their **Results**...

The results are **published** — scientists need to let others know about their work. Scientists publish their results in **scientific journals**. These are just like normal magazines, only they contain **scientific reports** (called papers) instead of the latest celebrity gossip.

1) Scientific reports are similar to the **lab write-ups** you do in school. And just as a lab write-up is **reviewed** (marked) by your teacher, reports in scientific journals undergo **peer review** before they're published.
   Scientists use standard terminology when writing their reports. This way they know that other scientists will understand them. For instance, there are internationally agreed rules for naming organic compounds, so that scientists across the world will know exactly what substance is being referred to. See page 49.
2) The report is sent out to **peers** — other scientists who are experts in the **same area**. They go through it bit by bit, examining the methods and data, and checking it's all clear and logical. When the report is approved, it's **published**. This makes sure that work published in scientific journals is of a **good standard**.
3) But peer review **can't guarantee** the science is **correct** — other scientists still need to **reproduce** it.
4) Sometimes **mistakes** are made and bad work is published. Peer review **isn't perfect** but it's probably the best way for scientists to self-regulate their work and to publish **quality reports**.

...Then **Other Scientists** Will **Test** the Theory Too

1) Other scientists read the published theories and results, and try to **test the theory** themselves. This involves:
   - Repeating the **exact same experiments**.
   - Using the theory to make **new predictions** and then testing them with **new experiments**.
2) If all the experiments in the world provide evidence to back it up, the theory is thought of as **scientific 'fact'** (for now).
3) If **new evidence** comes to light that **conflicts** with the current evidence the theory is questioned all over again. More rounds of **testing** will be carried out to try to find out where the theory **falls down**.

This is how the **scientific process works** — **evidence** supports a theory, loads of other scientists read it and test it for themselves, eventually all the scientists in the world **agree** with it and then bingo, you get to **learn** it.

This is exactly how scientists arrived at the structure of the atom (see pages 6-7) — and how they came to the conclusion that electrons are arranged in shells and orbitals (see page 22). It took years and years for these models to be developed and accepted — this is often the case with the scientific process.

The Scientific Process

*If the **Evidence** Supports a Theory, It's **Accepted — for Now***

Our currently accepted theories have survived this '**trial by evidence**'. They've been tested **over and over again** and each time the results have backed them up. **BUT**, and this is a big but (teehee), they never become totally indisputable fact. Scientific **breakthroughs or advances** could provide new ways to question and test the theory, which could lead to **changes and challenges** to it. Then the testing starts all over again...

And this, my friend, is the **tentative nature of scientific knowledge** — it's always **changing** and **evolving**.

When CFCs were first used in fridges in the 1930s, scientists thought they were problem-free — well, why not? There was no evidence to say otherwise. It was decades before anyone found out that CFCs were actually making a whopping great hole in the ozone layer. See page 94.

Evidence** Comes From **Lab Experiments...

1) Results from **controlled experiments** in **laboratories** are **great**.
2) A lab is the easiest place to **control variables** so that they're all **kept constant** (except for the one you're investigating).
3) This means you can draw meaningful **conclusions**.

For example, if you're investigating how temperature affects the rate of a reaction you need to keep everything but the temperature constant, e.g. the pH of the solution, the concentration of the solution, etc.

*...But You **Can't** Always do a Lab Experiment*

There are things you **can't** study in a lab. And outside the lab controlling the variables is tricky, if not impossible.

- *Are increasing CO_2 emissions causing climate change?*
  There are other variables which may have an effect, such as changes in solar activity. You can't easily rule out every possibility. Also, climate change is a very **gradual process**. Scientists won't be able to tell if their predictions are correct for donkey's years.
- *Does drinking chlorinated tap water increase the risk of developing certain cancers?*
  There are always differences between groups of people. The best you can do is to have a **well-designed study** using **matched groups** — **choose two groups** of people (those who drink tap water and those who don't) which are **as similar as possible** (same mix of ages, same mix of diets etc). But you still can't rule out every possibility. Taking new-born identical twins and treating them identically, except for making one drink gallons of tap water and the other only pure water, might be a fairer test, but it would present huge **ethical problems**.

See pages 92-93 for more on climate change.

Samantha thought her study was very well designed — especially the fitted bookshelf.

*Science Helps to Inform **Decision-Making***

Lots of scientific work eventually leads to **important discoveries** that **could** benefit humankind — but there are often **risks** attached (and almost always **financial costs**).

Society (that's you, me and everyone else) must weigh up the information in order to **make decisions** — about the way we live, what we eat, what we drive, and so on. Information is also be used by **politicians** to devise policies and laws.

- **Chlorine** is added to water in **small quantities** to disinfect it. Some studies link drinking chlorinated water with certain types of cancer (see page 47). But the risks from drinking water contaminated by nasty bacteria are far, far greater. There are other ways to get rid of bacteria in water, but they're heaps **more expensive**.
- Scientific advances mean that **non-polluting hydrogen-fuelled cars** can be made. They're better for the environment, but are really expensive. Also, it'd cost a fortune to adapt the existing filling stations to store hydrogen.
- Pharmaceutical drugs are really expensive to develop, and drug companies want to make money. So they put most of their efforts into developing drugs that they can sell for a good price. Society has to consider the **cost** of buying new drugs — the **NHS** can't afford the most expensive drugs without **sacrificing** something else.

So there you have it — how science works...

Hopefully these pages have given you a nice intro to how science works, e.g. what scientists do to provide you with 'facts'. You need to understand this, as you're expected to know how science works yourselves — for the exam and for life.

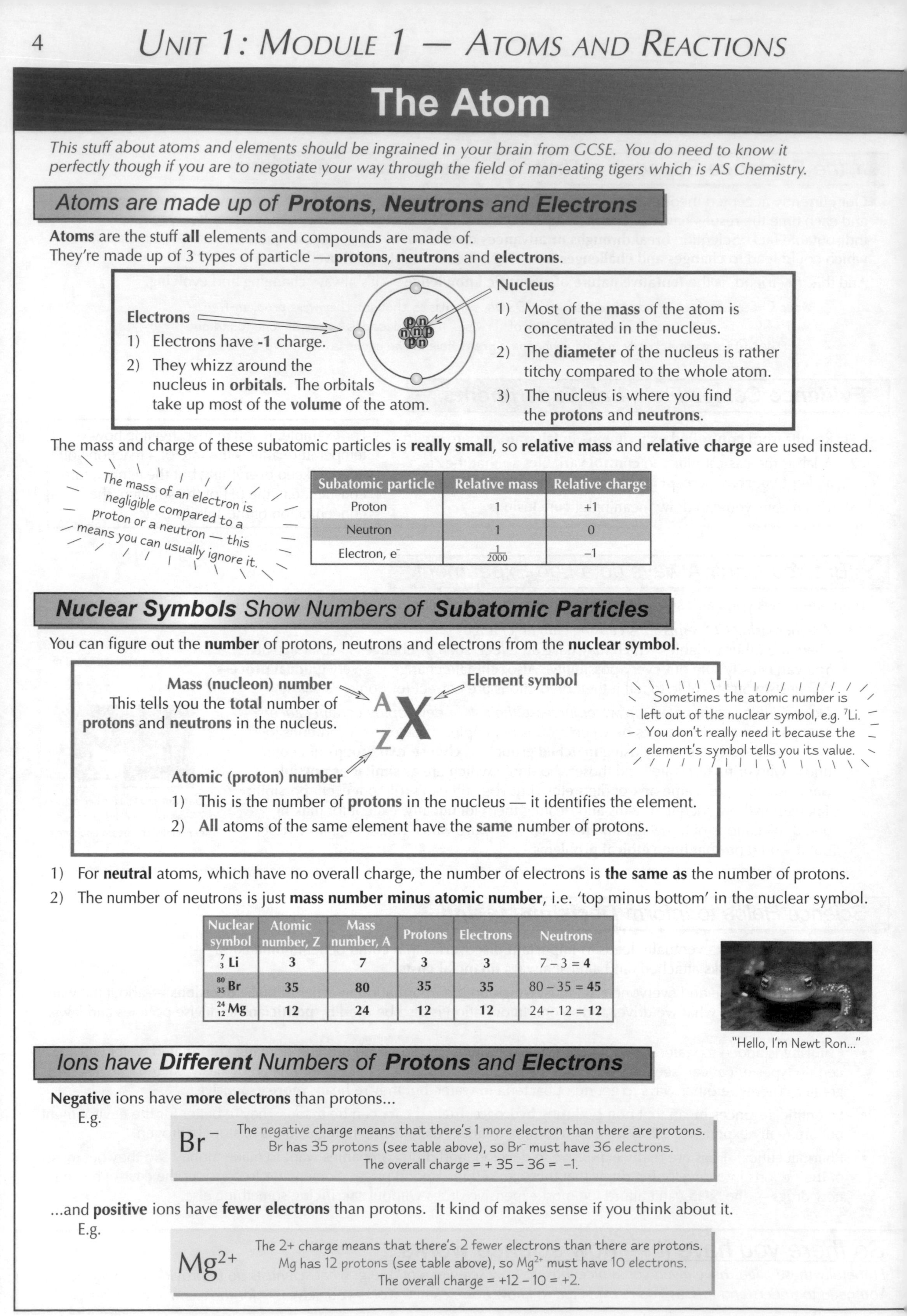

The Atom

This stuff about atoms and elements should be ingrained in your brain from GCSE. You do need to know it perfectly though if you are to negotiate your way through the field of man-eating tigers which is AS Chemistry.

Atoms are made up of Protons, Neutrons and Electrons

Atoms are the stuff **all** elements and compounds are made of.
They're made up of 3 types of particle — **protons**, **neutrons** and **electrons**.

Electrons
1) Electrons have **-1** charge.
2) They whizz around the nucleus in **orbitals**. The orbitals take up most of the **volume** of the atom.

Nucleus
1) Most of the **mass** of the atom is concentrated in the nucleus.
2) The **diameter** of the nucleus is rather titchy compared to the whole atom.
3) The nucleus is where you find the **protons** and **neutrons**.

The mass and charge of these subatomic particles is **really small**, so **relative mass** and **relative charge** are used instead.

The mass of an electron is negligible compared to a proton or a neutron — this means you can usually ignore it.

Subatomic particle	Relative mass	Relative charge
Proton	1	+1
Neutron	1	0
Electron, e^-	$\frac{1}{2000}$	−1

Nuclear Symbols Show Numbers of Subatomic Particles

You can figure out the **number** of protons, neutrons and electrons from the **nuclear symbol**.

$$^{A}_{Z}\mathrm{X}$$

Mass (nucleon) number
This tells you the **total** number of **protons** and **neutrons** in the nucleus.

Element symbol

Atomic (proton) number
1) This is the number of **protons** in the nucleus — it identifies the element.
2) **All** atoms of the same element have the **same** number of protons.

Sometimes the atomic number is left out of the nuclear symbol, e.g. $^{7}\mathrm{Li}$. You don't really need it because the element's symbol tells you its value.

1) For **neutral** atoms, which have no overall charge, the number of electrons is **the same as** the number of protons.
2) The number of neutrons is just **mass number minus atomic number**, i.e. 'top minus bottom' in the nuclear symbol.

Nuclear symbol	Atomic number, Z	Mass number, A	Protons	Electrons	Neutrons
$^{7}_{3}\mathrm{Li}$	3	7	3	3	7 − 3 = **4**
$^{80}_{35}\mathrm{Br}$	**35**	**80**	**35**	**35**	80 − 35 = **45**
$^{24}_{12}\mathrm{Mg}$	**12**	**24**	**12**	**12**	24 − 12 = **12**

"Hello, I'm Newt Ron..."

Ions have Different Numbers of Protons and Electrons

Negative ions have **more electrons** than protons...
E.g.

Br^- The negative charge means that there's 1 more electron than there are protons. Br has 35 protons (see table above), so Br^- must have 36 electrons. The overall charge = + 35 − 36 = −1.

...and **positive** ions have **fewer electrons** than protons. It kind of makes sense if you think about it.
E.g.

Mg^{2+} The 2+ charge means that there's 2 fewer electrons than there are protons. Mg has 12 protons (see table above), so Mg^{2+} must have 10 electrons. The overall charge = +12 − 10 = +2.

The Atom

Isotopes are Atoms of the Same Element with Different Numbers of Neutrons

Make sure you **learn** this definition and totally **understand** what it means —

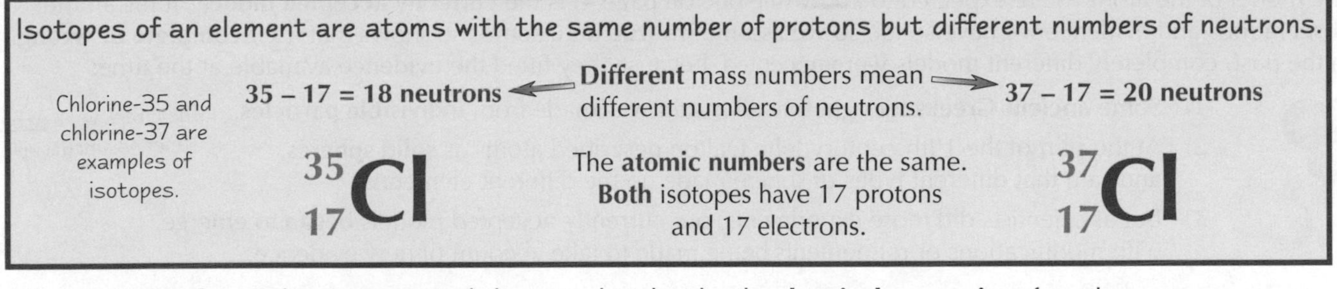

1) It's the **number** and **arrangement** of electrons that decides the **chemical properties** of an element. Isotopes have the **same configuration of electrons**, so they've got the **same** chemical properties.
2) Isotopes of an element do have slightly different **physical properties** though, such as different densities, rates of diffusion, etc. This is because **physical properties** tend to depend more on the **mass** of the atom.

Here's another example — naturally occurring **magnesium** consists of 3 isotopes.

^{24}Mg (79%)	^{25}Mg (10%)	^{26}Mg (11%)
12 protons	12 protons	12 protons
12 neutrons	**13** neutrons	**14** neutrons
12 electrons	12 electrons	12 electrons

The periodic table gives the atomic number for each element. The other number isn't the mass number — it's the relative atomic mass (see page 8). They're a bit different, but you can often assume they're equal — it doesn't matter unless you're doing really accurate work.

Practice Questions

Q1 Draw a diagram showing the structure of an atom, labelling each part.

Q2 Define the term 'isotope' and give an example.

Q3 Draw a table showing the relative charge and relative mass of the three subatomic particles found in atoms.

Q4 Using an example, explain the terms 'atomic number' and 'mass number'.

Q5 Where is the mass concentrated in an atom, and what makes up most of the volume of an atom?

Exam Questions

Q1 Hydrogen, deuterium and tritium are all isotopes of each other.

a) Identify one similarity and one difference between these isotopes. [2 marks]

b) Deuterium can be written as ^{2}H. Determine the number of protons, neutrons and electrons in a neutral deuterium atom. [3 marks]

c) Write a nuclear symbol for tritium, given that it has 2 neutrons. [1 mark]

Q2 This question relates to the atoms or ions A to D: A. $^{32}_{16}S^{2-}$, B. $^{40}_{18}Ar$, C. $^{30}_{16}S$, D. $^{42}_{20}Ca$

a) Identify the similarity for each of the following pairs, justifying your answer in each case.

(i) A and B. [2 marks]

(ii) A and C. [2 marks]

(iii) B and D. [2 marks]

b) Which two of the atoms or ions are isotopes of each other? Explain your reasoning. [2 marks]

Got it learned yet? — Isotope so...

This is a nice straightforward page just to ease you in to things. Remember that positive ions have fewer electrons than protons, and negative ions have more electrons than protons. Get that straight in your mind or you'll end up in a right mess. There's nowt too hard about isotopes neither. They're just the same element with different numbers of neutrons.

Atomic Models

Things ain't how they used to be, you know. Take atomic structure, for starters.

*The **Accepted Model** of the **Atom** Has **Changed** Throughout History*

The model of the atom you're expected to know (the one on page 4) is the currently **accepted model**. It fits all the observations and evidence we have so far, so we **assume it's true** until someone shows that it's **incomplete or wrong**. In the past, completely different models were accepted, because they fitted the evidence available at the time:

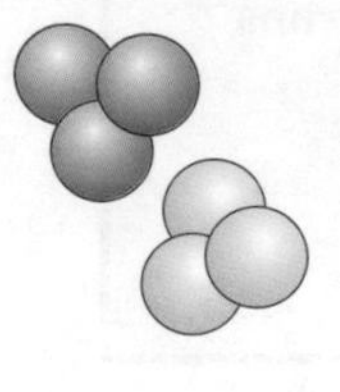

1) Some **ancient Greeks** thought that all matter was made from **indivisible particles**.
2) At the start of the 19th century John Dalton described atoms as **solid spheres**, and said that different types of sphere made up the different elements.
3) But as scientists did more experiments, our currently accepted models began to emerge, with modifications or refinements being made to take account of new evidence.

The Greek word atomos means 'uncuttable'.

Experimental Evidence** Showed that Atoms **Weren't Solid Spheres

In 1897 J J Thompson did a whole series of experiments and concluded that atoms **weren't** solid and indivisible.

1) His measurements of **charge** and **mass** showed that an atom must contain even smaller, negatively charged particles. He called these particles 'corpuscles' — we call them **electrons**.
2) The 'solid sphere' idea of atomic structure had to be changed. The new model was known as the **'plum pudding model'** — a positively charged sphere with negative electrons embedded in it.

Rutherford** Showed that the **Plum Pudding** Model Was **Wrong

1) In 1909 Ernest Rutherford and his students Hans Geiger and Ernest Marsden conducted the famous **gold foil experiment**. They fired **alpha particles** (which are positively charged) at an extremely thin sheet of gold.
2) From the plum pudding model, they were expecting **most** of the alpha particles to be deflected **very slightly** by the positive 'pudding' that made up most of an atom.
3) In fact, most of the alpha particles passed **straight through** the gold atoms, and a very small number were deflected **backwards** (through more than 90°). This showed that the plum pudding model **couldn't be right**.
4) So Rutherford came up with a model that **could** explain this new evidence — the **nuclear model** of the atom:

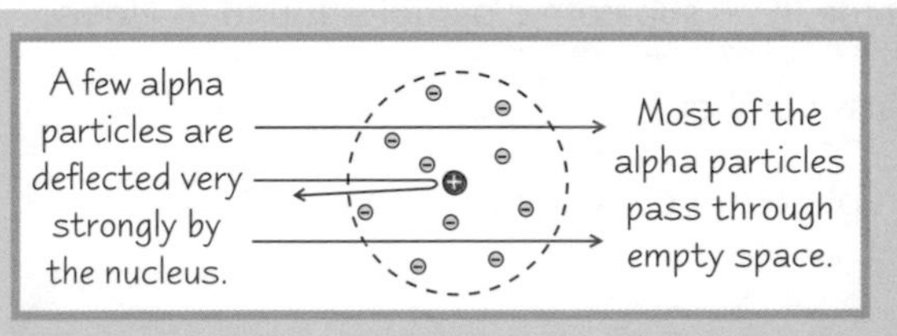

1) There is a **tiny, positively charged nucleus** at the centre of the atom, where most of the atom's mass is concentrated.
2) The nucleus is surrounded by a **'cloud'** of **negative electrons**.
3) Most of the atom is **empty space**.

*Rutherford's **Nuclear Model** Was **Modified** Several Times*

Rutherford's model seemed pretty convincing, but (there's always a but)... the scientists of the day didn't just say, "Well done Ernest old chap, you've got it", then all move to Patagonia to farm goats. No, they stuck at their experiments, wanting to be sure of the truth. (And it's just conceivable they wanted some fame and fortune too.)

1) Henry Moseley discovered that the charge of the nucleus **increased** from one element to another in units of one.
2) This led Rutherford to investigate the nucleus further. He finally discovered that it contained **positively charged** particles that he called **protons**. The charges of the nuclei of different atoms could then be explained — the atoms of **different elements** have a **different number of protons** in their nucleus.
3) There was still one problem with the model — the nuclei of atoms were **heavier** than they would be if they just contained protons. Rutherford predicted that there were other particles in the nucleus, that had **mass but no charge** — and the **neutron** was eventually discovered by James Chadwick.

This is nearly always the way scientific knowledge develops — **new evidence** prompts people to come up with **new, improved ideas**. Then other people go through each new, improved idea with a fine-tooth comb as well — modern '**peer review**' (see p2) is part of this process.

Atomic Models

The **Bohr Model** Was a Further Improvement

1) Scientists realised that electrons in a '**cloud**' around the nucleus of an atom would **spiral down** into the nucleus, causing the atom to **collapse**. Niels Bohr proposed a new model of the atom with four basic principles:

   1) Electrons can only exist in **fixed orbits**, or **shells**, and not anywhere in between.
   2) Each shell has a **fixed energy**.
   3) When an electron moves between shells **electromagnetic radiation** is **emitted** or **absorbed**.
   4) Because the energy of shells is fixed, the radiation will have a **fixed frequency**.

2) The frequencies of radiation emitted and absorbed by atoms were already known from experiments. The Bohr model fitted these observations — it looked good.

3) The Bohr model also explained why some elements (the noble gases) are **inert**. He said that the shells of an atom can only hold **fixed numbers of electrons**, and that an element's reactivity is due to its electrons. When an atom has **full shells** of electrons it is **stable** and does not react.

There's **More Than One** Model of Atomic Structure in Use Today

1) We now know that the Bohr model is **not perfect** — but it's still widely used to describe atoms because it's simple and explains many **observations** from experiments, like bonding and ionisation energy trends.
2) The most accurate model we have today involves complicated quantum mechanics. Basically, you can never **know** where an electron is or which direction it's going in at any moment, but you can say **how likely** it is to be at any particular point in the atom. Oh, and electrons can act as **waves** as well as particles (but you don't need to worry about the details).
3) This model might be **more accurate**, but it's a lot harder to get your head round and visualise. It **does** explain some observations that can't be accounted for by the Bohr model though. So scientists use whichever model is most relevant to whatever they're investigating.

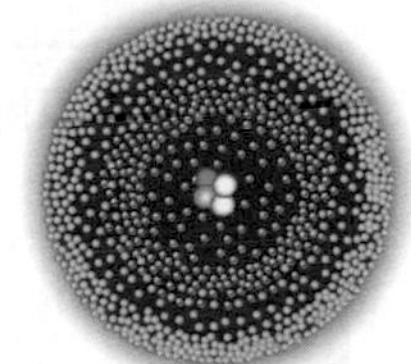
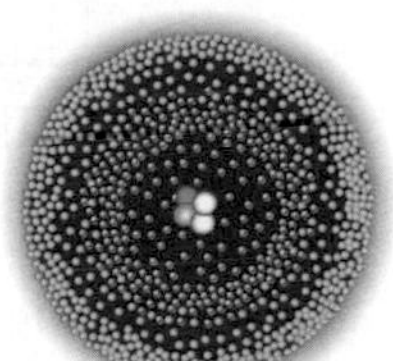

The quantum model of an atom with two shells of electrons. The denser the dots, the more likely an electron is to be there.

Practice Questions

Q1 What particle did J J Thompson discover?

Q2 Describe the model of the atom that was adopted because of Thompson's work.

Q3 Who developed the 'nuclear' model of the atom? What evidence did they have for it?

Q4 What are the names of the two particles in the nucleus of an atom?

Exam Question

Q1 Scientific theories are constantly being revised in the light of new evidence. New theories are accepted if they have been successfully tested by experiments or because they help to explain certain observations.

a) Niels Bohr thought that the model of the atom proposed by Ernest Rutherford did not describe the electrons in an atom correctly. Why did he think this and how was his model of the atom different from Rutherford's? [2 marks]

b) What happens when electrons in an atom move from one shell to another? [1 mark]

c) How did Bohr explain the lack of reactivity of the noble gases? [2 marks]

These models are tiny — even smaller than size zero, I reckon...

The process of developing a model to fit the evidence available, looking for more evidence to show if it's correct or not, then revising the model if necessary is really important. It happens with all new scientific ideas. Remember, scientific 'facts' are only accepted as true because no one's proved yet that they aren't. It might all be bunkum.

Relative Mass

Relative mass... What? Eh?...Read on...

Relative Masses** are Masses of Atoms Compared to **Carbon-12

The actual mass of an atom is **very, very tiny**. Don't worry about exactly how tiny for now, but it's far **too small** to weigh. So, the mass of one atom is compared to the mass of a different atom. This is its **relative mass**.
Here are some **definitions** for you to learn:

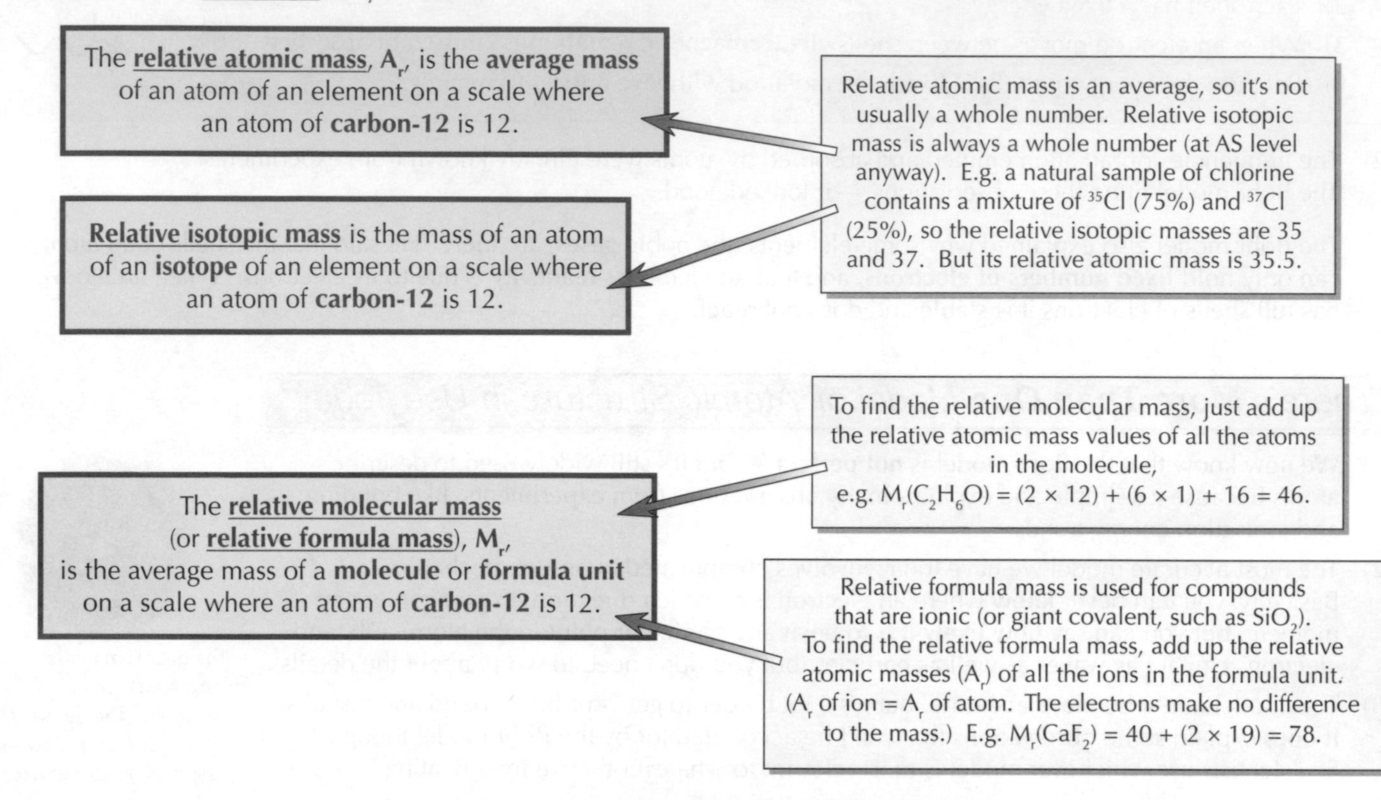

*A_r Can Be Worked Out from a **Isotopic Abundances***

You need to know how to calculate the **relative atomic mass** (A_r) of an element from its **isotopic abundances**.

1) Different isotopes of an element occur in different quantities, or **isotopic abundances**. For example, **76%** of the chlorine atoms found on Earth have a relative isotopic mass of 35, while **24%** have a relative isotopic mass of 37.
2) The relative atomic mass of chlorine is the **average** mass of all chlorine atoms.
   If you've got the isotopic abundances as **percentages**, the easiest way to do this is to imagine you have 100 chlorine atoms, and then find the average mass. Here's the method...

   Step 1: **Multiply** each **relative isotopic mass** by its **% relative isotopic abundance**, and **add up** the results: $(76 \times 35) + (24 \times 37) = 2660 + 888 = \mathbf{3548}$

   Step 2: **Divide** by **100**: $3548 \div 100 = \mathbf{35.5}$ (to 1 decimal place)

   So the relative atomic mass of chlorine is **35.5** (just like your textbook says).
3) You might be given your isotopic abundances in the form of a **graph**, such as a **mass spectrum**. (Mass spectra are produced by mass spectrometers — devices which are used to find out what samples are made up of by measuring the masses of their components, see page 76.)
   For example, using the data from this mass spectrum for neon...

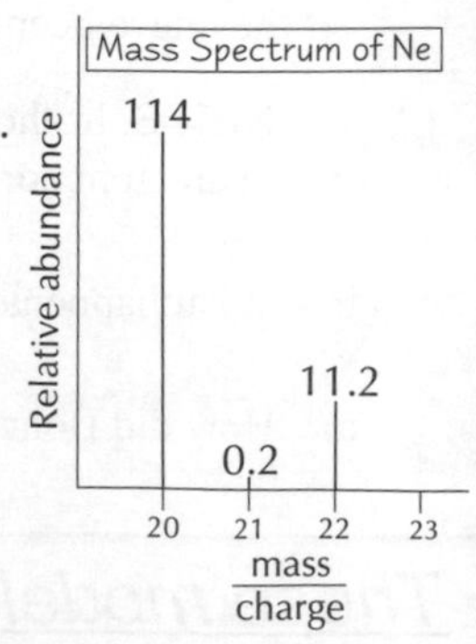

   This time the isotopic abundances aren't given as percentages. But just like before...

   Step 1: **Multiply** each **relative isotopic mass** by its **relative isotopic abundance**, and **add up** the results: $(20 \times 114) + (21 \times 0.2) + (22 \times 11.2) = \mathbf{2530.6}$

   Step 2: **Divide** by the **sum** of the isotopic abundances $(114 + 0.2 + 11.2 = 125.4)$.
   $2530.6 \div 125.4 = \mathbf{20.2}$ (to 1 decimal place)
4) This process of finding an **average** explains why the relative atomic mass is not usually a **whole number**.

Relative Mass

Spreadsheets Can Help You Find Relative Atomic Mass

This method is particularly useful when a large amount of data is being used.
You can put the **relative isotopic abundance** data into a spreadsheet, then get the spreadsheet to work out the **A_r** for you.
This example uses data from the mass spectrum of zirconium.

	A	B	C
1	Relative Isotopic Mass	Relative Abundance	Relative Mass × Abundance
2	90	100.00	9000.00
3	91	21.81	1984.71
4	92	33.33	3066.36
5	94	33.78	3175.32
6	96	5.44	522.24
7		**194.36**	**17748.63**
8			
9	Relative Atomic Mass =		91.31833
10			

In this column, the spreadsheet has multiplied the numbers in column A by the numbers in column B to give the **product** of each **relative isotopic mass** and its **relative abundance**.

In these cells, the spreadsheet has **added** up the numbers in the cells above to give the two totals you need to calculate the **A_r**.

To work out the **relative atomic mass**, the sum in cell C7 is **divided** by the sum in B7. So the relative atomic mass of zirconium is 91.3, to 1 decimal place.

Practice Questions

Q1 Explain what relative atomic mass (A_r) and relative isotopic mass mean.

Q2 Explain the difference between relative molecular mass and relative formula mass.

Q3 Explain what relative isotopic abundance means.

Q4 Explain why the relative atomic mass is rarely a whole number.

Exam Questions

Q1 Copper exists in two main isotopic forms, ^{63}Cu and ^{65}Cu.
a) Calculate the relative atomic mass of copper using the information from the mass spectrum. [2 marks]
b) Explain why the relative atomic mass of copper is not a whole number. [2 marks]

Mass Spectrum of Cu
Relative abundance
120.8
54.0
61
63
65
67
mass/charge

Q2 The percentage make-up of naturally occurring potassium is 93.11 % ^{39}K, 0.12 % ^{40}K and 6.77 % ^{41}K.
Use the information to determine the relative atomic mass of potassium. [2 marks]

You can't pick your relatives, you just have to learn them...

Working out A_r is dead easy — and using a spreadsheet makes it even easier. It'll really help if you know the mass numbers for the first 20 elements or so, or you'll spend half your time looking back at the periodic table. I hope you've done the Practice Questions, cos they pretty much cover the rest of the stuff, and if you can get them right, you've nailed it.

The Mole

It'd be handy to be able to count out atoms — but they're way too tiny. You can't even see them, never mind get hold of them with tweezers. But not to worry — using the idea of relative mass, you can figure out how many atoms you've got.

A **Mole** is Just a (Very Large) **Number of Particles**

Chemists often talk about 'amount of substance'. Basically, all they mean is 'number of particles'.

1) Amount of substance is measured using a unit called the **mole** (**mol** for short) and given the symbol ***n***.
2) One mole is roughly **6.02×10^{23} particles** (**the Avogadro constant, N_A**).
3) It **doesn't matter** what the particles are.
   They can be atoms, molecules, penguins — **anything**.
4) Here's a nice simple formula for finding the number of moles from the number of atoms or molecules:

$$\text{Number of moles} = \frac{\text{Number of particles you have}}{\text{Number of particles in a mole}}$$

Example: I have 1.5×10^{24} carbon atoms. How many moles of carbon is this?

$$\text{Number of moles} = \frac{1.5 \times 10^{24}}{6.02 \times 10^{23}} \approx 2.49 \text{ moles}$$

Molar Mass is the Mass of **One Mole**

Molar mass, **M**, is the mass of **one mole** of something.

But the main thing to remember is:

Molar mass is just the same as the relative molecular mass, M_r

That's why the mole is such a ridiculous number of particles (6.02×10^{23}) — it's the number of particles for which the weight in g is the same as the relative molecular mass.

The only difference is you stick a 'g mol^{-1}' on the end...

Example: Find the molar mass of $CaCO_3$.

Relative formula mass, M_r, of $CaCO_3$ = 40 + 12 + (3 × 16) = 100
So the molar mass, M, is **100 g mol^{-1}**. — i.e. 1 mole of $CaCO_3$ weighs 100 g.

Here's another formula. This one's really important — you need it **all the time**:

$$\text{Number of moles} = \frac{\text{mass of substance}}{\text{molar mass}}$$

Example: How many moles of aluminium oxide are present in 5.1 g of Al_2O_3?

Molar mass of Al_2O_3 = (2 × 27) + (3 × 16)
= 102 g mol^{-1}

Number of moles of Al_2O_3 = $\frac{5.1}{102}$ = **0.05 moles**

Example: How many moles of chlorine molecules are present in 71 g of chlorine gas?

We're talking chlorine **molecules** (not chlorine atoms), so it's Cl_2 we're interested in.
Molar mass of Cl_2 = (2 × 35.5) = 71 g mol^{-1}

Number of moles of Cl_2 = $\frac{71}{71}$ = **1 mole**

But note that it would be **2 moles** of chlorine **atoms**, since chlorine **atoms** have a molar mass of 35.5 g mol^{-1}.

The Mole

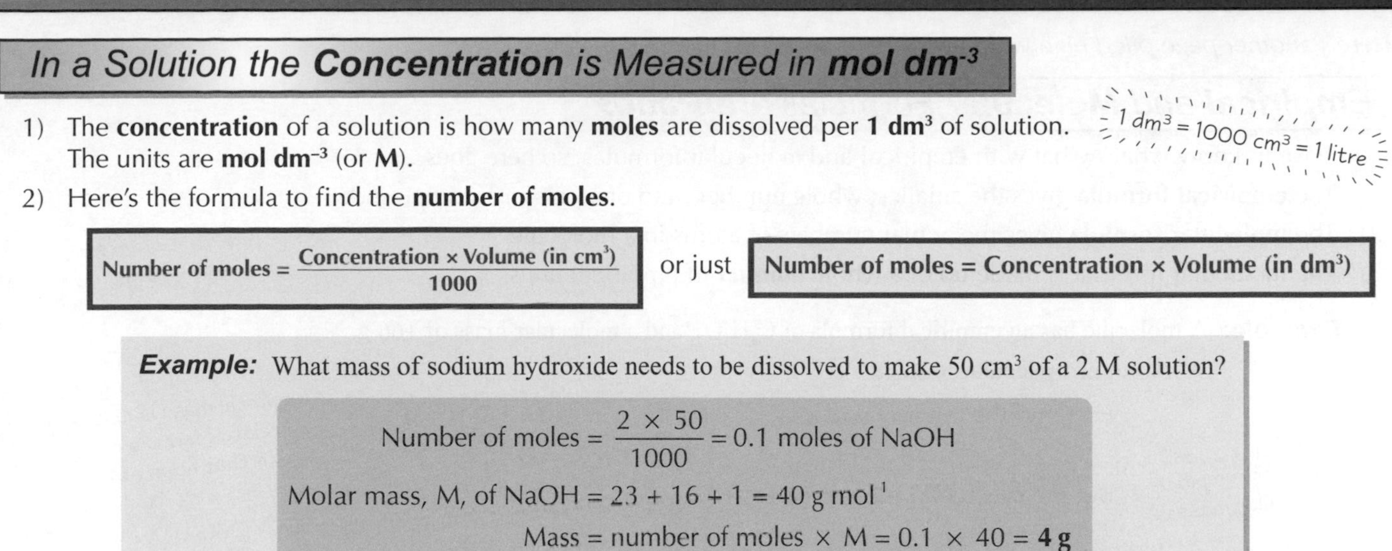

In a Solution the **Concentration** is Measured in **mol dm^{-3}**

1) The **concentration** of a solution is how many **moles** are dissolved per **1 dm³** of solution. The units are **mol dm^{-3}** (or **M**).

1 dm³ = 1000 cm³ = 1 litre

2) Here's the formula to find the **number of moles**.

$$\text{Number of moles} = \frac{\text{Concentration} \times \text{Volume (in cm}^3\text{)}}{1000}$$

or just

$$\text{Number of moles} = \text{Concentration} \times \text{Volume (in dm}^3\text{)}$$

Example: What mass of sodium hydroxide needs to be dissolved to make 50 cm³ of a 2 M solution?

$$\text{Number of moles} = \frac{2 \times 50}{1000} = 0.1 \text{ moles of NaOH}$$

Molar mass, M, of NaOH = 23 + 16 + 1 = 40 g mol^{-1}

Mass = number of moles × M = 0.1 × 40 = **4 g**

3) A solution that has **more moles per dm³** than another is **more concentrated**. A solution that has **fewer moles per dm³** than another is **more dilute**.

All Gases Take Up the **Same Volume** under the Same Conditions

If temperature and pressure stay the same, **one mole** of **any** gas always has the **same volume**.
At **room temperature and pressure** (r.t.p.), this happens to be **24 dm³**, (r.t.p is 298 K (25 °C) and 101.3 kPa).
Here are 2 formulas for working out the number of moles in a volume of gas. Don't forget — **ONLY** use them for r.t.p.

$$\text{Number of moles} = \frac{\text{Volume in dm}^3}{24} \quad \text{OR} \quad \text{Number of moles} = \frac{\text{Volume in cm}^3}{24\,000}$$

Example: How many moles are there in 6 dm³ of oxygen gas at r.t.p.?

$$\text{Number of moles} = \frac{6}{24} = \mathbf{0.25 \text{ moles of oxygen molecules}}$$

Practice Questions

Q1 How many molecules are there in one mole of ethane molecules?

Q2 Which has the most particles, a solution of concentration 0.1 mol dm^{-3} or an equal volume of one that is 0.1 M?

Q3 What volume does 1 mole of gas occupy at r.t.p.?

Exam Questions

Q1 Calculate the mass of 0.36 moles of ethanoic acid, CH_3COOH. [2 marks]

Q2 What mass of H_2SO_4 is needed to produce 60 cm³ of 0.25 M solution? [2 marks]

Q3 What volume will be occupied by 88 g of propane gas (C_3H_8) at r.t.p.? [2 marks]

Put your back teeth on the scale and find out your molar mass...

You need this stuff for loads of the calculation questions you might get, so make sure you know it inside out. Before you start plugging numbers into formulas, make sure they're in the right units. If they're not, you need to know how to convert them or you'll be tossing marks out the window. Learn all the definitions and formulas, then have a bash at the questions.

Empirical and Molecular Formulas

Here's another page piled high with numbers — it's all just glorified maths really.

Empirical and **Molecular** Formulas are **Ratios**

You have to know what's what with empirical and molecular formulas, so here goes...

1) The **empirical formula** gives the smallest whole number ratio of atoms in a compound.
2) The **molecular formula** gives the **actual** numbers of atoms in a molecule.
3) The molecular formula is made up of a whole **number** of empirical units.

Example: A molecule has an empirical formula of $C_4H_3O_2$, and a molecular mass of 166 g. Work out its molecular formula.

Compare the empirical and molecular masses.

First find the **empirical mass** — $(4 \times 12) + (3 \times 1) + (2 \times 16)$
$= 48 + 3 + 32 = 83$ g

Empirical mass is just like the relative formula mass... (if that helps at all...).

But the **molecular mass** is 166 g,

so there are $\frac{166}{83} = 2$ empirical units in the molecule.

The molecular formula must be the **empirical formula × 2**,
so the molecular formula = $\mathbf{C_8H_6O_4}$. So there you go.

Empirical Formulas are Calculated from **Experiments**

You need to be able to work out empirical formulas from **experimental results** too.

Example: When a hydrocarbon is burnt in excess oxygen, 4.4 g of carbon dioxide and 1.8 g of water are made. What is the empirical formula of the hydrocarbon?

First work out how many moles of the products you have.

No. of moles of $CO_2 = \frac{\text{mass}}{M} = \frac{4.4}{12+(16\times2)} = \frac{4.4}{44} = 0.1$ moles

1 mole of CO_2 contains 1 mole of carbon atoms, so you must have started with **0.1 moles of carbon atoms.**

No. of moles of $H_2O = \frac{1.8}{(2\times1)+16} = \frac{1.8}{18} = 0.1$ moles

1 mole of H_2O contains 2 moles of hydrogen atoms (H), so you must have started with **0.2 moles of hydrogen atoms.**

Ratio C : H = 0.1 : 0.2 . Now you divide both numbers by the **smallest** — here it's 0.1.
So, the ratio C : H = 1 : 2. So the empirical formula must be $\mathbf{CH_2}$.

This works because the only place the carbon in the carbon dioxide and the hydrogen in the water could have come from is the hydrocarbon.

As if that's not enough, you also need to know how to work out empirical formulas from the **percentages** of the different elements.

Example: A compound is found to have percentage composition 56.5% potassium, 8.7% carbon and 34.8% oxygen by mass. Calculate its empirical formula.

If you assume you've got 100 g of the compound, you can turn the % straight into mass, and then work out the number of moles as normal.

In **100 g** of compound there are:

Use $n = \frac{\text{mass}}{M}$

$\frac{56.5}{39} = 1.449$ moles of K $\frac{8.7}{12} = 0.725$ moles of C $\frac{34.8}{16} = 2.175$ moles of O

Divide each number of moles by the **smallest number** — in this case it's 0.725.

K: $\frac{1.449}{0.725} = 2.0$ C: $\frac{0.725}{0.725} = 1.0$ O: $\frac{2.175}{0.725} = 3.0$

The ratio of K : C : O = 2 : 1 : 3. So you know the empirical formula's got to be $\mathbf{K_2CO_3}$.

Empirical and Molecular Formulas

Molecular Formulas are Calculated from Experimental Data Too

Once you know the empirical formula, you just need a bit more info and you can work out the **molecular formula** too.

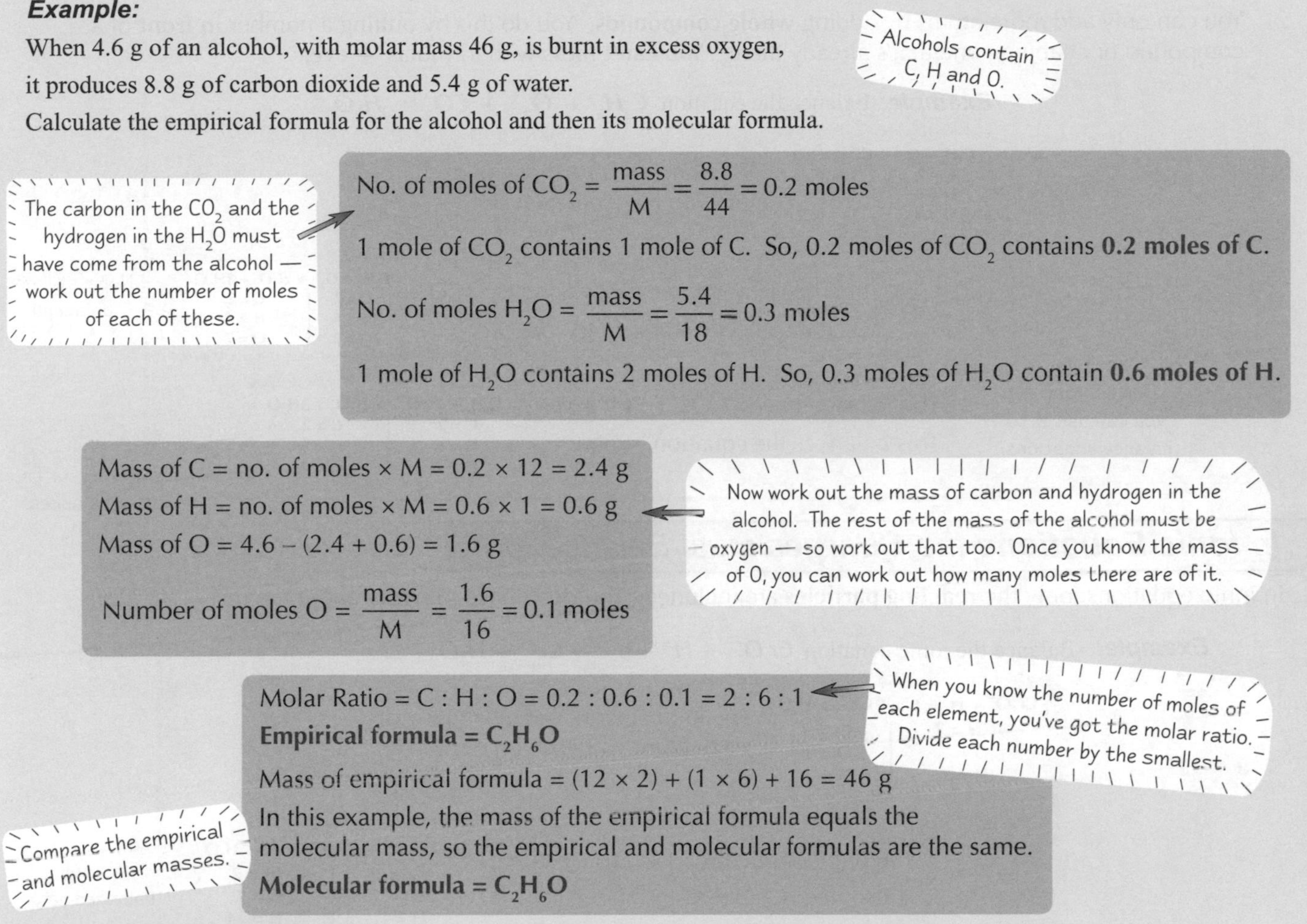

Practice Questions

Q1 Define 'empirical formula'.

Q2 What is the difference between a molecular formula and an empirical formula?

Exam Questions

Q1 Hydrocarbon X has a molecular mass of 78 g. It is found to have 92.3% carbon and 7.7% hydrogen by mass. Calculate the empirical and molecular formulae of X. [3 marks]

Q2 When 1.2 g of magnesium ribbon is heated in air, it burns to form a white powder, which has a mass of 2 g. What is the empirical formula of the powder? [2 marks]

Q3 When 19.8 g of an organic acid, A, is burnt in excess oxygen, 33 g of carbon dioxide and 10.8 g of water are produced.
Calculate the empirical formula for A and hence its molecular formula, if M_r(A) = 132. [4 marks]

Hint: organic acids contain C, H and O.

The Empirical Strikes Back...

With this stuff, it's not enough to learn a few facts parrot-fashion, to regurgitate in the exam — you've gotta know how to use them. The only way to do that is to practise. Go through all the examples on these two pages again, this time working the answers out for yourself. Then test yourself on the practice exam questions. It'll help you sleep at night — honest.

Equations and Calculations

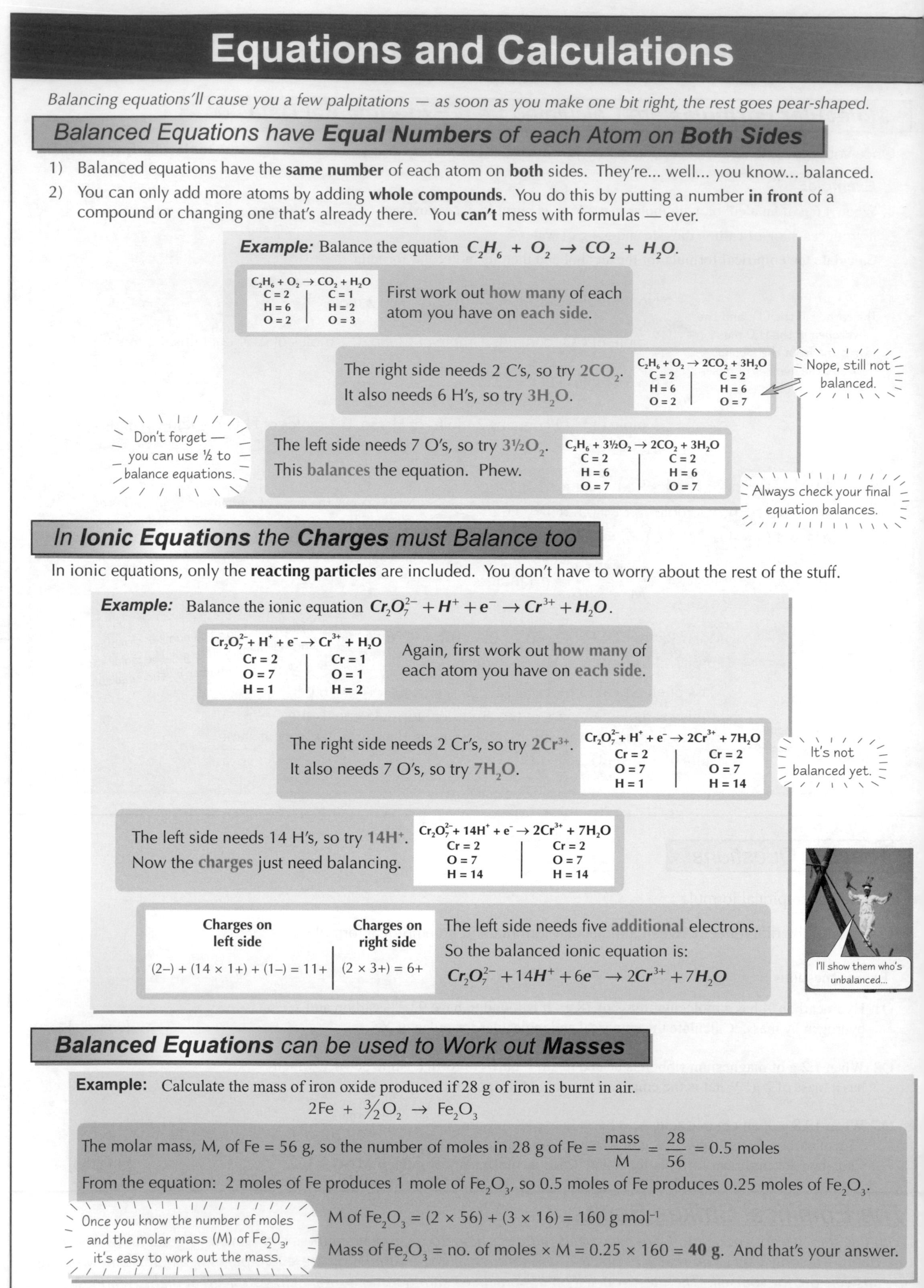

Balancing equations'll cause you a few palpitations — as soon as you make one bit right, the rest goes pear-shaped.

Balanced Equations have **Equal Numbers** of each Atom on **Both Sides**

1) Balanced equations have the **same number** of each atom on **both** sides. They're... well... you know... balanced.
2) You can only add more atoms by adding **whole compounds**. You do this by putting a number **in front** of a compound or changing one that's already there. You **can't** mess with formulas — ever.

Example: Balance the equation $C_2H_6 + O_2 \rightarrow CO_2 + H_2O$.

$C_2H_6 + O_2 \rightarrow$	$CO_2 + H_2O$
C = 2	C = 1
H = 6	H = 2
O = 2	O = 3

First work out **how many** of each atom you have on **each side**.

The right side needs 2 C's, so try $2CO_2$. It also needs 6 H's, so try $3H_2O$.

$C_2H_6 + O_2 \rightarrow$	$2CO_2 + 3H_2O$
C = 2	C = 2
H = 6	H = 6
O = 2	O = 7

Nope, still not balanced.

The left side needs 7 O's, so try $3\frac{1}{2}O_2$. This **balances** the equation. Phew.

$C_2H_6 + 3\frac{1}{2}O_2 \rightarrow$	$2CO_2 + 3H_2O$
C = 2	C = 2
H = 6	H = 6
O = 7	O = 7

Don't forget — you can use ½ to balance equations.

Always check your final equation balances.

In **Ionic Equations** the **Charges** must **Balance** too

In ionic equations, only the **reacting particles** are included. You don't have to worry about the rest of the stuff.

Example: Balance the ionic equation $Cr_2O_7^{2-} + H^+ + e^- \rightarrow Cr^{3+} + H_2O$.

$Cr_2O_7^{2-} + H^+ + e^- \rightarrow$	$Cr^{3+} + H_2O$
Cr = 2	Cr = 1
O = 7	O = 1
H = 1	H = 2

Again, first work out **how many** of each atom you have on **each side**.

The right side needs 2 Cr's, so try $2Cr^{3+}$. It also needs 7 O's, so try $7H_2O$.

$Cr_2O_7^{2-} + H^+ + e^- \rightarrow$	$2Cr^{3+} + 7H_2O$
Cr = 2	Cr = 2
O = 7	O = 7
H = 1	H = 14

It's not balanced yet.

The left side needs 14 H's, so try $14H^+$. Now the **charges** just need balancing.

$Cr_2O_7^{2-} + 14H^+ + e^- \rightarrow$	$2Cr^{3+} + 7H_2O$
Cr = 2	Cr = 2
O = 7	O = 7
H = 14	H = 14

Charges on left side	Charges on right side
(2–) + (14 × 1+) + (1–) = 11+	(2 × 3+) = 6+

The left side needs five **additional** electrons. So the balanced ionic equation is:

$$Cr_2O_7^{2-} + 14H^+ + 6e^- \rightarrow 2Cr^{3+} + 7H_2O$$

I'll show them who's unbalanced...

Balanced Equations can be used to **Work out Masses**

Example: Calculate the mass of iron oxide produced if 28 g of iron is burnt in air.

$$2Fe + \tfrac{3}{2}O_2 \rightarrow Fe_2O_3$$

The molar mass, M, of Fe = 56 g, so the number of moles in 28 g of Fe = $\frac{\text{mass}}{M} = \frac{28}{56} = 0.5$ moles

From the equation: 2 moles of Fe produces 1 mole of Fe_2O_3, so 0.5 moles of Fe produces 0.25 moles of Fe_2O_3.

M of $Fe_2O_3 = (2 \times 56) + (3 \times 16) = 160\ \text{g mol}^{-1}$

Mass of Fe_2O_3 = no. of moles × M = 0.25 × 160 = **40 g**. And that's your answer.

Once you know the number of moles and the molar mass (M) of Fe_2O_3, it's easy to work out the mass.

Equations and Calculations

That's not all... Balanced Equations can be used to Work Out Gas Volumes

It's pretty handy to be able to work out **how much gas** a reaction will produce, so that you can use **large enough apparatus**. Or else there might be a rather large bang.

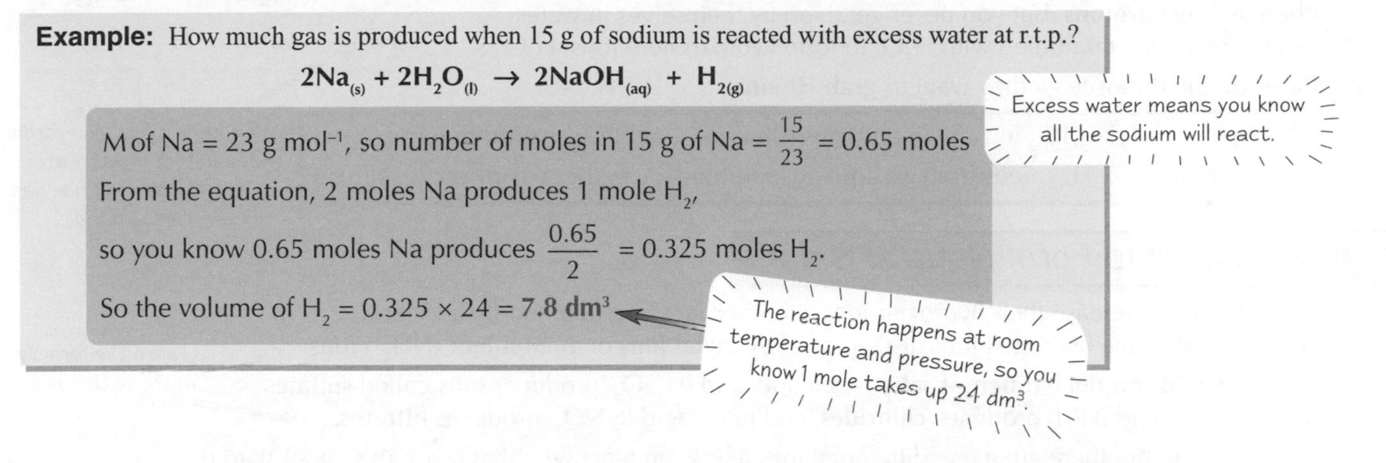

Example: How much gas is produced when 15 g of sodium is reacted with excess water at r.t.p.?

$$2Na_{(s)} + 2H_2O_{(l)} \rightarrow 2NaOH_{(aq)} + H_{2(g)}$$

M of Na = 23 g mol^{-1}, so number of moles in 15 g of Na = $\frac{15}{23}$ = 0.65 moles

From the equation, 2 moles Na produces 1 mole H_2,

so you know 0.65 moles Na produces $\frac{0.65}{2}$ = 0.325 moles H_2.

So the volume of H_2 = 0.325 × 24 = **7.8 dm³**

State Symbols Give a bit More Information about the Substances

State symbols are put after each compound in an equation. They tell you what **state of matter** things are in.

s = solid
l = liquid
g = gas
aq = aqueous (solution in water)

To show you what I mean, here's an example —

$$CaCO_{3\,(s)} + 2HCl_{(aq)} \rightarrow CaCl_{2\,(aq)} + H_2O_{(l)} + CO_{2\,(g)}$$

solid — aqueous — aqueous — liquid — gas

Practice Questions

Q1 What is the state symbol for a solution of hydrochloric acid?

Q2 What is the difference between a balanced equation and an ionic equation?

Exam Questions

Q1 Calculate the mass of ethene required to produce 258 g of chloroethane, C_2H_5Cl.

$C_2H_4 + HCl \rightarrow C_2H_5Cl$ [4 marks]

Q2 15 g of calcium carbonate is heated strongly so that it fully decomposes. $CaCO_{3(s)} \rightarrow CaO_{(s)} + CO_{2(g)}$

a) Calculate the mass of calcium oxide produced. [3 marks]

b) Calculate the volume of gas produced. [3 marks]

Q3 Balance this equation: $KI + Pb(NO_3)_2 \rightarrow PbI_2 + 2KNO_3$ [1 mark]

Don't get in a state about equations...

You're probably completely fed up with all these equations, calculations, moles and whatnot... well hang in there — there are just a few more pages coming up. I've said it once, and I'll say it again — practise, practise, practise... it's the only road to salvation (by the way, where is salvation anyway?). Keep going... you're nearly there.

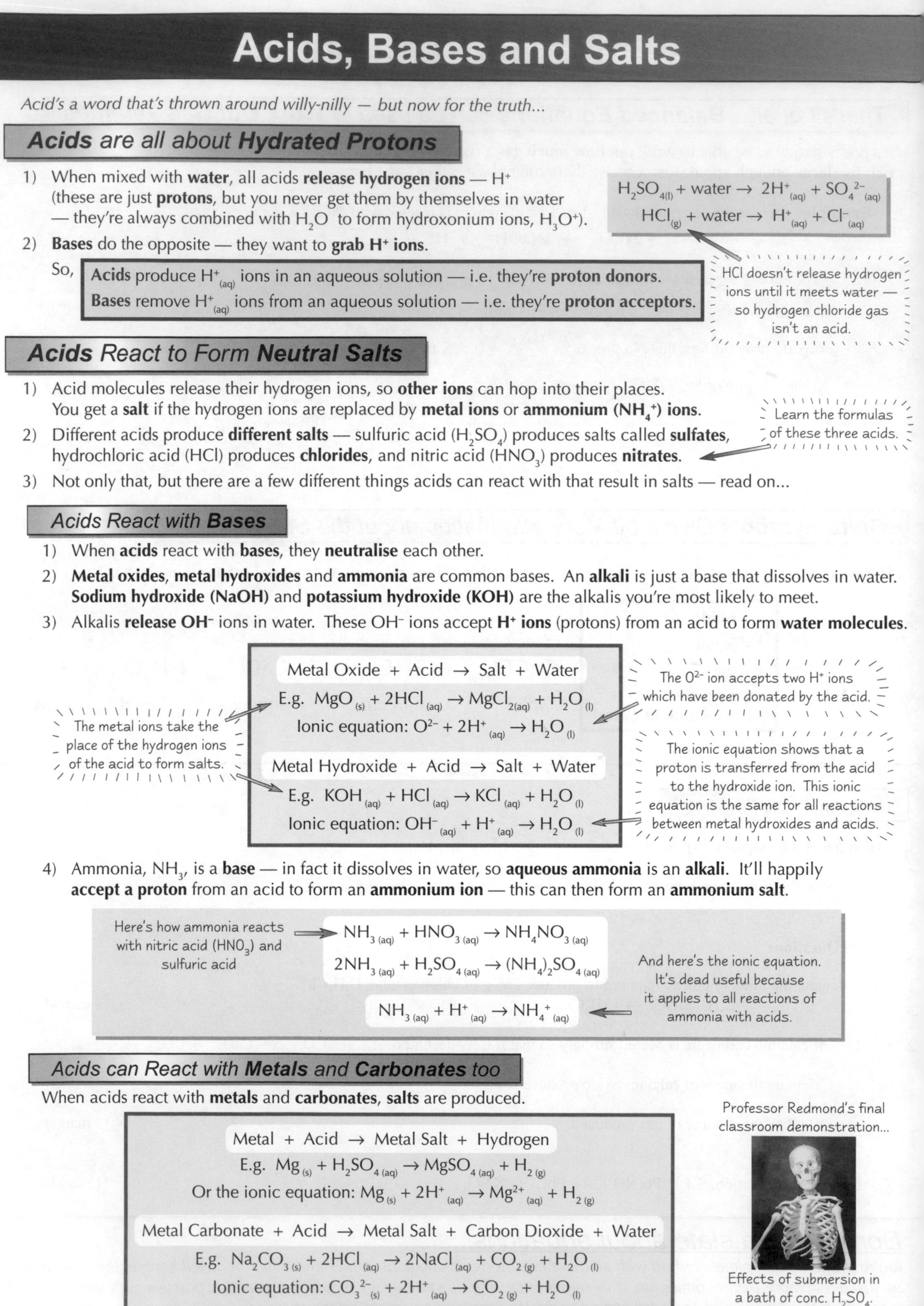

Acids, Bases and Salts

Acid's a word that's thrown around willy-nilly — but now for the truth...

Acids are all about Hydrated Protons

1) When mixed with **water**, all acids **release hydrogen ions** — H^+ (these are just **protons**, but you never get them by themselves in water — they're always combined with H_2O to form hydroxonium ions, H_3O^+).

$$H_2SO_{4(l)} + \text{water} \rightarrow 2H^+_{(aq)} + SO_4^{2-}{}_{(aq)}$$
$$HCl_{(g)} + \text{water} \rightarrow H^+_{(aq)} + Cl^-_{(aq)}$$

HCl doesn't release hydrogen ions until it meets water — so hydrogen chloride gas isn't an acid.

2) **Bases** do the opposite — they want to **grab H^+ ions.**

So,

> **Acids** produce $H^+_{(aq)}$ ions in an aqueous solution — i.e. they're **proton donors.**
> **Bases** remove $H^+_{(aq)}$ ions from an aqueous solution — i.e. they're **proton acceptors.**

Acids React to Form Neutral Salts

1) Acid molecules release their hydrogen ions, so **other ions** can hop into their places. You get a **salt** if the hydrogen ions are replaced by **metal ions** or **ammonium (NH_4^+) ions.**
2) Different acids produce **different salts** — sulfuric acid (H_2SO_4) produces salts called **sulfates**, hydrochloric acid (HCl) produces **chlorides**, and nitric acid (HNO_3) produces **nitrates.**
3) Not only that, but there are a few different things acids can react with that result in salts — read on...

Learn the formulas of these three acids.

Acids React with Bases

1) When **acids** react with **bases**, they **neutralise** each other.
2) **Metal oxides**, **metal hydroxides** and **ammonia** are common bases. An **alkali** is just a base that dissolves in water. **Sodium hydroxide (NaOH)** and **potassium hydroxide (KOH)** are the alkalis you're most likely to meet.
3) Alkalis **release OH^-** ions in water. These OH^- ions accept **H^+ ions** (protons) from an acid to form **water molecules.**

Metal Oxide + Acid → Salt + Water

E.g. $MgO_{(s)} + 2HCl_{(aq)} \rightarrow MgCl_{2(aq)} + H_2O_{(l)}$

Ionic equation: $O^{2-} + 2H^+_{(aq)} \rightarrow H_2O_{(l)}$

Metal Hydroxide + Acid → Salt + Water

E.g. $KOH_{(aq)} + HCl_{(aq)} \rightarrow KCl_{(aq)} + H_2O_{(l)}$

Ionic equation: $OH^-_{(aq)} + H^+_{(aq)} \rightarrow H_2O_{(l)}$

The metal ions take the place of the hydrogen ions of the acid to form salts.

The O^{2-} ion accepts two H^+ ions which have been donated by the acid.

The ionic equation shows that a proton is transferred from the acid to the hydroxide ion. This ionic equation is the same for all reactions between metal hydroxides and acids.

4) Ammonia, NH_3, is a **base** — in fact it dissolves in water, so **aqueous ammonia** is an **alkali**. It'll happily **accept a proton** from an acid to form an **ammonium ion** — this can then form an **ammonium salt.**

Here's how ammonia reacts with nitric acid (HNO_3) and sulfuric acid

$$NH_{3\,(aq)} + HNO_{3\,(aq)} \rightarrow NH_4NO_{3\,(aq)}$$
$$2NH_{3\,(aq)} + H_2SO_{4\,(aq)} \rightarrow (NH_4)_2SO_{4\,(aq)}$$
$$NH_{3\,(aq)} + H^+_{(aq)} \rightarrow NH_4^+{}_{(aq)}$$

And here's the ionic equation. It's dead useful because it applies to all reactions of ammonia with acids.

Acids can React with Metals and Carbonates too

When acids react with **metals** and **carbonates**, **salts** are produced.

Metal + Acid → Metal Salt + Hydrogen

E.g. $Mg_{(s)} + H_2SO_{4\,(aq)} \rightarrow MgSO_{4\,(aq)} + H_{2\,(g)}$

Or the ionic equation: $Mg_{(s)} + 2H^+_{(aq)} \rightarrow Mg^{2+}_{(aq)} + H_{2\,(g)}$

Metal Carbonate + Acid → Metal Salt + Carbon Dioxide + Water

E.g. $Na_2CO_{3\,(s)} + 2HCl_{(aq)} \rightarrow 2NaCl_{(aq)} + CO_{2\,(g)} + H_2O_{(l)}$

Ionic equation: $CO_3^{2-}{}_{(s)} + 2H^+_{(aq)} \rightarrow CO_{2\,(g)} + H_2O_{(l)}$

Professor Redmond's final classroom demonstration...

Effects of submersion in a bath of conc. H_2SO_4.

Acids, Bases and Salts

Salts Can Be Anhydrous or Hydrated

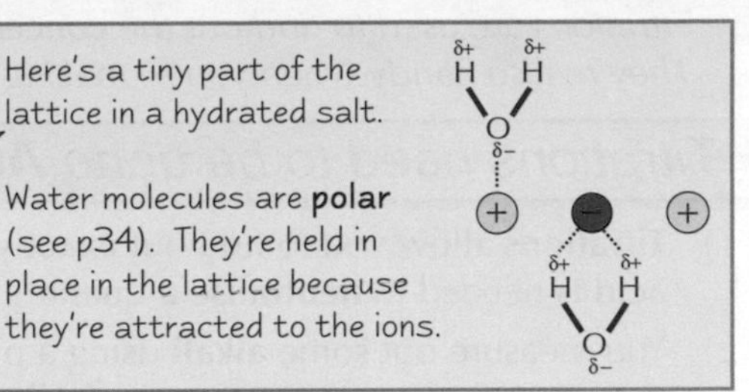

1) All solid salts consist of a **lattice** of positive and negative ions. In some salts, **water molecules** are incorporated in the lattice too.
2) The water in a lattice is called **water of crystallisation**. A solid salt containing water of crystallisation is **hydrated**. A salt is **anhydrous** if it doesn't contain water of crystallisation.
3) **One mole** of a particular hydrated salt always has the **same number of moles** of water of crystallisation — its **formula** shows **how many** (it's always a whole number).
4) For example, **hydrated copper sulfate** has **five** moles of water for every mole of the salt. So its formula is **$CuSO_4.5H_2O$**. ⟸ Notice that there's a dot between $CuSO_4$ and $5H_2O$.
5) Many hydrated salts **lose** their water of crystallisation **when heated**, to become **anhydrous**. If you know the mass of the salt when hydrated and anhydrous, you can work its formula out like this:

Example: Heating 3.210 g of hydrated magnesium sulfate, $MgSO_4.XH_2O$, forms 1.567 g of anhydrous magnesium sulfate. Find the value of **X** and write the formula of the hydrated salt.

First you find the number of moles of water lost.

Mass of water lost: 3.210 – 1.567 = 1.643 g
Number of moles of water lost: mass ÷ molar mass = 1.643 g ÷ 18 = **0.0913 moles**

Then you find the number of moles of anhydrous salt.

Molar mass of $MgSO_4$: 24 + 32 + (4 × 16) = 120 g mol^{-1}
Number of moles (in 1.567 g): mass ÷ molar mass = 1.567 ÷ 120 = **0.0131 moles**

Now you work out the ratio of moles of anhydrous salt to moles of water in the form 1 : n.

From the experiment, **0.0131 moles of salt : 0.0913 moles of water,**

So, **1 mole of salt : $\frac{0.0913}{0.0131}$ = 6.97 moles of water.**

You might be given the percentage of the mass that is water — use the method on p12.

X must be a whole number, and some errors are to be expected in any experiment, so you can safely round off your result — so the formula of the hydrated salt is **$MgSO_4.7H_2O$**.

Practice Questions

Q1 Write an ionic equation for the reaction between a metal hydroxide and an acid.

Q2 Explain what an alkali is.

Q3 Why can water molecules become fixed in an ionic lattice?

Exam Questions

Q1 Chloric(VII) acid, $HClO_4$, and sulfuric acid, H_2SO_4, are both strong acids.

a) Write a balanced equation, including state symbols, for the reaction between chloric(VII) acid and calcium carbonate, $CaCO_3$. [3 marks]

b) Sulfuric acid reacts with lithium metal, potassium hydroxide and ammonia.
(i) Write an ionic equation for the reaction with lithium. [2 marks]
(ii) Write an equation for the reaction with potassium hydroxide. [2 marks]
(iii) Write an equation for the reaction with aqueous ammonia. [2 marks]

Q2 A sample of hydrated calcium sulfate, $CaSO_4.XH_2O$, was prepared by reacting calcium hydroxide with sulfuric acid. 1.883 g of hydrated salt was produced. This was then heated until all the water of crystallisation was driven off and the product was then reweighed. Its mass was 1.133 g.

a) How many moles of anhydrous calcium sulfate were produced? [2 marks]
b) What mass of water was present in the hydrated salt? [1 mark]
c) Calculate the value of **X** in the formula $CaSO_4.XH_2O$. (**X** is an integer.) [3 marks]

It's a stick-up — your protons or your life...

Remember — all acids have protons to give away and bases just love to take them. It's what makes them acids and bases. It's like how bus drivers drive buses... it's what makes them bus drivers. Learn the formulas for the common acids — hydrochloric, sulfuric and nitric, and the common alkalis — sodium hydroxide, potassium hydroxide and aqueous ammonia.

Titrations

*Titrations are used to find out the **concentrations** of acid or alkali solutions. They're also handy when you're making salts of soluble bases.*

*Titrations need to be done **Accurately***

1) **Titrations** allow you to find out **exactly** how much acid is needed to **neutralise** a quantity of alkali.
2) You measure out some **alkali** using a pipette and put it in a flask, along with some **indicator**, e.g. **phenolphthalein**.
3) First of all, do a rough titration to get an idea where the **end point** is (the point where the alkali is **exactly neutralised** and the indicator changes colour). Add the **acid** to the alkali using a **burette** — giving the flask a regular **swirl**.
4) Now do an **accurate** titration. Run the acid in to within 2 cm^3 of the end point, then add the acid **dropwise**. If you don't notice exactly when the solution changed colour you've **overshot** and your result won't be accurate.
5) **Record** the amount of acid used to **neutralise** the alkali. It's best to **repeat** this process a few times, making sure you get the same answer each time.

Pipette
Pipettes measure only one volume of solution. Fill the pipette just above the line, then take the pipette out of the solution (or the water pressure will hold up the level). Now drop the level down carefully to the line.

Burette
Burettes measure different volumes and let you add the solution drop by drop.

acid

scale

alkali and indicator

You can also do titrations the other way round — adding alkali to acid.

Indicators** Show you when the Reaction's **Just Finished

Indicators change **colour**, as if by magic. In titrations, indicators that change colour quickly over a **very small pH range** are used so you know **exactly** when the reaction has ended.

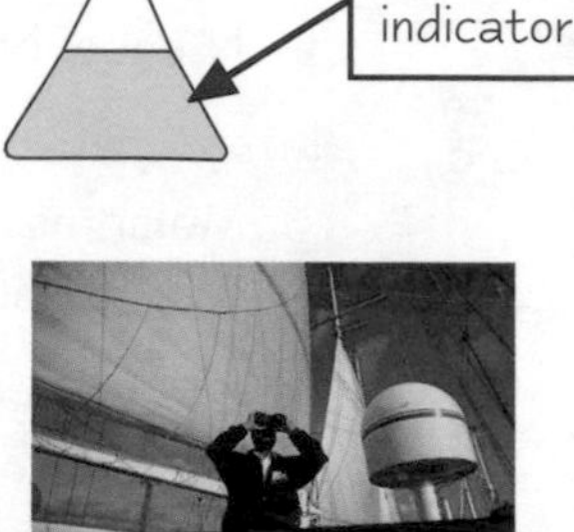

Choppy seas made it difficult for Captain Blackbird to read the burette accurately.

The main two indicators for **acid/alkali reactions** are —

methyl orange — turns **yellow** to **red** when adding acid to alkali.
phenolphthalein — turns **red** to **colourless** when adding acid to alkali.

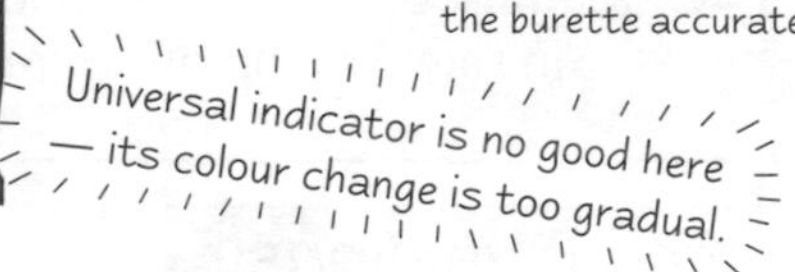

*You can Calculate **Concentrations** from Titrations*

Now for the calculations...

Example: 25 cm^3 of 0.5 M HCl was used to neutralise 35 cm^3 of NaOH solution.
Calculate the concentration of the sodium hydroxide solution in mol dm^{-3}.

First write a **balanced equation** and decide **what you know** and what you **need to know**:

HCl	+ NaOH	→	NaCl + H_2O
25 cm^3	**35 cm^3**		
0.5 M	**?**		

Now work out how many **moles of HCl** you have:

It's just the formula from page 11.

$$\text{Number of moles HCl} = \frac{\text{concentration} \times \text{volume (cm}^3)}{1000} = \frac{0.5 \times 25}{1000} = 0.0125 \text{ moles}$$

From the equation, you know 1 mole of HCl neutralises 1 mole of NaOH.
So 0.0125 moles of HCl must neutralise **0.0125** moles of NaOH.

Now it's a doddle to work out the **concentration of NaOH**.

$$\text{Concentration of NaOH}_{(aq)} = \frac{\text{moles of NaOH} \times 1000}{\text{volume (cm}^3)} = \frac{0.0125 \times 1000}{35} = \mathbf{0.36 \text{ mol dm}^{-3}}$$

Titrations

You use a *Pretty Similar Method* to Calculate *Volumes* for Reactions

This is usually used for **planning experiments**.

You need to use this formula again, but this time **rearrange** it to find the volume.

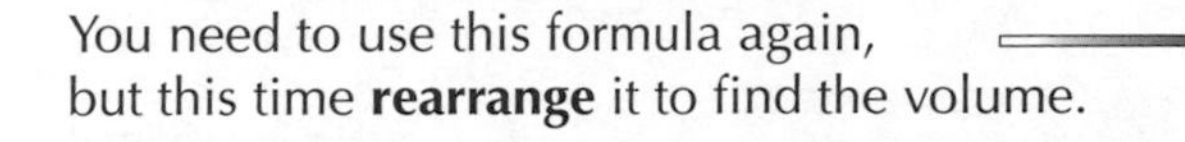

$$\text{number of moles} = \frac{\text{concentration} \times \text{volume (cm}^3\text{)}}{1000}$$

Example: 20.4 cm³ of a 0.5 M solution of sodium carbonate reacts with 1.5 M nitric acid. Calculate the volume of nitric acid required to neutralise the sodium carbonate.

First write a **balanced equation** for the reaction and decide **what you know** and what you **want to know**:

$$Na_2CO_3 + 2HNO_3 \rightarrow 2NaNO_3 + H_2O + CO_2$$

Na_2CO_3	$2HNO_3$
20.4 cm³	**?**
0.5 M	**1.5 M**

Now work out how many **moles** of Na_2CO_3 you've got:

$$\text{No. of moles of } Na_2CO_3 = \frac{\text{concentration} \times \text{volume (cm}^3\text{)}}{1000} = \frac{0.5 \times 20.4}{1000} = 0.0102 \text{ moles}$$

1 mole of Na_2CO_3 neutralises 2 moles of HNO_3,
so 0.0102 moles of Na_2CO_3 neutralises **0.0204 moles of HNO_3**.

Now you know the number of moles of HNO_3 and the concentration, you can work out the **volume**:

$$\text{Volume of } HNO_3 = \frac{\text{number of moles} \times 1000}{\text{concentration}} = \frac{0.0204 \times 1000}{1.5} = \mathbf{13.6 \text{ cm}^3}$$

Practice Questions

Q1 Describe the procedure for doing a titration.

Q2 What colour change would you expect to see if you added enough hydrochloric acid to a conical flask containing sodium hydroxide and methyl orange?

Exam Questions

Q1 Calculate the concentration (in M) of a solution of ethanoic acid, CH_3COOH, if 25.4 cm³ of it is neutralised by 14.6 cm³ of 0.5 M sodium hydroxide solution. $CH_3COOH + NaOH \rightarrow CH_3COONa + H_2O$ [3 marks]

Q2 You are supplied with 0.75 g of calcium carbonate and a solution of 0.25 M sulfuric acid.
What volume of acid will be needed to neutralise the calcium carbonate?
$CaCO_3 + H_2SO_4 \rightarrow CaSO_4 + H_2O + CO_2$ [4 marks]

Burettes and pipettes — big glass things, just waiting to be dropped...

Titrations are annoyingly fiddly. But you do get to use big, impressive-looking equipment and feel like you're doing something important. It's really tempting to rush it and let half the acid gush into the alkali first. But it's totally not worth it, cos you'll just have to do it again. Yep, this is definitely one of those slow-and-steady-wins-the-race situations.

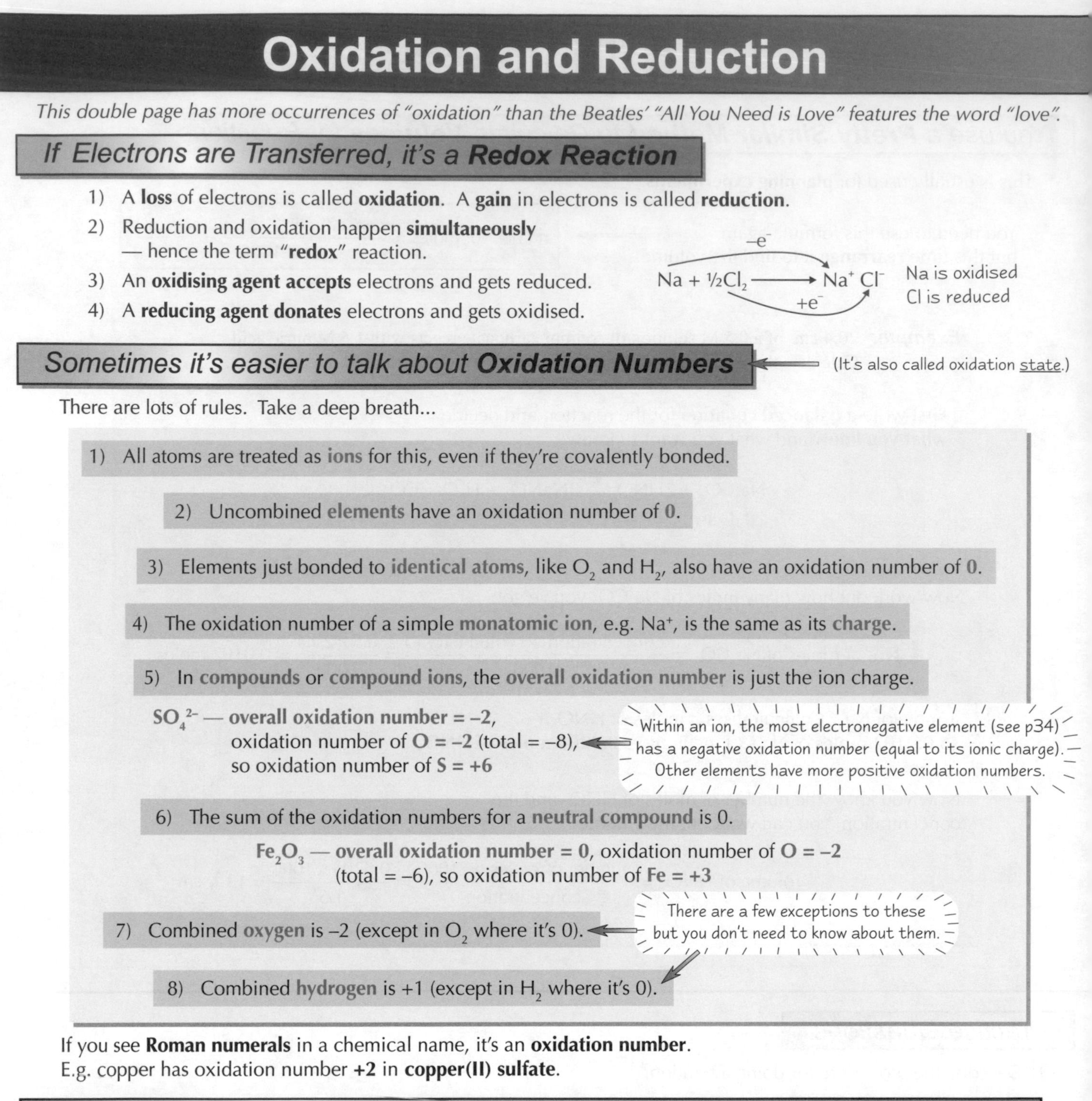

Oxidation and Reduction

This double page has more occurrences of "oxidation" than the Beatles' "All You Need is Love" features the word "love".

If Electrons are Transferred, it's a **Redox Reaction**

1) A **loss** of electrons is called **oxidation**. A **gain** in electrons is called **reduction**.
2) Reduction and oxidation happen **simultaneously** — hence the term "**redox**" reaction.
3) An **oxidising agent accepts** electrons and gets reduced.
4) A **reducing agent donates** electrons and gets oxidised.

$$Na + \tfrac{1}{2}Cl_2 \longrightarrow Na^+ \, Cl^-$$

Sometimes it's easier to talk about **Oxidation Numbers**

(It's also called oxidation state.)

There are lots of rules. Take a deep breath...

1) All atoms are treated as **ions** for this, even if they're covalently bonded.

2) Uncombined **elements** have an oxidation number of **0**.

3) Elements just bonded to **identical atoms**, like O_2 and H_2, also have an oxidation number of **0**.

4) The oxidation number of a simple **monatomic ion**, e.g. Na^+, is the same as its **charge**.

5) In **compounds** or **compound ions**, the **overall oxidation number** is just the ion charge.

SO_4^{2-} — overall oxidation number = –2,
oxidation number of **O = –2** (total = –8),
so oxidation number of **S = +6**

Within an ion, the most electronegative element (see p34) has a negative oxidation number (equal to its ionic charge). Other elements have more positive oxidation numbers.

6) The sum of the oxidation numbers for a **neutral compound** is 0.

Fe_2O_3 — overall oxidation number = 0, oxidation number of **O = –2** (total = –6), so oxidation number of **Fe = +3**

7) Combined **oxygen** is –2 (except in O_2 where it's 0).

There are a few exceptions to these but you don't need to know about them.

8) Combined **hydrogen** is +1 (except in H_2 where it's 0).

If you see **Roman numerals** in a chemical name, it's an **oxidation number**.
E.g. copper has oxidation number **+2** in **copper(II) sulfate**.

You Can Work Out **Oxidation Numbers** from **Formulas** or **Systematic Names**

Systematic names make the oxidation numbers of all the atoms making up a compound ion clear.

1) Ions with names ending in **-ate** (e.g. sulfate, nitrate, carbonate) contain **oxygen**, as well as another element. For example, **sulfates** contain **sulfur** and **oxygen**, **nitrates** contain **nitrogen** and **oxygen**... and so on.
2) But sometimes the 'other' element in the ion can exist with different oxidation numbers, and so form different '-ate ions'. In these cases, the oxidation number is attached as a Roman numeral.
E.g., **sulfate(VI)** tells you that the **sulfur** has oxidation number **+6** — this is the SO_4^{2-} ion.
However, in the **sulfate(IV)** ion, **sulfur** has oxidation number **+4** — this is the SO_3^{2-} ion.
Several ions have widely used common names that are different from their correct systematic names. For example, the sulfate(IV) ion (SO_3^{2-}) is often called the sulfite ion.

The oxidation number applies to the sulfur not the oxygen, because oxygen is always –2.

3) You might have to work out the systematic name for a compound, given its formula — e.g. KNO_3.
Okay... it's potassium nitrate, but you need to give the **oxidation number** of the **nitrogen**. Here's how...

You know that potassium **always** forms K^+ ions, so the charge on the **nitrate** ion must be **–1**. Each **oxygen** atom in the NO_3^- ion has oxidation number **–2**. This gives 3 × –2 = **–6**. Then, since the ion has an overall number of –1, the nitrogen must be in the **+5** state. So the compound is **potassium nitrate(V)**.

NO_3^- is called the nitrate(V) ion.

Oxidation and Reduction

*Oxidation Numbers go **Up** or **Down** as Electrons are **Lost** or **Gained***

1) The oxidation number for an atom will **increase by 1** for each **electron lost**.
2) The oxidation number will **decrease by 1** for each **electron gained**.
3) When **metals** form compounds, they generally **donate** electrons to form **positive ions** — meaning they usually have **positive oxidation numbers**.
4) When **non-metals** form compounds, they generally **gain** electrons — meaning they usually have **negative oxidation numbers**.
5) In a **redox** reaction, some oxidation numbers will **change** — like in this reaction between iron(III) oxide and **carbon(II) oxide** (aka carbon monoxide). The products are the element iron and **carbon(IV) oxide** (more commonly known as carbon dioxide).

Fe oxidation number reduced from +3 to 0 — reduction

$$Fe_2O_3 + 3CO \rightarrow 2Fe + 3CO_2$$

C oxidation number increased from +2 to +4 — oxidation

*Many **Metals Reduce Dilute Acids***

1) On page 16 you saw how metals react with acids to produce a salt and hydrogen gas. Well this is a redox reaction:
   - The metal atoms are **oxidised**, losing electrons and forming soluble metal ions.
   - The hydrogen ions in solution are **reduced**, gaining electrons and forming hydrogen molecules.
2) For example, magnesium reacts with dilute hydrochloric acid like this:

Mg oxidation number increased from 0 to +2 — oxidation

$$Mg_{(s)} + 2HCl_{(aq)} \rightarrow MgCl_{2\,(aq)} + H_{2\,(g)}$$

H oxidation number decreased from +1 to 0 — reduction

Notice that the chloride ions don't change oxidation number — they're still chloride ions, with oxidation number −1.

Hands up if you like Roman numerals...

3) If you use **sulfuric acid** instead of hydrochloric acid, exactly the same processes of **oxidation** and **reduction** take place. For example, potassium is oxidised to K^+ ions:

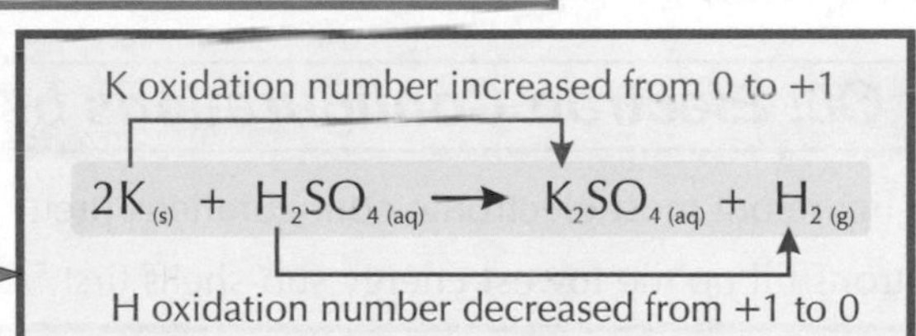

Practice Questions

Q1 What is a reducing agent?

Q2 What is the usual oxidation number for oxygen combined with another element?

Q3 Explain why the systematic name of $NaNO_2$ is sodium nitrate(III).

Q4 Explain which element is oxidised and which is reduced in the reaction: $Ca + 2HCl \rightarrow CaCl_2 + H_2$

Exam Question

Q1 When hydrogen iodide gas is bubbled through warm concentrated sulfuric acid, hydrogen sulfide and iodine are produced.

a) Balance the equation below for the reaction. [1 mark]

$$H_2SO_{4\,(aq)} + HI_{(g)} \rightarrow H_2S_{(g)} + I_{2\,(s)} + H_2O_{(l)}$$

b) State the oxidation number of sulfur in H_2SO_4 and in H_2S. [2 marks]

c) In this reaction, which is the reducing agent? Give a reason. [2 marks]

Redox — relax in a lovely warm bubble bath...

The thing here is to take your time. Questions on oxidation numbers aren't usually that hard, but they are easy to get wrong. So don't panic, take it easy, and get all the marks.

And while we're on the oxidation page, I suppose you ought to learn the most famous memory aid thingy in the world...

OIL RIG
- **Oxidation Is Loss**
- **Reduction Is Gain**

(of electrons)

Electronic Structure

Those little electrons prancing about like mini bunnies decide what'll react with what — it's what chemistry's all about.

Electron Shells are Made Up of Sub-Shells and Orbitals

1) Electrons move around the nucleus in **shells** (sometimes called **energy levels**). These shells are all given numbers known as **principal quantum numbers**.
2) Shells **further** from the nucleus have a greater energy level than shells closer to the nucleus.
3) The shells contain different types of **sub-shell**. These sub-shells have different numbers of **orbitals**, which can each hold up to **2 electrons**.

This table shows the number of electrons that fit in each type of sub-shell.

Sub-shell	Number of orbitals	Maximum electrons
s	1	1 × 2 = 2
p	3	3 × 2 = 6
d	5	5 × 2 = 10
f	7	7 × 2 = 14

And this one shows the sub-shells and electrons in the first four energy levels.

Shell	Sub-shells	Total number or electrons	
1st	1s	2	= 2
2nd	2s 2p	2 + (3 × 2)	= 8
3rd	3s 3p 3d	2 + (3 × 2) + (5 × 2)	= 18
4th	4s 4p 4d 4f	2 + (3 × 2) + (5 × 2) + (7 × 2)	= 32

Orbitals Have Characteristic Shapes

There are a few things you need to know about orbitals... like what they are —

1) An orbital is the **bit of space** that an electron moves in. Orbitals within the same sub-shell have the **same energy**.
2) The electrons in the orbitals have to 'spin' in **opposite** directions — this is called **spin-pairing**.
3) s orbitals are **spherical** — p orbitals have **dumbbell shapes**. There are 3 p orbitals and they're at right angles to one another.

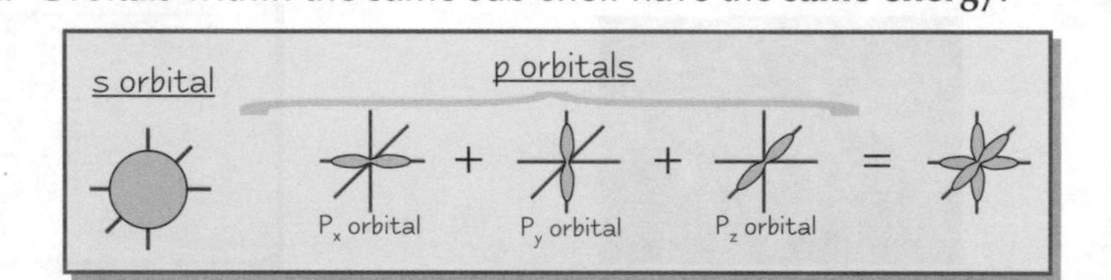

Work Out Electron Configurations by Filling the Lowest Energy Levels First

You can figure out most electronic configurations pretty easily, so long as you know a few simple rules —

1) Electrons fill up the **lowest** energy sub-shells first.

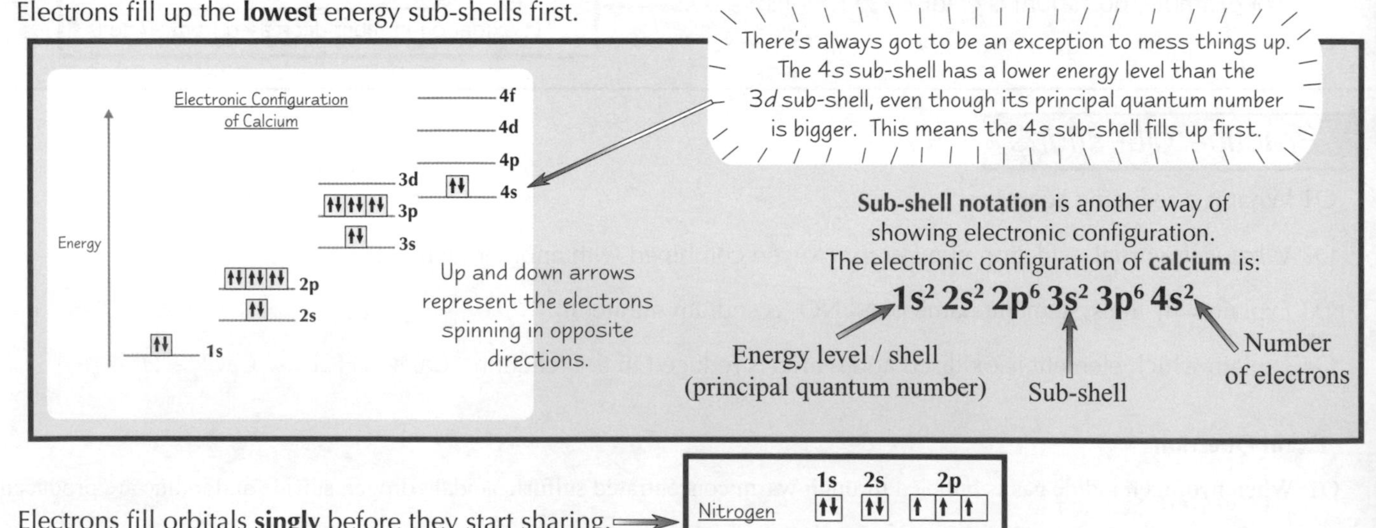

2) Electrons fill orbitals **singly** before they start sharing.

	1s	2s	2p
Nitrogen	↑↓	↑↓	↑ ↑ ↑
Oxygen	↑↓	↑↓	↑↓ ↑ ↑

See the next page for more on the s and p block.

3) For the configuration of **ions** from the **s** and **p** blocks of the periodic table, just **remove or add** the electrons to or from the highest-energy occupied sub-shell. E.g. $Mg^{2+} = 1s^2\ 2s^2\ 2p^6$, $Cl^- = 1s^2\ 2s^2\ 2p^6\ 3s^2\ 3p^6$

Watch out — **noble gas symbols**, like that of argon (Ar), are sometimes used in electron configurations. For example, calcium ($1s^2\ 2s^2\ 2p^6\ 3s^2\ 3p^6\ 4s^2$) can be written as $[Ar]4s^2$, where $[Ar] = 1s^2\ 2s^2\ 2p^6\ 3s^2\ 3p^6$.

Electronic Structure

Electronic Structure Decides the **Chemical Properties** of an Element

The number of **outer shell electrons** decides the chemical properties of an element.

1) The **s block** elements (Groups 1 and 2) have 1 or 2 outer shell electrons. These are easily **lost** to form positive ions with an **inert gas configuration**.

E.g. Na: $1s^2\ 2s^2\ 2p^6\ 3s^1 \rightarrow$ Na^+: $1s^2\ 2s^2\ 2p^6$ ← This is the electron configuration of neon.

2) The elements in Groups 5, 6 and 7 (in the p block) can **gain** 1, 2 or 3 electrons to form negative ions with an **inert gas configuration**.

E.g. O: $1s^2\ 2s^2\ 2p^4 \rightarrow O^{2-}$: $1s^2\ 2s^2\ 2p^6$

Groups 4 to 7 can also **share** electrons when they form covalent bonds.

3) Group 0 (the inert gases) have **completely filled** s and p sub-shells and don't need to bother gaining, losing or sharing electrons — their full sub-shells make them **inert**.
4) The **d block elements** (which include the transition metals) tend to **lose** s and d electrons to form positive ions.

Sub-shells and the periodic table

1s, 2s, 3s, 4s, 5s, 6s, 7s, 3d, 4d, 5d, 1s, 2p, 3p, 4p, 5p, 6p

Practice Questions

Q1 Write down the sub-shells in order of increasing energy up to 4p.

Q2 How many electrons do full s, p and d sub-shells contain?

Q3 Draw diagrams to show the shapes of an s and a p orbital.

Q4 What does the term 'spin-pairing' mean?

Exam Questions

Q1 Potassium reacts with oxygen to form potassium oxide, K_2O.

a) Give the electron configurations of the K atom and K^+ ion. [2 marks]

b) Give the electron configuration of the oxygen atom. [1 mark]

c) Explain why it is the outer shell electrons, not those in the inner shells, which determine the chemistry of potassium and oxygen. [2 marks]

Q2 This question concerns electron configurations in atoms and ions.

a) What is the electron configuration of a manganese atom? [1 mark]

b) Identify the element with the 4th shell configuration of $4s^2\,4p^2$. [1 mark]

c) Suggest the identity of an atom, a positive ion and a negative ion with the configuration $1s^2\ 2s^2\ 2p^6\ 3s^2\ 3p^6$. [3 marks]

d) Give the electron configuration of the Al^{3+} ion. [1 mark]

She shells sub-sells on the shesore...

The way electrons fill up the orbitals is kind of like how strangers fill up seats on a bus. Everyone tends to sit in their own seat till they're forced to share. Except for the huge, scary man who comes and sits next to you. Make sure you learn the order that the sub-shells are filled up in, so you can write electron configurations for any atom or ion they throw at you.

Ionisation Energies

This page gets a trifle brain-boggling, so I hope you've got a few aspirin handy...

Ionisation** is the **Removal** of One or More **Electrons

When electrons have been removed from an atom or molecule, it's been **ionised**.
The energy you need to remove the first electron is called the **first ionisation energy**:

> The **first ionisation energy** is the energy needed to remove 1 electron from **each atom** in **1 mole** of **gaseous** atoms to form 1 mole of gaseous 1+ ions.

You have to put energy **in** to ionise an atom or molecule, so it's an **endothermic process**.

You can write **equations** for this process — here's the equation for the **first ionisation of oxygen**:

$$O_{(g)} \rightarrow O^{+}_{(g)} + e^{-} \qquad \textbf{1st ionisation energy = +1314 kJ mol}^{-1}$$

Here are a few rather important points about ionisation energies:

1) You **must** use the gas state symbol, **(g)**, because ionisation energies are measured for gaseous atoms.
2) Always refer to **1 mole** of atoms, as stated in the definition, rather than to a single atom.
3) The **lower** the ionisation energy, the **easier** it is to form an ion.

*The **Factors** Affecting Ionisation Energy are...*

You need to know all about these...

The **more protons** there are in the nucleus, the more positively charged the nucleus is and the **stronger the attraction** for the electrons.

Attraction falls off very **rapidly with distance**. An electron **close** to the nucleus will be **much more** strongly attracted than one further away.

As the number of electrons **between** the outer electrons and the nucleus **increases**, the outer electrons feel less attraction towards the nuclear charge. This lessening of the pull of the nucleus by inner shells of electrons is called **shielding (or screening)**.

A **high ionisation energy** means there's a **high attraction** between the **electron** and the **nucleus**.

***Successive Ionisation Energies** Involve Removing **Additional** Electrons*

1) You can remove **all** the electrons from an atom, leaving only the nucleus.
   Each time you remove an electron, there's a **successive ionisation energy**.
2) The definition for the **second ionisation energy** is:

> The **second ionisation energy** is the energy needed to remove 1 electron from **each ion** in **1 mole** of **gaseous** 1+ ions to form 1 mole of gaseous 2+ ions.

And here's the equation for the **second ionisation of oxygen** :

$$O^{+}_{(g)} \rightarrow O^{2+}_{(g)} + e^{-} \qquad \textbf{2nd ionisation energy = +3388 kJ mol}^{-1}$$

Ionisation Energies

Successive Ionisation Energies Show **Shell Structure**

If you have the successive ionisation energies of an element you can work out the number of electrons in each shell of the atom and which element the group is in.

A **graph** of successive ionisation energies (like this one for sodium) provides evidence for the **shell structure** of atoms.

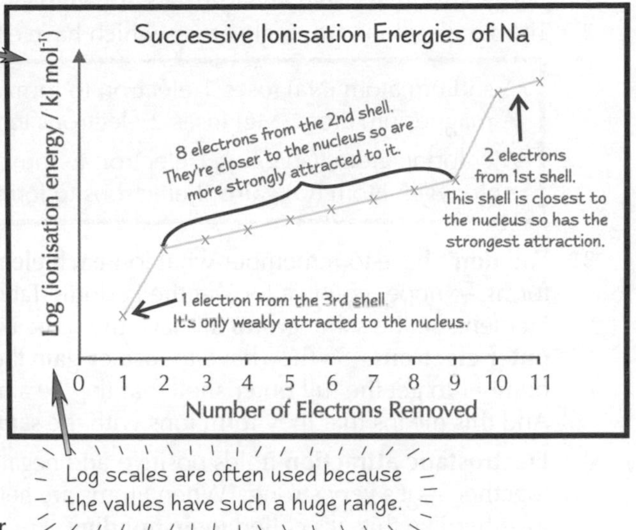

Log scales are often used because the values have such a huge range.

1) **Within each shell**, successive ionisation energies **increase.** This is because electrons are being removed from an **increasingly positive ion** — there's **less repulsion** amongst the remaining electrons, so they're **held more strongly** by the nucleus.
2) The **big jumps** in ionisation energy happen when a new shell is broken into — an electron is being removed from a shell **closer** to the nucleus.

1) Graphs like this can tell you which **group** of the periodic table an element belongs to. Just count **how many electrons are removed** before the first big jump to find the group number.
2) These graphs can be used to predict the **electronic structure** of an element. Working from **right to left**, count how many points there are before each big jump to find how many electrons are in each shell, starting with the first.

E.g. In the graph for sodium, **one electron** is removed before the first big jump — sodium is in **group 1**.
The graph has **2 points** on the right-hand side, then a jump, then **8 points**, a jump, and **1 final point**.
Sodium has **2 electrons** in the first shell, **8** in the second and **1** in the third.

Practice Questions

Q1 Why is the (g) state symbol always used when writing ionisation energy equations?

Q2 Name three factors which affect the size of an ionisation energy.

Q3 Write the definition of the first and second ionisation energies.

Q4 Explain how you can tell that it's chlorine's successive ionisation energies that are shown on the graph on the right.

Log (ionisation energies kJ mol⁻¹)
Number of electrons removed

Exam Question

Q1 This graph shows the successive ionisation energies of a certain element.

a) To which group of the periodic table does this element belong? [1 mark]

b) Explain why it takes more energy to remove each successive electron. [2 marks]

c) What causes the sudden increases in ionisation energy? [1 mark]

d) What is the total number of shells of electrons in this element? [1 mark]

Log (Ionisation energies (kJ mol⁻¹))
0 1 2 3 4 5 6 7 8 9 10 11 12 13
Number of electrons removed

Shirt crumpled — ionise it...

When you're talking about ionisation energies in exams, always use the three main factors — shielding, nuclear charge and distance from nucleus. Make sure you're comfortable interpreting the jumps in those graphs without getting stressed. And recite the definitions of the first and second ionisation energies to yourself until the men in white coats get to you. Then stop.

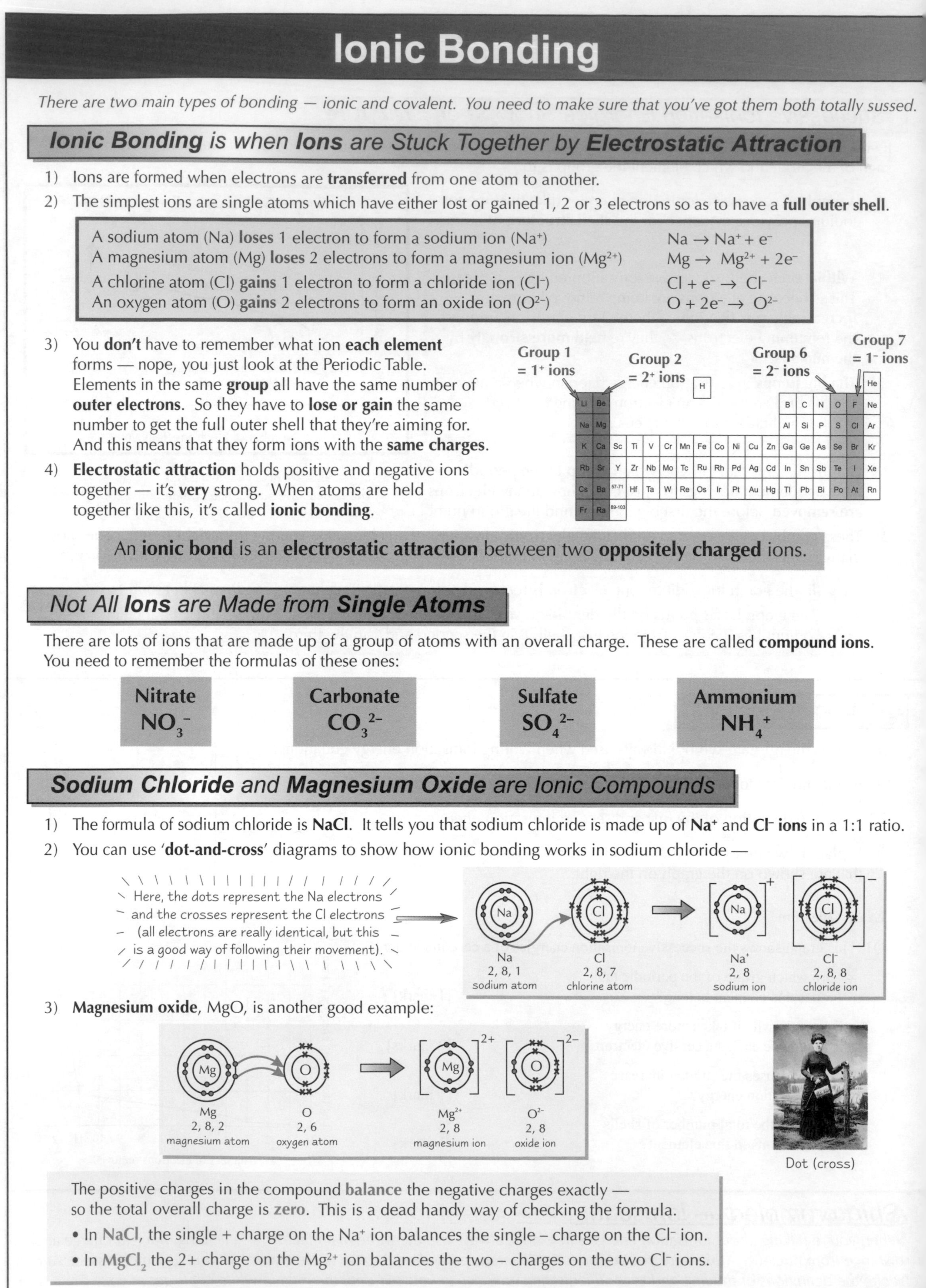

Ionic Bonding

There are two main types of bonding — ionic and covalent. You need to make sure that you've got them both totally sussed.

Ionic Bonding is when Ions are Stuck Together by Electrostatic Attraction

1) Ions are formed when electrons are **transferred** from one atom to another.
2) The simplest ions are single atoms which have either lost or gained 1, 2 or 3 electrons so as to have a **full outer shell**.

A sodium atom (Na) **loses** 1 electron to form a sodium ion (Na^+)	$Na \rightarrow Na^+ + e^-$
A magnesium atom (Mg) **loses** 2 electrons to form a magnesium ion (Mg^{2+})	$Mg \rightarrow Mg^{2+} + 2e^-$
A chlorine atom (Cl) **gains** 1 electron to form a chloride ion (Cl^-)	$Cl + e^- \rightarrow Cl^-$
An oxygen atom (O) **gains** 2 electrons to form an oxide ion (O^{2-})	$O + 2e^- \rightarrow O^{2-}$

3) You **don't** have to remember what ion **each element** forms — nope, you just look at the Periodic Table. Elements in the same **group** all have the same number of **outer electrons**. So they have to **lose or gain** the same number to get the full outer shell that they're aiming for. And this means that they form ions with the **same charges**.
4) **Electrostatic attraction** holds positive and negative ions together — it's **very** strong. When atoms are held together like this, it's called **ionic bonding**.

An **ionic bond** is an **electrostatic attraction** between two **oppositely charged** ions.

Not All Ions are Made from Single Atoms

There are lots of ions that are made up of a group of atoms with an overall charge. These are called **compound ions**. You need to remember the formulas of these ones:

Nitrate	Carbonate	Sulfate	Ammonium
NO_3^-	CO_3^{2-}	SO_4^{2-}	NH_4^+

Sodium Chloride and Magnesium Oxide are Ionic Compounds

1) The formula of sodium chloride is **NaCl**. It tells you that sodium chloride is made up of **Na^+** and **Cl^- ions** in a 1:1 ratio.
2) You can use '**dot-and-cross**' diagrams to show how ionic bonding works in sodium chloride —

3) **Magnesium oxide**, MgO, is another good example:

Dot (cross)

The positive charges in the compound **balance** the negative charges exactly — so the total overall charge is **zero**. This is a dead handy way of checking the formula.
- In **NaCl**, the single + charge on the Na^+ ion balances the single – charge on the Cl^- ion.
- In **$MgCl_2$** the 2+ charge on the Mg^{2+} ion balances the two – charges on the two Cl^- ions.

Ionic Bonding

Sodium Chloride has a **Giant Ionic Lattice** Structure

1) In **sodium chloride**, the Na^+ and Cl^- ions are packed together in a regular structure called a **lattice**.
2) The structure's called **'giant'** because it's made up of the same basic unit repeated over and over again.
3) The sodium chloride lattice is **cube** shaped — different ionic compounds have different shaped structures, but they're all still giant lattices.
4) Sodium chloride's got very strong **ionic bonds**, so it takes loads of **energy** to break up the lattice. This gives it a high melting point (801 °C).

The lines show the ionic bonds between the ions.

The Na^+ and Cl^- ions alternate.

But it's not just melting points — the structure decides other **physical properties** too...

Ionic Structure Explains the **Behaviour** of Ionic Compounds

1) **Ionic compounds conduct electricity when they're molten or dissolved — but not when they're solid.**
   The ions in a liquid are free to move (and they carry a charge).
   In a solid they're fixed in position by the strong ionic bonds.
2) **Ionic compounds have high melting points.**
   The giant ionic lattices are held together by strong electrostatic forces. It takes loads of energy to overcome these forces, so melting points are very high (801 °C for sodium chloride).
3) **Ionic compounds tend to dissolve in water.**
   Water molecules are polar — part of the molecule has a small negative charge, and the other bits have small positive charges (see p34). The water molecules pull the ions away from the lattice and cause it to dissolve.

Practice Questions

Q1 State the formula of the carbonate ion.

Q2 Draw a dot-and-cross diagram showing the bonding between magnesium and oxygen.

Q3 What type of force holds ionic substances together?

Q4 Why do many ionic compounds dissolve in water?

Exam Questions

Q1 a) Draw a labelled diagram to show the structure of sodium chloride. [3 marks]

b) What is the name of this type of structure? [1 mark]

c) Would you expect sodium chloride to have a high or a low melting point?
Explain your answer. [4 marks]

Q2 a) Ions can be formed by electron transfer. Explain this and give an example of a positive and a negative ion. [3 marks]

b) Solid lead bromide does not conduct electricity, but molten lead bromide does.
Explain this with reference to ionic bonding. [4 marks]

A black fly in your Chardonnay — isn't it ionic...

This stuff's easy marks in exams. Just make sure you can draw dot-and-cross diagrams showing the bonding in ionic compounds, and you're sorted. Remember — atoms are lazy. It's easier to lose two electrons to get a full shell than it is to gain six, so that's what an atom's going to do. Practise drawing sodium chloride too, and don't stop till you're perfect.

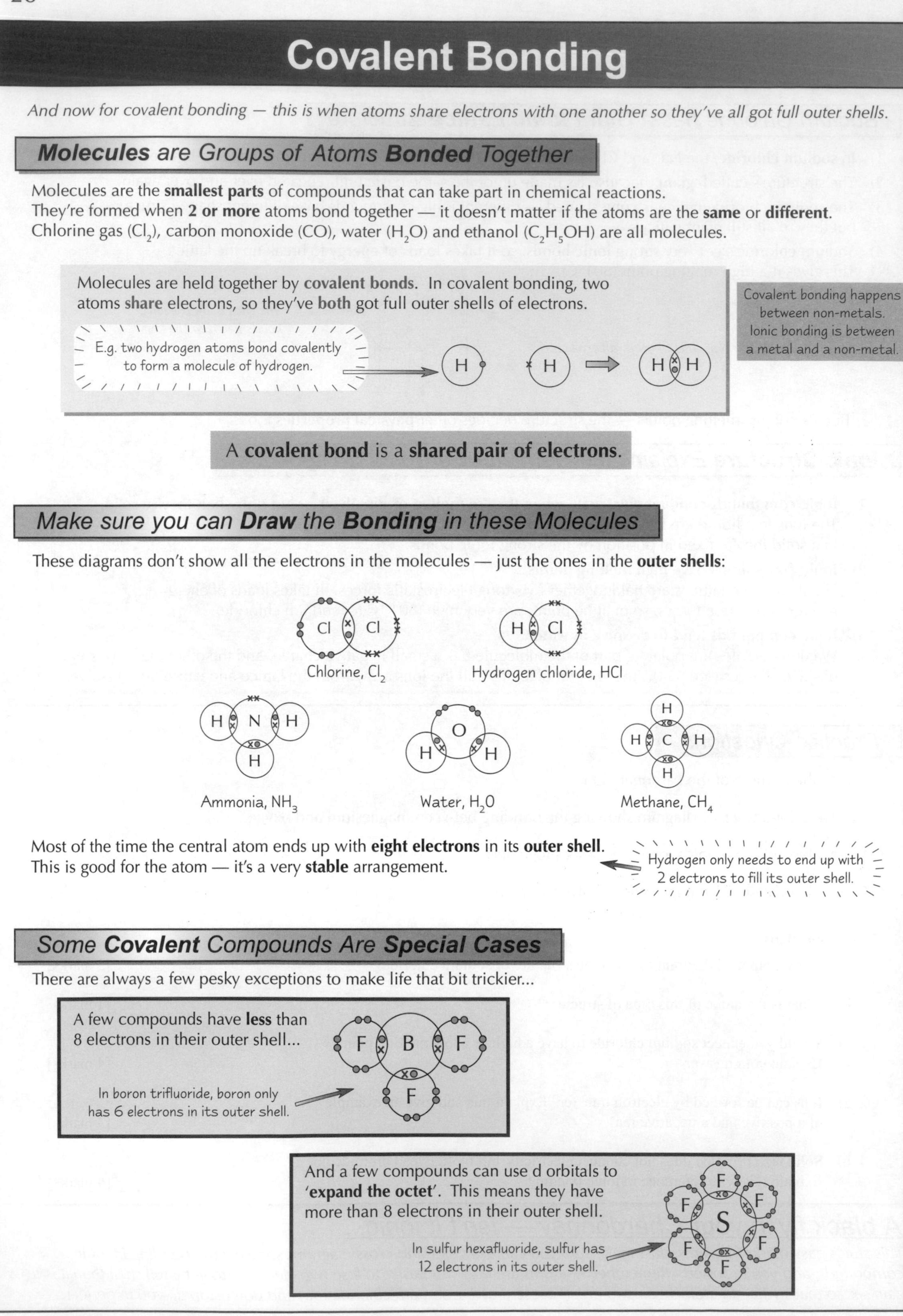

Covalent Bonding

And now for covalent bonding — this is when atoms share electrons with one another so they've all got full outer shells.

Molecules are Groups of Atoms *Bonded* Together

Molecules are the **smallest parts** of compounds that can take part in chemical reactions.
They're formed when **2 or more** atoms bond together — it doesn't matter if the atoms are the **same** or **different**.
Chlorine gas (Cl_2), carbon monoxide (CO), water (H_2O) and ethanol (C_2H_5OH) are all molecules.

Molecules are held together by **covalent bonds.** In covalent bonding, two atoms **share** electrons, so they've **both** got full outer shells of electrons.

E.g. two hydrogen atoms bond covalently to form a molecule of hydrogen.

H H → H H

Covalent bonding happens between non-metals. Ionic bonding is between a metal and a non-metal.

A **covalent bond** is a **shared pair of electrons.**

Make sure you can *Draw* the *Bonding* in these Molecules

These diagrams don't show all the electrons in the molecules — just the ones in the **outer shells**:

Cl Cl

Chlorine, Cl_2

H Cl

Hydrogen chloride, HCl

H N H H

Ammonia, NH_3

O H H

Water, H_2O

H H C H H

Methane, CH_4

Most of the time the central atom ends up with **eight electrons** in its **outer shell**.
This is good for the atom — it's a very **stable** arrangement.

Hydrogen only needs to end up with 2 electrons to fill its outer shell.

Some *Covalent* Compounds Are *Special Cases*

There are always a few pesky exceptions to make life that bit trickier...

A few compounds have **less** than 8 electrons in their outer shell...

F B F F

In boron trifluoride, boron only has 6 electrons in its outer shell.

And a few compounds can use d orbitals to '**expand the octet**'. This means they have more than 8 electrons in their outer shell.

F F F S F F F

In sulfur hexafluoride, sulfur has 12 electrons in its outer shell.

Covalent Bonding

Some Atoms Share **More Than One Pair** of **Electrons**

Atoms don't just form single bonds — some can form **double** or even **triple covalent bonds.** An example of a molecule that has a double bond is **oxygen,** O_2.

O O

You can show oxygen's bonding as a **dot-and-cross diagram** too.

Nitrogen can triple bond, and carbon dioxide has two double bonds:

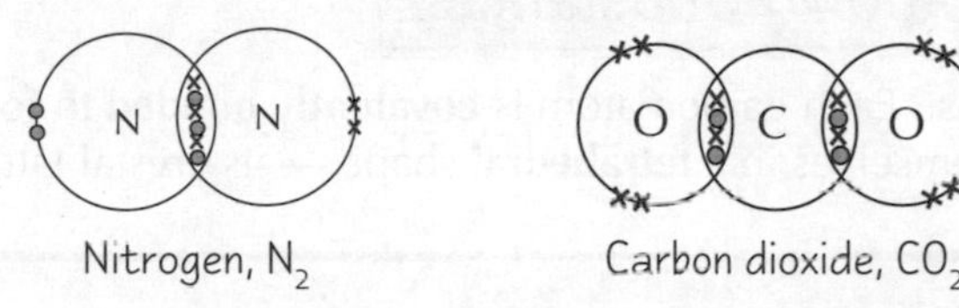

Nitrogen, N_2 Carbon dioxide, CO_2

Dative Covalent Bonding is where **Both Electrons** come from **One Atom**

The **ammonium ion** (NH_4^+) is formed by dative (or coordinate) covalent bonding — it's an example the examiners love. It forms when the nitrogen atom in an ammonia molecule **donates a pair of electrons** to a proton (H^+):

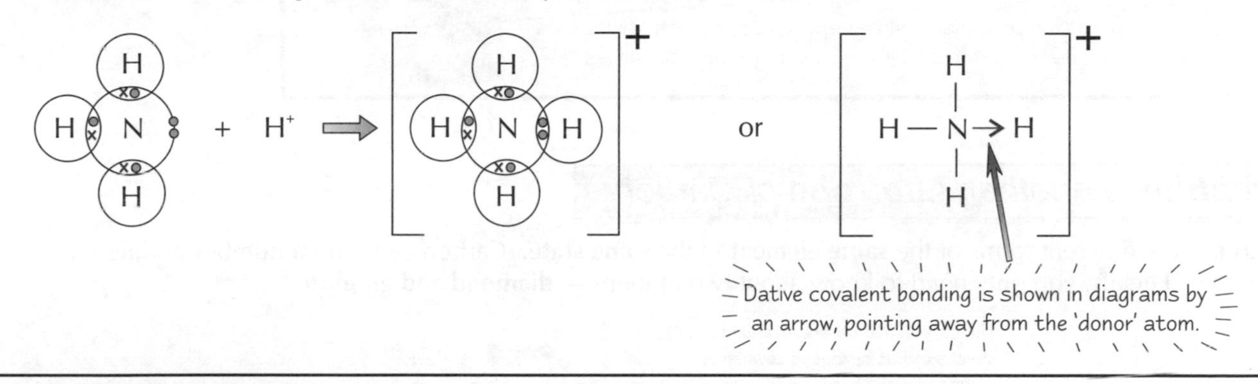

Practice Questions

Q1 Draw a dot-and-cross diagram to show the arrangement of the outer electrons in a molecule of hydrogen chloride.

Q2 What's special about the bonding in boron trifluoride?
Draw a diagram showing the outer electrons in a molecule of boron trifluoride.

Q3 Name a molecule with a double covalent bond. Draw a diagram showing the outer electrons in this molecule.

Exam Questions

Q1 Methane, CH_4, contains atoms of two non-metals.

a) What type of bonding would you expect it to have? [1 mark]

b) Draw a dot-and-cross diagram to show the **full** electronic arrangement in a molecule of methane. [2 marks]

Q2 a) What type of bonding is present in the ammonium ion (NH_4^+)? [1 mark]

b) Explain how this type of bonding occurs. [2 marks]

Interesting fact #795 — $TiCl_4$ is known as 'tickle' in the chemical industry...

More pretty diagrams to learn here folks — practise till you get every single dot and cross in the right place. It's totally amazing to think of these titchy little atoms sorting themselves out so they've got full outer shells of electrons. Remember — covalent bonding happens between two non-metals, whereas ionic bonding happens between a metal and a non-metal.

Giant Covalent Lattices and Metallic Bonding

Atoms can form giant structures as well as piddling little molecules — well... 'giant' in molecular terms anyway. Compared to structures like the Eiffel Tower or even your granny's carriage clock, they're still unbelievably tiny.

Diamond *and* ***Graphite*** *are Giant Covalent Lattices*

1) **Giant covalent lattices** are huge networks of **covalently** bonded atoms. (They're sometimes called **macromolecular structures** too.)
2) **Carbon** atoms can form this type of structure because they can each form four strong, covalent bonds.

Diamond *is the* ***Hardest*** *known Substance*

Diamond is made up of **carbon atoms**. Each carbon atom is **covalently bonded** to **four** other carbon atoms. The atoms arrange themselves in a **tetrahedral** shape — its crystal lattice structure.

Diamond

Because of its **strong covalent** bonds:

1) Diamond has a **very high melting point** — it actually sublimes at over 3800 K.
2) Diamond is extremely **hard** — it's used in diamond-tipped drills and saws.
3) **Vibrations** travel easily through the stiff lattice, so it's a **good thermal conductor**.
4) It **can't conduct** electricity — all the outer electrons are held in localised bonds.
5) It won't dissolve in **any** solvent.

You can 'cut' diamond to form gemstones. Its structure makes it refract light a lot, which is why it sparkles.

'Sublimes' means it changes straight from a solid to a gas, skipping out the liquid stage.

Graphite *is another* ***Allotrope*** *of Carbon*

Allotropes are different forms of the **same element** in the **same state**. Carbon can form a number of different allotropes. Luckily, you only need to know about two of them — **diamond** and **graphite**.

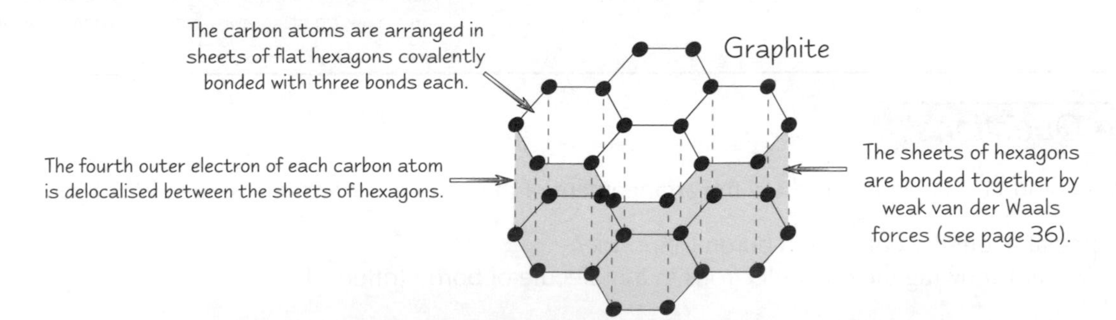

Graphite's **structure** means it has some **different properties** from diamond.

1) The weak bonds **between** the layers in graphite are easily broken, so the sheets can slide over each other — graphite feels **slippery** and is used as a **dry lubricant** and in **pencils**.
2) The **'delocalised'** electrons in graphite aren't attached to any particular carbon atom and are **free to move** along the sheets, so an **electric current** can flow.
3) The layers are quite **far apart** compared to the length of the covalent bonds, so graphite is **less dense** than diamond and is used to make **strong**, **lightweight** sports equipment.
4) Because of the **strong covalent bonds** in the hexagon sheets, graphite also has a **very high melting point** (it sublimes at over 3900 K).
5) Like diamond, graphite is **insoluble** in any solvent. The covalent bonds in the sheets are **too difficult** to break.

Giant Covalent Lattices and Metallic Bonding

Metals have Giant Structures Too

Metal elements exist as **giant metallic lattice structures**.

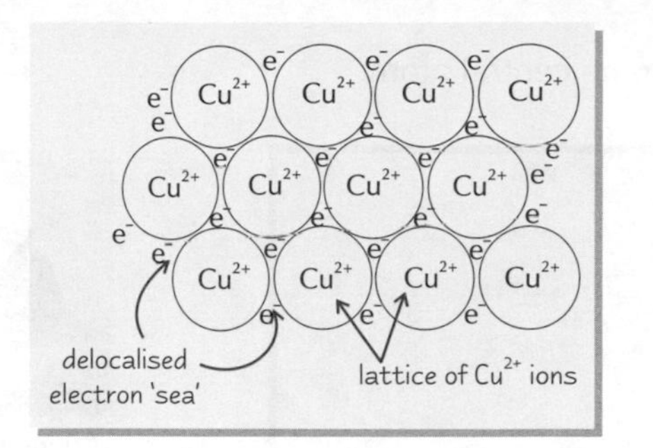

1) The electrons in the outermost shell of a metal atom are **delocalised** — the electrons are free to move about the metal. This leaves a **positive metal ion**, e.g. Na^+, Mg^{2+}, Al^{3+}.
2) The positive metal ions are **attracted** to the delocalised negative electrons. They form a lattice of closely packed positive ions in a **sea** of delocalised electrons — this is **metallic bonding**.

Metallic bonding explains why metals do what they do —

1) The **number of delocalised electrons per atom** affects the melting point. The **more** there are, the **stronger** the bonding will be and the **higher** the melting point. $\mathbf{Mg^{2+}}$ has **two** delocalised electrons per atom, so it's got a **higher melting point** than $\mathbf{Na^+}$, which only has **one**. The **size** of the metal ion and the **lattice structure** also affect the melting point.
2) As there are **no bonds** holding specific ions together, the metal ions can slide past each other when the structure is pulled, so metals are **malleable** (can be hammered into sheets) and **ductile** (can be drawn into a wire).
3) The delocalised electrons can pass **kinetic energy** to each other, making metals **good thermal conductors**.
4) Metals are **good electrical conductors** because the **delocalised electrons** can carry a **current**.
5) Metals are **insoluble**, except in **liquid metals**, because of the **strength** of the metallic bonds.

Practice Questions

Q1 How are the carbon sheets in graphite held together?

Q2 Diamond has a giant covalent lattice structure. Give two properties that it has as a result of this.

Q3 Why are metals malleable?

Exam Questions

Q1 Carbon can be found as the allotropes diamond and graphite.

a) What type of structure do diamond and graphite display? [1 mark]

b) Draw diagrams to illustrate the structures of diamond and graphite. [2 marks]

c) Compare and explain the electrical conductivities of diamond and graphite in terms of their structure and bonding. [4 marks]

Q2 Illustrate with a suitable labelled diagram the structure of copper and explain what is meant by metallic bonding. [4 marks]

Carbon is a girl's best friend...

Examiners love giving you questions on diamond and graphite. Close the book and do a quick sketch of each allotrope, together with a list of their properties — then look back at the page and see what you missed. It might be less fun than ironing your underwear, but it's much more useful and the only way to make sure you sparkle in the exam.

Shapes of Molecules

Chemistry would be heaps more simple if all molecules were flat. But they're not.

Molecular Shape** depends on **Electron Pairs** around the **Central Atom

Molecules and molecular ions come in loads of **different shapes**.
The shape depends on the **number of pairs** of electrons in the outer shell of the central atom.

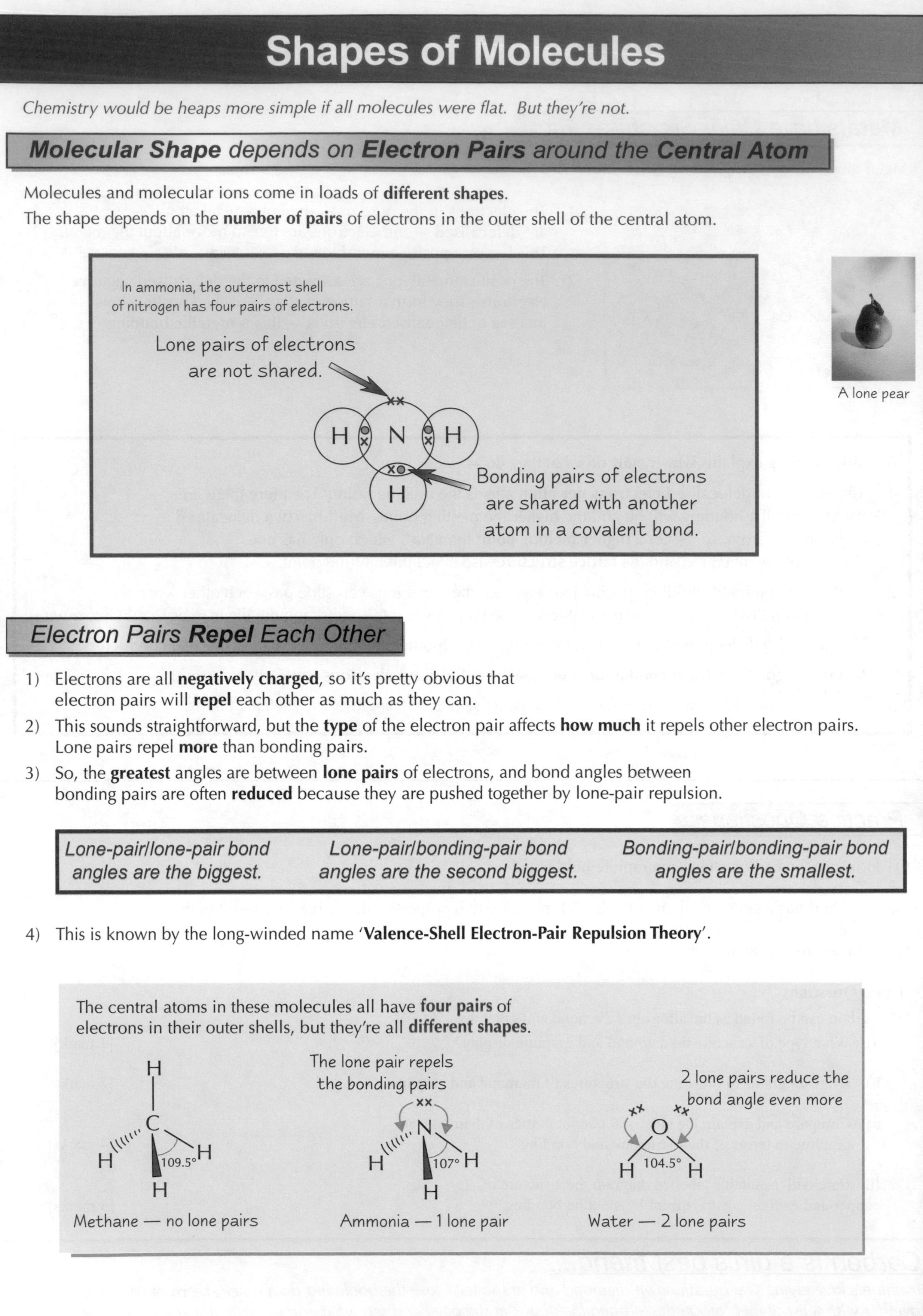

*Electron Pairs **Repel** Each Other*

1) Electrons are all **negatively charged**, so it's pretty obvious that electron pairs will **repel** each other as much as they can.
2) This sounds straightforward, but the **type** of the electron pair affects **how much** it repels other electron pairs. Lone pairs repel **more** than bonding pairs.
3) So, the **greatest** angles are between **lone pairs** of electrons, and bond angles between bonding pairs are often **reduced** because they are pushed together by lone-pair repulsion.

Lone-pair/lone-pair bond angles are the biggest.	*Lone-pair/bonding-pair bond angles are the second biggest.*	*Bonding-pair/bonding-pair bond angles are the smallest.*

4) This is known by the long-winded name '**Valence-Shell Electron-Pair Repulsion Theory**'.

The central atoms in these molecules all have **four pairs** of electrons in their outer shells, but they're all **different shapes**.

The lone pair repels the bonding pairs

2 lone pairs reduce the bond angle even more

109.5° 107° 104.5°

Methane — no lone pairs

Ammonia — 1 lone pair

Water — 2 lone pairs

Shapes of Molecules

Practise Drawing these Molecules

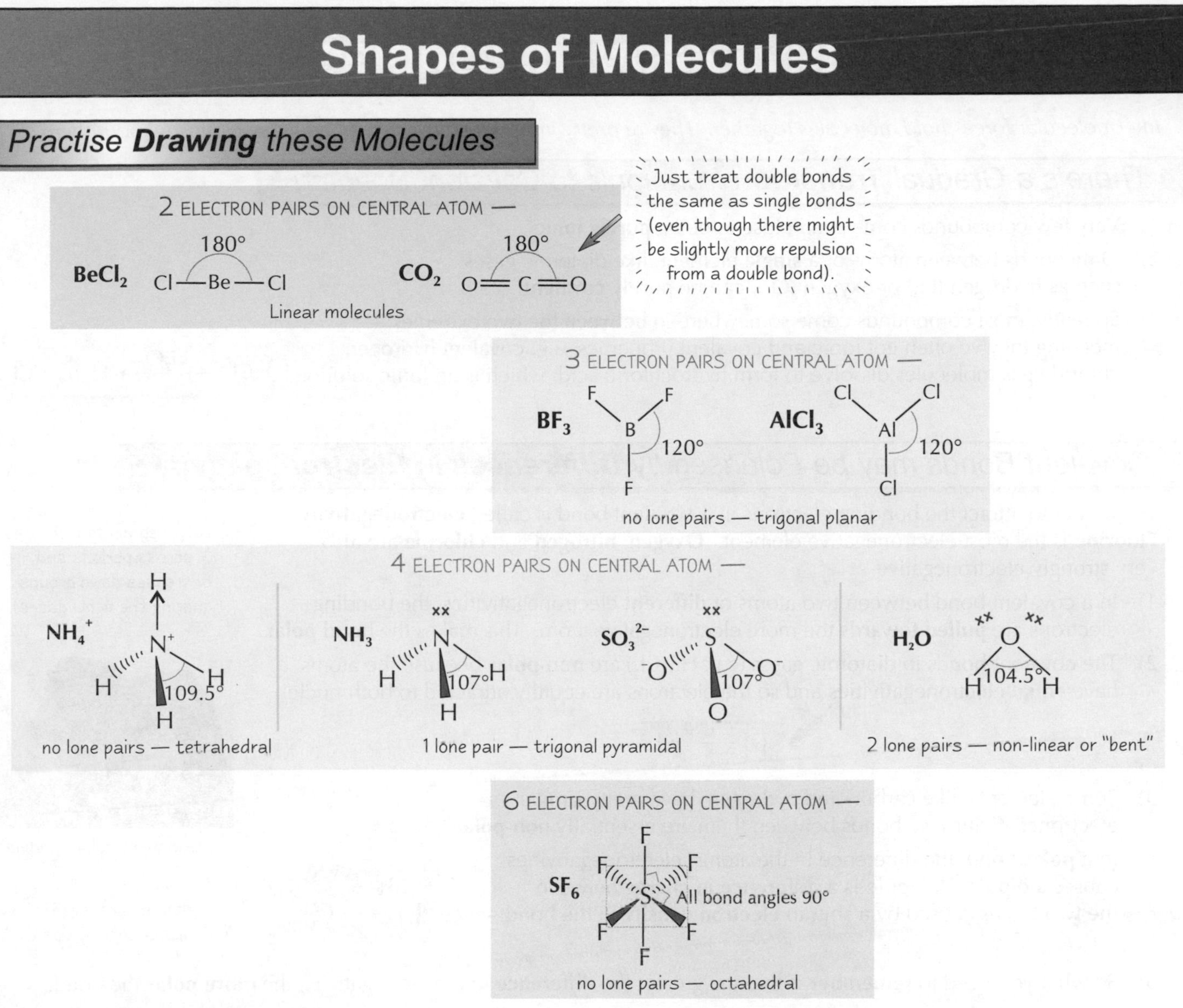

Practice Questions

Q1 What is a lone pair of electrons?

Q2 Explain why a water molecule is not linear.

Q3 Write down the order of the strength of repulsion between different kinds of electron pair.

Q4 Draw a tetrahedral molecule.

Exam Question

Q1 Nitrogen and boron can form the chlorides NCl_3 and BCl_3.

a) Draw 'dot and cross' diagrams to show the bonding in NCl_3 and BCl_3. [2 marks]

b) Draw the shapes of the molecules NCl_3 and BCl_3.
Show the approximate values of the bond angles on the diagrams and name each shape. [6 marks]

c) Explain why the shapes of NCl_3 and BCl_3 are different. [3 marks]

These molecules ain't square...

In the exam, those evil examiners might try to throw you by asking you to predict the shape of an unfamiliar molecule. Don't panic — it'll be just like one you do know, e.g. PH_3 is the same shape as NH_3. Make sure you can draw every single molecule on this page. Yep, that's right — from memory. And you need to know what the shapes are called too.

Electronegativity and Intermolecular Forces

Intermolecular forces hold molecules together. They're pretty important, cos we'd all be gassy clouds without them.

There's a Gradual **Transition** from Ionic to Covalent Bonding

1) Very few compounds come even close to being **purely ionic**.
2) Only bonds between atoms of a **single** element, like diatomic gases such as hydrogen (H_2) or oxygen (O_2), can be **purely covalent**.
3) So really, most compounds come somewhere **in between** the two extremes — meaning they've often got ionic **and** covalent properties, e.g. covalent hydrogen chloride gas molecules dissolve to form hydrochloric acid, which is an ionic solution.

$$HCl_{(g)} \xrightarrow{H_2O} H^+_{(aq)} + Cl^-_{(aq)}$$

Covalent Bonds may be Polarised by **Differences** in **Electronegativity**

The ability to attract the bonding electrons in a covalent bond is called **electronegativity**. **Fluorine** is the most electronegative element. **Oxygen**, **nitrogen** and **chlorine** are also very strongly electronegative.

Electronegativity increases across periods and decreases down groups (ignoring the noble gases).

1) In a covalent bond between two atoms of **different** electronegativities, the bonding electrons are **pulled towards** the more electronegative atom. This makes the bond **polar**.
2) The covalent bonds in diatomic gases (e.g. H_2, Cl_2) are **non-polar** because the atoms have **equal** electronegativities and so the electrons are equally attracted to both nuclei.

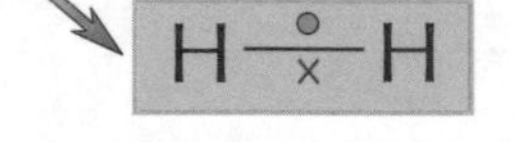

Permanent polar bonding

3) Some elements, like carbon and hydrogen, have pretty **similar** electronegativities, so bonds between them are essentially **non-polar**.
4) In a **polar bond**, the difference in the atoms' electronegativities causes a **dipole**. A dipole is a **difference in charge** between the two atoms caused by a shift in **electron density** in the bond.

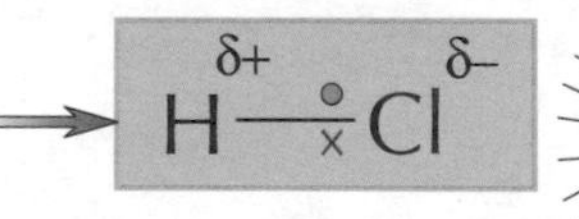

'δ' (delta) means 'slightly', so 'δ+' means 'slightly positive'.

5) So what you need to **remember** is that the greater the **difference** in electronegativity, the **more polar** the bond.

Polar Molecules Mean **Intermolecular** Attraction

Intermolecular forces are forces between molecules. They're much weaker than covalent, ionic or metallic bonds.

The **δ+** and **δ-** charges on **polar molecules** cause **weak electrostatic forces** of attraction **between** molecules. These are called **permanent dipole-dipole interactions**.

E.g. hydrogen chloride gas has polar molecules.

$$\overset{\delta+}{H}-\overset{\delta-}{Cl}\cdots\cdots\overset{\delta+}{H}-\overset{\delta-}{Cl}\cdots\cdots\overset{\delta+}{H}-\overset{\delta-}{Cl}$$

Now this is pretty cool:

If you put an **electrostatically charged rod** next to a polar liquid, like water, the liquid will **move** towards the rod. I wouldn't believe me either, but it's true. It's because **polar liquids** contain molecules with **permanent dipoles**. It doesn't matter if the rod is **positively** or **negatively** charged. The polar molecules in the liquid can **turn around** so the oppositely charged end is attracted towards the rod.

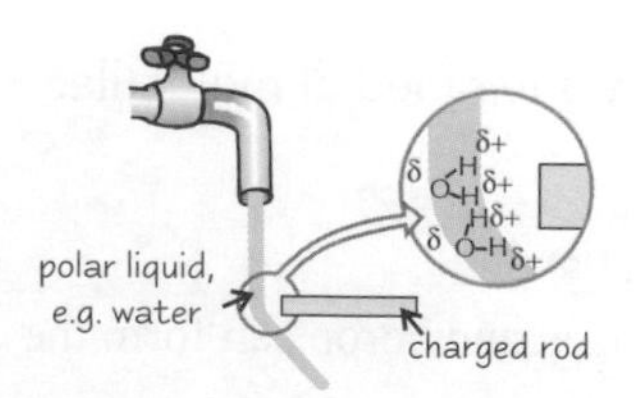

Intermolecular Forces are **Very Weak**

There are three types of **intermolecular force** you need to know about, but they're all very **weak** compared to the bonds **within** the molecule.

Sometimes the term 'van der Waals forces' is considered to include all three types of intermolecular force.

1) **Permanent dipole-dipole interactions** (see previous page)
2) **Hydrogen bonding** (this is the strongest type)
3) **Temporary dipole-induced dipole** or **van der Waals** forces — this is the weakest type

See page 36 for more info.

Electronegativity and Intermolecular Forces

Hydrogen Bonding is the Strongest Intermolecular Force

1) Hydrogen bonding can **only** happen when **hydrogen** is covalently bonded to **fluorine**, **nitrogen** or **oxygen**. Hydrogen has a **high charge density** because it's so small and fluorine, nitrogen and oxygen are very **electronegative**. The bond is so **polarised** that the hydrogen of one molecule forms a weak bond with the fluorine, nitrogen or oxygen of **another molecule**.
2) Molecules which have hydrogen bonding are usually **organic**, containing **-OH** or **-NH** groups.

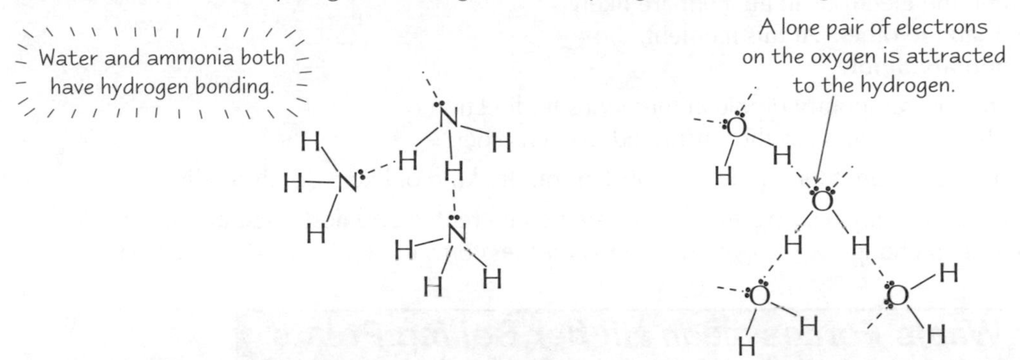

3) Hydrogen bonding has a huge effect on the properties of substances. They are **soluble** in water and have **higher boiling and freezing points** than non-polar molecules of a similar size.

Water, ammonia and **hydrogen fluoride** generally have the **highest boiling points** if you compare them with other hydrides in their groups, because of the **extra energy** needed to break the H bonds.

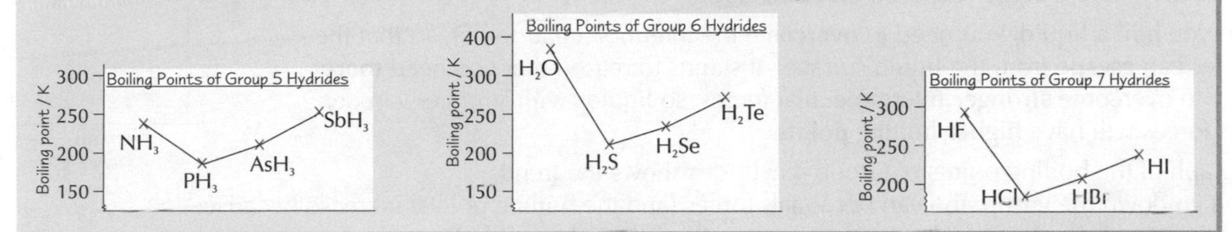

4) In ice, molecules of H_2O are held together in a **lattice** by hydrogen bonds. And because hydrogen bonds are relatively **long**, ice is **less dense** than liquid water.

This is unusual... most substances get denser when they freeze.

Practice Questions

Q1 What are the only bonds which can be purely covalent?

Q2 What is the most electronegative element?

Q3 What is a dipole?

Q4 What atoms must be covalently bonded together for hydrogen bonding to exist?

Exam Questions

Q1 Many covalent molecules have a permanent dipole, due to differences in electronegativities.

a) Define the term electronegativity. [2 marks]

b) Draw the shapes and marking any bond polarities clearly on your diagram: (i) Br_2 (ii) H_2O (iii) NH_3 [5 marks]

Q2 a) Name three types of intermolecular force. [3 marks]

b) Water, H_2O, boils at 373 K.

(i) Explain why water's boiling point is higher than expected in comparison to other similar molecules. [2 marks]

(ii) Draw a labelled diagram showing the intermolecular bonding that takes place in water. [2 marks]

Enough of this chemistry rubbish. Here are some interesting facts...

If you chop the head off a beetle, it wouldn't die of being beheaded, but actually starvation. It's true. If you ate 14 lbs of almonds, you'd die of cyanide poisoning. It's true! Daddy-long-legs are actually the most poisonous insects in the world, but they can't pierce the skin... it's TRUE. Every night, the human body sweats enough to fill a swimming pool. It's true...

Van der Waals Forces

Hang on in there — you're almost at the end of the section. If you can just hold out for another two pages...

Van der Waals Forces are Found Between All Atoms and Molecules

Van der Waals forces cause **all** atoms and molecules to be **attracted** to each other.
Even **noble gas atoms** are affected, despite not being at all interested in forming other types of bond.

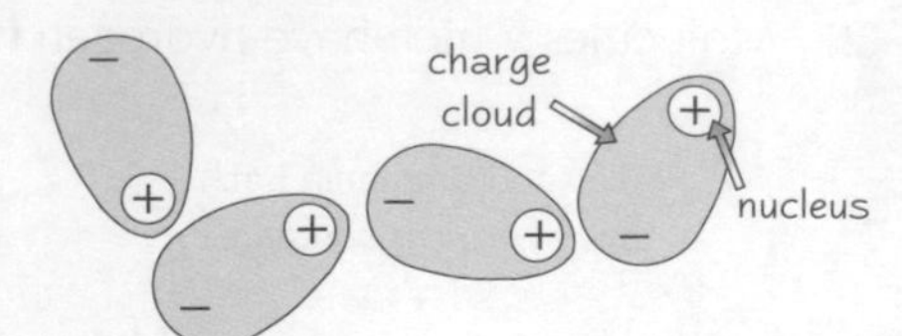

1) **Electrons** in charge clouds are always **moving** really quickly. At any particular moment, the electrons in an atom are likely to be more to one side than the other. At this moment, the atom would have a **temporary dipole**.
2) This dipole can cause **another** temporary dipole in the opposite direction on a neighbouring atom. The two dipoles are then **attracted** to each other.
3) The second dipole can cause yet another dipole in a **third atom**. It's kind of like a domino rally.
4) Because the electrons are constantly moving, the dipoles are being **created** and **destroyed** all the time. Even though the dipoles keep changing, the **overall effect** is for the atoms to be **attracted** to each other.

Stronger Van der Waals Forces mean Higher Boiling Points

1) Not all van der Waals forces are the same strength — larger molecules have **larger electron clouds**, meaning **stronger** van der Waals forces. Molecules with greater **surface areas** also have stronger van der Waals forces because they have a **bigger exposed electron cloud**.

Van der Waals forces affect other physical properties, such as melting point and viscosity, too.

2) When you **boil** a liquid, you need to **overcome** the intermolecular forces, so that the particles can **escape** from the liquid surface. It stands to reason that you need **more energy** to overcome **stronger** intermolecular forces, so liquids with stronger van der Waals forces will have **higher boiling points**.

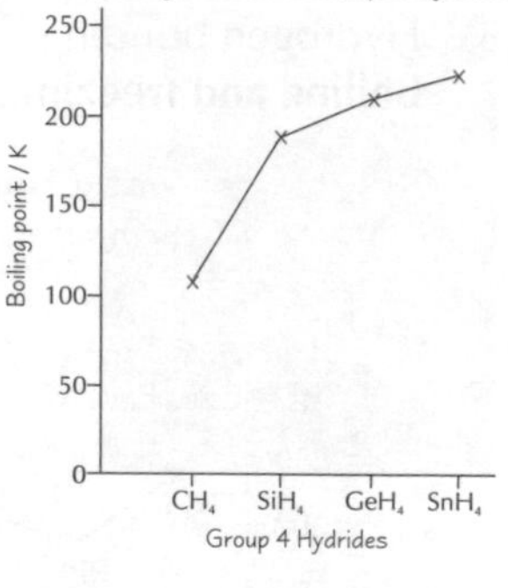

3) This graph of the boiling points of Group 4 hydrides shows the trend. As you go down the group, the van der Waals forces (and the boiling points) increase because:
   (i) the **atomic/molecular size** increases, (ii) the number of **shells** of electrons increases.

Van der Waals Forces Can Hold Molecules in a Lattice

Van der Waals forces are responsible for holding **iodine** molecules together in a **lattice**.

1) Iodine atoms are held together in pairs by **strong** covalent bonds to form molecules of $\mathbf{I_2}$.
2) But the molecules are then held together in a **molecular lattice** arrangement by **weak** van der Waals attractions.

Covalent Bonds Don't Break during Melting and Boiling

Except for giant molecular substances, like diamond.

This is something that confuses loads of people — get it sorted in **your** head now...

1) To **melt** or **boil** a simple covalent compound you only have to overcome the **van der Waals forces** or **hydrogen bonds** that hold the molecules together.
2) You **don't** need to break the much stronger covalent bonds that hold the atoms together in the molecules.
3) That's why simple covalent compounds have relatively **low melting** and **boiling points**.

When you boil water, you don't get hydrogen and oxygen.

For example:

Chlorine, Cl_2, has **stronger** covalent bonds than bromine, Br_2.
But under normal conditions, chlorine is a **gas** and bromine a **liquid**.
Bromine has the higher boiling point because its molecules are **bigger**, giving stronger van der Waals forces.

Van der Waals Forces

Learn the **Properties** of the Main Substance Types

Nearly finished... but not quite. Before you mentally clock off, make sure you know how all the various types of attraction between atoms and molecules affect a substance's **properties**. You need to know this table like the back of your spam...

Bonding	Examples	Melting and boiling points	Typical state at STP	Does solid conduct electricity?	Does liquid conduct electricity?	Is it soluble in water?
Ionic	NaCl $MgCl_2$	High	Solid	No (ions are held firmly in place)	Yes (ions are free to move)	Yes
Simple molecular (covalent)	CO_2 I_2 H_2O	Low (have to overcome van der Waals forces or hydrogen bonds, not covalent bonds)	Sometimes solid, usually liquid or gas (water is liquid because it has hydrogen bonds)	No	No	Depends on how polarised the molecule is
Giant covalent lattice	Diamond Graphite	High	Solid	No (except graphite)	— (will generally sublime)	No
Metallic	Fe Mg Al	High	Solid	Yes (delocalised electrons)	Yes (delocalised electrons)	No

Practice Questions

Q1 Explain why van der Waals attractions are present even in neutral atoms like argon.

Q2 Describe some of the effects of van der Waals forces on the physical properties of substances.

Q3 Describe two factors that affect the size of van der Waals forces.

Q4 What types of bond must be overcome in order for a simple molecular substance to boil or melt?

Q5 Do ionic compounds conduct electricity?

Q6 Why can metals conduct electricity?

Exam Questions

Q1

Substance	Melting point	Electrical conductivity of solid	Electrical conductivity of liquid	Solubility in water
A	High	Poor	Good	Soluble
B	Low	Poor	Poor	Insoluble
C	High	Good	Good	Insoluble
D	Very High	Poor	Poor	Insoluble

a) Identify the type of crystal structure present in each substance, A to D. [4 marks]

b) Which substance is most likely to be:
(i) diamond, (ii) aluminium, (iii) sodium chloride and (iv) iodine? [2 marks]

Q2 Explain the electrical conductivity of magnesium, sodium chloride and graphite.
In your answer you should consider the structure and bonding of each of these materials. [12 marks]

Van der Waal — a Dutch hit for Oasis...

You need to learn the info in the table above. With a quick glance in my crystal ball, I can almost guarantee you'll need a bit of it in your exam... let me look a bit closer and tell you which bit... mmm.... Nah — it's clouded over... you'll have to learn the lot. Sorry. Tell you what — close the book and see how much of the table you can scribble out from memory.

The Periodic Table

As far as Chemistry topics go, the Periodic Table is a bit of a biggie. So much so that they even want you to know the history of it. So make yourself comfortable and I'll tell you a story that began... oh, about 200 years ago...

*In the **1800s**, Elements Could Only Be Grouped by **Atomic Mass***

1) In the early 1800s, there were only two ways to categorise elements — by their **physical and chemical properties** and by their **relative atomic mass**. (The modern periodic table is arranged by **proton number**, but back then, they knew nothing about protons or electrons. The only thing they could measure was relative atomic mass.)
2) In 1817, Johann Döbereiner attempted to group similar elements — these groups were called **Döbereiner's triads**. He saw that **chlorine**, **bromine** and **iodine** had similar characteristics. He also realised that other properties of bromine (e.g. atomic weight) fell **halfway** between those of chlorine and iodine. He found other such groups of three elements (e.g. lithium, sodium and potassium), and called them **triads**. It was a start.
3) An English chemist called **John Newlands** had the first good stab at making a table of the elements in 1863. He noticed that if he arranged the elements in order of **mass**, similar elements appeared at regular intervals — every **eighth element** was similar. He called this the **law of octaves**, and he listed some known elements in rows of seven so that the similar elements lined up in columns.

Li	Be	B	C	N	O	F
Na	Mg	Al	Si	P	S	Cl

4) The problem was, the pattern broke down on the third row, with many transition metals like Fe, Cu and Zn messing it up completely.

Dmitri Mendeleev** Created the **First Accepted Version

1) In 1869, Russian chemist **Dmitri Mendeleev** produced a much better table, which wasn't far off the one we have today.
2) He arranged all the known elements by atomic mass (like Newlands did), but the clever thing he did was to leave **gaps** in the table where the next element didn't seem to fit. By putting in gaps, he could keep elements with similar chemical properties in the same group.
3) He also predicted the properties of **undiscovered elements** that would go in the gaps.
4) When elements were **later discovered** (e.g. germanium, scandium and gallium) with properties that matched Mendeleev's predictions, it showed that clever old Mendeleev had got it right.

	Group 1	Group 2	Group 3	Group 4	Group 5	Group 6	Group 7
Period 1	H						
Period 2	Li	Be	B	C	N	O	F
Period 3	Na	Mg	Al	Si	P	S	Cl
Period 4	K Cu	Ca Zn	* *	Ti *	V As	Cr Se	Mn Br
Period 5	Rb Ag	Sr Cd	Y In	Zr Sn	Nb Sb	Mo Te	* I

What do you think of the table? I made it myself...

Oh, Dmitri, I love it...

*The **Modern Periodic Table** Arranges Elements by **Proton Number***

The modern Periodic Table is pretty much the one produced by Henry Moseley in 1914.

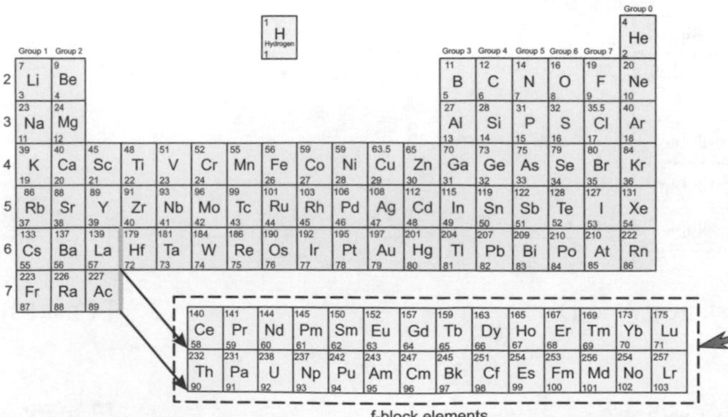

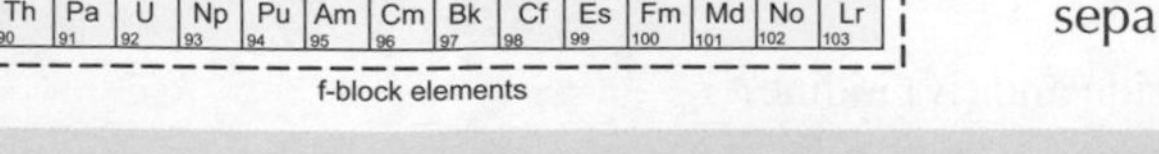

1) He arranged the elements according to **atomic number** rather than by mass.
2) This fixed a few elements that Mendeleev had put out of place using atomic mass.
3) He also added the **noble gases** (Group 0) which had been discovered in the 1890s.
4) The final big change was a result of the work of **Glenn Seaborg**. He suggested how the **f-block** elements fit into the Periodic Table (though they're usually shown separated from the main part of the table).

1) The modern Periodic Table is arranged into **periods** (rows) and **groups** (columns).
2) All the elements **within a period** have the same number of **electron shells** (if you don't worry about the sub-shells)
— the elements of Period 1 (hydrogen and helium) both have 1 electron shell.
— the elements in Period 2 have 2 electron shells. And so on down the table...
3) All the elements **within a group** have the same number of **electrons in their outer shell**. This means they have similar physical and chemical properties. The group number tells you the number of electrons in the outer shell, e.g. Group 1 elements have 1 electron in their outer shell, Group 4 elements have 4 electrons, and so on...

The Periodic Table

You can use the Periodic Table to work out **Electron Configurations**

The Periodic Table can be split into an **s block**, **d block** and **p block** like this: Doing this shows you which sub-shells all the electrons go into.

See page 22 if this sub-shell malarkey doesn't ring a bell.

1s: 1 H Hydrogen 1

s block

2s	7 Li 3	9 Be 4
3s	23 Na 11	24 Mg 12
4s	39 K 19	40 Ca 20
5s	86 Rb 37	88 Sr 38
6s	133 Cs 55	137 Ba 56
7s	223 Fr 87	226 Ra 88

d block

3d	45 Sc 21	48 Ti 22	51 V 23	52 Cr 24	55 Mn	56 Fe 26	59 Co 27	59 Ni 28	63.5 Cu 29	65 Zn 30
4d	89 Y 39	91 Zr 40	93 Nb 41	96 Mo 42	99 Tc 43	101 Ru 44	103 Rh 45	106 Pd 46	108 Ag 47	112 Cd 48
5d	57-71 Lanthanides	179 Hf 72	181 Ta 73	184 W 74	186 Re 75	190 Os 76	192 Ir 77	195 Pt 78	197 Au 79	201 Hg 80
6d	89-103 Actinides									

p block

						4 He 2
2p	11 B 5	12 C 6	14 N 7	16 O 8	19 F 9	20 Ne 10
3p	27 Al 13	28 Si 14	31 P 15	32 S 16	35.5 Cl 17	40 Ar 18
4p	70 Ga 31	73 Ge 32	75 As 33	79 Se 34	80 Br 35	84 Kr 36
5p	115 In 49	119 Sn 50	122 Sb 51	128 Te 52	127 I 53	131 Xe 54
6p	204 Tl 81	207 Pb 82	209 Bi 83	210 Po 84	210 At 85	222 Rn 86

1) The **s-block** elements have an outer shell electron configuration of s^1 or s^2.

   Examples: Lithium ($1s^2$ $\mathbf{2s^1}$) and magnesium ($1s^2$ $2s^2$ $2p^6$ $\mathbf{3s^2}$)

2) The **p-block** elements have an outer shell configuration of s^2p^1 to s^2p^6.

   Example: Chlorine ($1s^2$ $2s^2$ $2p^6$ $\mathbf{3s^2}$ $\mathbf{3p^5}$)

3) The **d-block** elements have electron configurations in which d sub-shells are being filled.

   Example: Cobalt ($1s^2$ $2s^2$ $2p^6$ $3s^2$ $3p^6$ $\mathbf{3d^7}$ $4s^2$)

   Even though the 3d sub-shell fills last in cobalt, it's not written at the end of the line.

When you've got the Periodic Table **labelled** with the **shells** and **sub-shells** like the one up there, it's pretty easy to read off the electron structure of any element by starting at the top and working your way across and down until you get to your element.

Example

Electron structure of phosphorus (P):

Period 1 — $1s^2$ ← Complete sub-shells
Period 2 — $2s^2$ $2p^6$ ←
Period 3 — $3s^2$ $3p^3$ ← Incomplete outer sub-shell

A wee apology...
This bit's really hard to explain clearly in words. If you're confused, just look at the examples until you get it...

Practice Questions

Q1 In what ways is Newlands' 'periodic table' not as good as Mendeleev's?

Q2 In what order did Mendeleev originally set out the elements?

Q3 In what order are the elements set out in the modern Periodic Table? Who was the first to do this?

Q4 What is the name given to the columns in the Periodic Table?

Q5 What is the name given to the rows in the Periodic Table? *(Err, hello — easy questions alert.)*

Exam Question

Q1 a) Complete the electronic configuration of sodium: $1s^2$ ____________________

b) State the block in the Periodic Table to which sodium belongs.

c) Complete the electronic configuration of bromine: $1s^2$ ____________________

d) State the block in the Periodic Table to which bromine belongs. [4 marks]

Periodic — probably the best table in the world...*

Dropped History for AS Chemistry, did you... Ha, bet you're regretting that now, aren't you. If so, you'll enjoy the free History lesson that you get here with the Periodic Table. Make sure you learn all the key details and particularly how to spell Mendeleev. This stuff's not here for fun — it's here because you're gonna get questions on it.

*Excluding Dinner and the Round, of course.

Periodic Trends

Periodicity is one of those words you hear a lot in Chemistry without ever really knowing what it means. Well it basically means the trends that occur (in physical and chemical properties) as you move across the periods. E.g. Metal to non-metal is a trend that occurs going left to right in each period...

Atomic Radius **Decreases** across a Period

1) As the number of protons increases, the **positive charge** of the nucleus increases. This means electrons are **pulled closer** to the nucleus, making the atomic radius smaller.
2) The extra electrons that the elements gain across a period are added to the **outer energy level** so they don't really provide any extra shielding effect (shielding works with inner shells mainly).

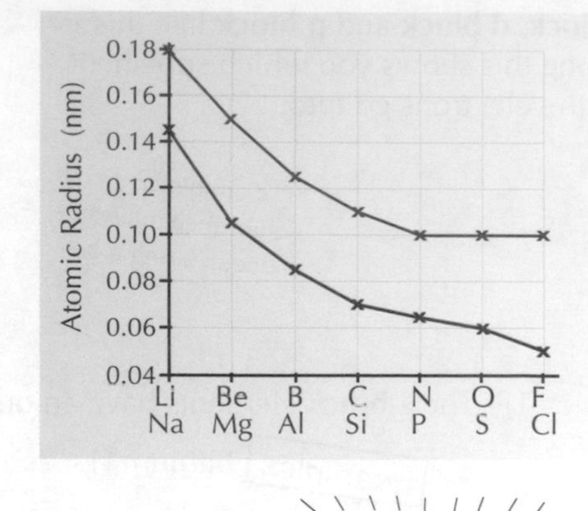

Ionisation Energy **Increases** across a Period...

Don't forget — there are **3 main things** that affect the size of ionisation energies:

1) **Atomic radius** — the further the outer shell electrons from the nucleus, the lower the ionisation energy.
2) **Nuclear charge** — the **more protons** in the nucleus, the higher the ionisation energy.
3) **Electron shielding** — the more inner shells there are, the more shielding there is, and the lower the ionisation energy.

See page 24 for more on ionisation energies.

The graph below shows the first ionisation energies of the elements in **Periods 2 and 3**.

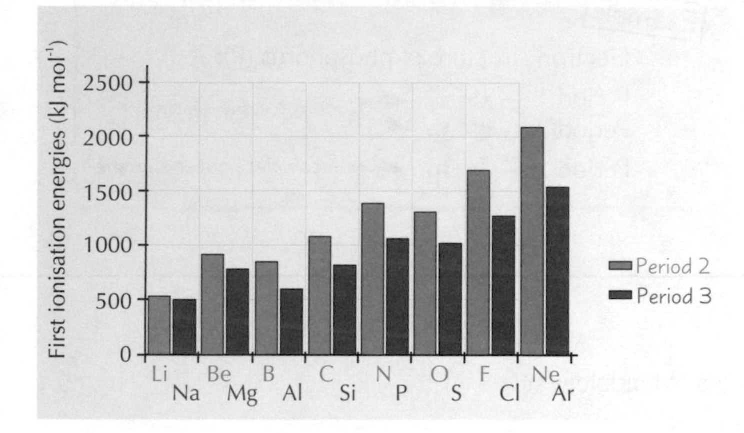

1) As you **move across** a period, the **general trend** is for the ionisation energies to **increase** — i.e. it gets harder to remove the outer electrons.
2) This is because the number of protons is increasing, which means a stronger **nuclear attraction**.
3) All the extra electrons are at **roughly the same** energy level, even if the outer electrons are in different orbital types.
4) This means there's generally little **extra shielding** effect or **extra distance** to lessen the attraction from the nucleus.

...and **Decreases** Down a Group

As you **go down** a group in the Periodic Table, ionisation energies generally **fall**, i.e. it gets **easier** to remove outer electrons. This is because:

- Elements further down a group have **extra electron shells** compared to ones above. The extra shells mean that the outer electrons are **further away** from the nucleus, which greatly reduces the attraction to the nucleus.
- The extra inner shells **shield** the outer electrons from the attraction of the nucleus.

The positive charge of the nucleus does increase as you go down a group (due to the extra protons), but this effect is overridden by the effect of the extra shells.

First ionisation energies of the first five elements of Group 1.

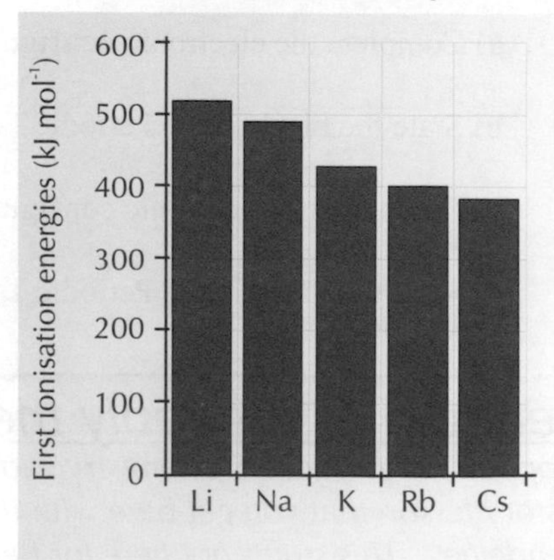

Periodic Trends

Melting and Boiling Points are linked to **Bond Strength** and **Structure**

Periods 2 and 3 show similar trends in their melting and boiling points. These trends are linked to changes in **structure** and **bond strength**.

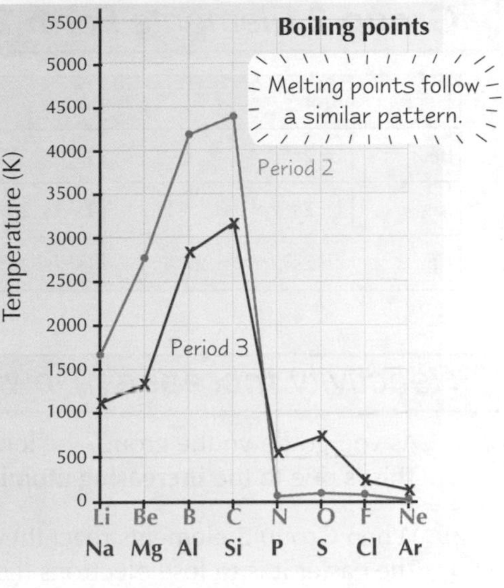

1) For the **metals** (Li and Be, Na, Mg and Al), melting and boiling points **increase** across the period because the **metal-metal bonds** get stronger. The bonds get stronger because the metal ions have an increasing number of **delocalised electrons** and a decreasing **ionic radius**. This leads to a higher charge density, which attracts the ions together more strongly.

2) The elements with **macromolecular** structures have **strong covalent bonds** linking all their atoms together. **A lot** of energy is needed to break these bonds. So, for example, carbon (as graphite or diamond) and silicon have the **highest** melting and boiling points in their periods. (The carbon data in the graph opposite is for graphite — diamond has an even higher boiling point. But neither of them actually melts or boils at atmospheric pressure, they sublime from solid to gas.)

3) Next come the **simple molecular substances** (N_2, O_2 and F_2, P_4, S_8 and Cl_2). Their melting and boiling points depend upon the strength of the **van der Waals forces** (see p36) between their molecules. Van der Waals forces are weak and easily overcome so these elements have **low** melting and boiling points.

4) More atoms in a molecule mean stronger van der Waals forces. For example, in Period 3 sulfur is the **biggest molecule** (S_8), so it's got higher melting and boiling points than phosphorus or chlorine.

5) The noble gases (neon and argon) have the **lowest** melting and boiling points because they exist as **individual atoms** (they're monatomic) resulting in **very weak** van der Waals forces.

Practice Questions

Q1 Name three factors that affect the size of ionisation energies.

Q2 How does the first ionisation energy change as you go across a period?

Q3 Which element in Period 3 has the highest melting point? Which has the highest boiling point?

Q4 Why does phosphorus have a lower melting point than magnesium?

Exam Questions

Q1 Explain why first ionisation energies show an overall tendency to increase across a period. [3 marks]

Q2 Explain why the melting point of magnesium is higher than that of sodium. [3 marks]

Q3 This table shows the melting points for the Period 3 elements.

Element	Na	Mg	Al	Si	P	S	Cl	Ar
Melting point / K	371	923	933	1680	317	392	172	84

In terms of structure and bonding explain why:

a) silicon has a high melting point. [2 marks]

b) the melting point of sulfur is higher than that of phosphorus. [2 marks]

Q4 State and explain the trend in atomic radius across Period 3. [4 marks]

Periodic trends — my mate Dom's always a decade behind...

*He still thinks Oasis, Blur and REM are the best bands around. The sad muppet. But not me. Oh no sirree, I'm up with the times — April Lavigne... Linkin' Pork... Christina Agorrilla. I'm hip, I'm with it. Da ga da ga da ga da ga... ooaarrr ooup **

** Obscure reference to Austin Powers: International Man of Mystery. You should watch it — it's better than doing Chemistry.*

Group 2 — The Alkaline Earth Metals

It would be easy for Group 2 elements to feel slightly inferior to those in Group 1. They're only in the second group, after all. That's why you should try to get to know and like them. They'd really appreciate it, I'm sure.

Group 2 Elements Form 2+ Ions

Element	Atom	Ion
Be	$1s^2 2s^2$	$1s^2$
Mg	$1s^2 2s^2 2p^6 3s^2$	$1s^2 2s^2 2p^6$
Ca	$1s^2 2s^2 2p^6 3s^2 3p^6 4s^2$	$1s^2 2s^2 2p^6 3s^2 3p^6$

Group 2 elements all have two electrons in their outer shell (s^2). They lose their two outer electrons to form **2+ ions**. Their ions then have every atom's dream electronic structure — that of a **noble gas**.

Reactivity Increases Down Group 2

1) As you go down the group, the **ionisation energies** decrease. This is due to the **increasing atomic radius** and **shielding effect** (see p40).
2) When Group 2 elements react they **lose electrons**, forming positive ions (**cations**). The easier it is to lose electrons (i.e. the lower the first and second ionisation energies), the more reactive the element, so **reactivity increases** down the group.

Mr Kelly has one final attempt at explaining electron shielding to his students...

Group 2 Elements React with Water and Oxygen

When Group 2 elements react, they are **oxidised** from a state of **0** to **+2**, forming M^{2+} ions. This is because Group 2 atoms contain 2 electrons in their outer shell.

	$M \rightarrow M^{2+} + 2e^-$	E.g. $Ca \rightarrow Ca^{2+} + 2e^-$
Oxidation number:	0 → +2	0 → +2

1) **GROUP 2 ELEMENTS REACT WITH WATER TO PRODUCE HYDROXIDES**

The Group 2 metals react with water to give a **metal hydroxide and hydrogen**.

$$M_{(s)} + 2H_2O_{(l)} \rightarrow M(OH)_{2\,(aq)} + H_{2\,(g)}$$

Oxidation number: M: 0 → +2

e.g. $Ca_{(s)} + 2H_2O_{(l)} \rightarrow Ca(OH)_{2\,(aq)} + H_{2\,(g)}$

Be	doesn't react
Mg	VERY slowly
Ca	steadily
Sr	fairly quickly
Ba	rapidly

These are redox reactions — see p20 for more info.

2) **THEY BURN IN OXYGEN TO FORM OXIDES**

When Group 2 metals burn in oxygen, you get solid white **oxides**.

$$2M_{(s)} + O_{2\,(g)} \rightarrow 2MO_{(s)}$$

Oxidation number of metal: 0 → +2

Oxidation number of oxygen: 0 → −2

e.g. $2Ca_{(s)} + O_{2\,(g)} \rightarrow 2CaO_{(s)}$ — Ca: 0 → +2; O: 0 → −2

Group 2 Oxides and Hydroxides are Bases

They form Alkaline Solutions in Water...

1) The **oxides** of the Group 2 metals react readily with **water** to form **metal hydroxides**, which dissolve. The **hydroxide ions, OH^-**, make these solutions **strongly alkaline** (e.g. pH 12 - 13).
2) Magnesium oxide is an exception — it only reacts slowly and the hydroxide isn't very soluble.
3) The oxides form **more strongly alkaline** solutions as you go down the group, because the hydroxides get more soluble.

$$CaO_{(s)} + H_2O_{(l)} \rightarrow Ca^{2+}_{(aq)} + 2OH^-_{(aq)}$$

Group 2 — The Alkaline Earth Metals

Thermal Stability of *Carbonates* Changes Down the Group

Thermal decomposition is when a substance **breaks down** (decomposes) when **heated**. The more thermally stable a substance is, the more heat it will take to break it down. Here's how it goes for **Group 2 carbonates...**

1) **Group 2 carbonates decompose to form the oxide and carbon dioxide.**

$$MCO_{3\,(s)} \rightarrow MO_{(s)} + CO_{2\,(g)}$$

e.g. $$CaCO_{3\,(s)} \rightarrow CaO_{(s)} + CO_{2\,(g)}$$

2) **Thermal stability increases down the group.**
So, it's take **more heat** to decompose, say, calcium carbonate than magnesium carbonate.

Group 2 Compounds are used to Neutralise Acidity

Group 2 elements are known as the **alkaline earth metals**, and many of their common compounds are used for neutralising acids. Here are a couple of common examples:

1) Calcium hydroxide (slaked lime, $Ca(OH)_2$) is used in **agriculture** to neutralise acid soils.

Daisy the cow*

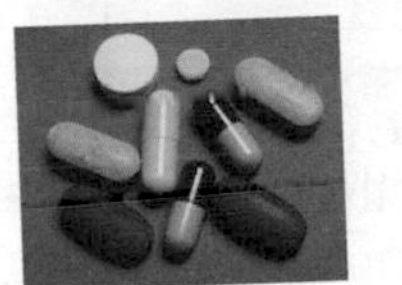

2) Magnesium hydroxide ($Mg(OH)_2$) is used in some indigestion tablets as an **antacid**.

In both cases, the ionic equation for the neutralisation is
$\mathbf{H^+_{(aq)} + OH^-_{(aq)} \rightarrow H_2O_{(l)}}$

*She wanted to be in the book. I said OK.

Practice Questions

Q1 Which is the least reactive metal in Group 2?

Q2 Why does reactivity with water increase down Group 2?

Q3 Which of the following increases in size down Group 2? **atomic radius, first ionisation energy**

Q4 Give a use of magnesium hydroxide.

Q5 Write an equation for the thermal decomposition of calcium carbonate.

Exam Questions

Q1 The reactivity of an element depends on its ionisation energies. Explain the difference in first ionisation energies of magnesium and calcium. [4 marks]

Q2 Calcium (Ca) can be burned in oxygen.
a) Write an equation for the reaction. [1 mark]
b) Show the change in oxidation state of calcium. [1 mark]
c) Predict the appearance of the product. [2 marks]
d) What type of bonding does the product have? [1 mark]

Q3 The table shows the atomic radii of three elements from Group 2.

Element	Atomic radius (nm)
X	0.089
Y	0.198
Z	0.176

a) Predict which element would react most rapidly with water. [1 mark]
b) Explain your answer. [2 marks]

I'm not gonna make it. You've gotta get me out of here, Doc...

We're deep in the dense jungle of Inorganic Chemistry now. Those carefree days of Section Two are well behind us. It's now an endurance test and you've just got to keep going. By now, all the facts are probably blurring into one. It's tough, but you've got to stay awake, stay focused and keep learning. That's all you can do.

Group 7 — The Halogens

Now you can wave goodbye to those pesky s-block elements. Here come the halogens.

Halogens are the **Highly Reactive Non-Metals** of Group 7

The table below gives some of the main properties of the first 4 halogens.

halogen	formula	colour	physical state	electronic structure
fluorine	F_2	pale yellow	gas	$1s^2\ 2s^2\ 2p^5$
chlorine	Cl_2	green	gas	$1s^2\ 2s^2\ 2p^6\ 3s^2\ 3p^5$
bromine	Br_2	red-brown	liquid	$1s^2\ 2s^2\ 2p^6\ 3s^2\ 3p^6\ 3d^{10}\ 4s^2\ 4p^5$
iodine	I_2	grey	solid	$1s^2\ 2s^2\ 2p^6\ 3s^2\ 3p^6\ 3d^{10}\ 4s^2\ 4p^6\ 4d^{10}\ 5s^2\ 5p^5$

The boiling and melting points of the halogens increase down the group. This is due to the increasing strength of the **van der Waals forces** as the size and relative mass of the atoms increases. This trend is shown in the changes of **physical state** from chlorine (gas) to iodine (solid). (A substance is said to be **volatile** if it has a low boiling point. So volatility **decreases** down the group.)

The word halogen should be used when describing the atom (X) or molecule (X_2), but the word halide is used to describe the negative ion (X^-).

Halogens get **Less Reactive** Down the Group

$$X + e^- \rightarrow X^-$$
ox. number: 0 → −1

1) Halogen atoms react by **gaining an electron** in their outer shell. This means they're **reduced**. As they're reduced, they **oxidise** another substance (it's a redox reaction) — so they're **oxidising agents**.
2) As you go down the group, the atoms become **larger** so the outer electrons are **further** from the nucleus. The outer electrons are also **shielded** more from the attraction of the positive nucleus, because there are more inner electrons. This makes it harder for larger atoms to attract the electron needed to form an ion (despite the increased charge on the nucleus), so larger atoms are less reactive.
3) Another way of saying that the halogens get **less reactive** down the group is to say that they become **less oxidising**.

Halogens **Displace** Less Reactive Halide Ions from Solution

1) The halogens' **relative oxidising strengths** can be seen in their **displacement reactions** with halide ions. For example, if you mix bromine water, $Br_{2\,(aq)}$, with potassium iodide solution, the bromine displaces the iodide ions (it oxidises them), giving iodine (I_2) and potassium bromide solution, $KBr_{(aq)}$.

$$Br_{2(aq)} + 2I^-_{(aq)} \rightarrow 2Br^-_{(aq)} + I_{2(aq)}$$

Oxidation number of Br: **0** → **−1**
Oxidation number of I: **−1** → **0**

2) When these displacement reactions happen, there are **colour changes** — you can see what happens by following them. Iodine water ($I_{2\,(aq)}$) is **brown** and bromine water ($Br_{2\,(aq)}$) is **orange**.
3) You can make the changes easier to see by shaking the reaction mixture with an **organic solvent** like hexane. The halogen that's present will dissolve readily in the organic solvent, which settles out as a distinct layer above the aqueous solution. A **violet/pink** colour shows the presence of **iodine**. An **orange/red** colour shows **bromine**, and a **very pale yellow/green** shows **chlorine**.

hexane layer
aqueous layer

4) Here are the colour changes you'll see:

	Potassium chloride solution **$KCl_{(aq)}$** – colourless	Potassium bromide solution **$KBr_{(aq)}$** – colourless	Potassium iodide solution **$KI_{(aq)}$** – colourless
Chlorine water **$Cl_{2\,(aq)}$** – colourless	no reaction	orange/red solution (Br_2) formed with organic solvent	violet/pink solution (I_2) formed with organic solvent
Bromine water **$Br_{2\,(aq)}$** – orange	no reaction	no reaction	violet/pink solution (I_2) formed with organic solvent
Iodine solution **$I_{2\,(aq)}$** – brown	no reaction	no reaction	no reaction

Group 7 — The Halogens

Displacement Reactions Can Help to Identify Solutions

These displacement reactions can be used to help **identify** which halogen (or halide) is present in a solution.

A **halogen** will **displace a halide** from solution if the halide is **below it** in the Periodic Table, e.g.

You can also say a halogen will **oxidise** a halide if the halide is below it in the Periodic Table.

Periodic table	Displacement reaction	Ionic equation
Cl	chlorine (Cl_2) will displace bromide (Br^-) and iodide (I^-)	$Cl_{2(aq)} + 2Br^-_{(aq)} \rightarrow 2Cl^-_{(aq)} + Br_{2(aq)}$ $Cl_{2(aq)} + 2I^-_{(aq)} \rightarrow 2Cl^-_{(aq)} + I_{2(aq)}$
Br	bromine (Br_2) will displace iodide (I^-)	$Br_{2(aq)} + 2I^-_{(aq)} \rightarrow 2Br^-_{(aq)} + I_{2(aq)}$
I	no reaction with F^-, Cl^-, Br^-	

Silver Nitrate Solution is used to Test for Halides

The test for halides is dead easy. First you add **dilute nitric acid** to remove ions that might interfere with the test. Then you just add **silver nitrate solution** ($AgNO_{3\ (aq)}$). A **precipitate** is formed (of the silver halide).

$$Ag^+_{(aq)} + X^-_{(aq)} \rightarrow AgX_{(s)}$$ ...where X is Cl, Br or I

1) The **colour** of the precipitate identifies the halide.
2) Then to be extra sure, you can test your results by adding **ammonia solution**. (Each silver halide has a different solubility in ammonia.)

SILVER NITRATE TEST FOR HALIDE IONS...	
Chloride Cl^-:	white precipitate, dissolves in dilute $NH_{3(aq)}$
Bromide Br^-:	cream precipitate, dissolves in conc. $NH_{3(aq)}$
Iodide I^-:	yellow precipitate, insoluble in conc. $NH_{3(aq)}$

Practice Questions

Q1 Describe the trend in boiling points as you go down Group 7.

Q2 What do you see when potassium iodide is added to bromine water?

Q3 Write the ionic equation for the reaction that happens when chlorine is added to a solution of iodide ions.

Q4 How would you test whether an aqueous solution contained chloride ions?

Exam Questions

Q1 a) Write an ionic equation for the reaction between iodine solution and sodium astatide (NaAt). [1 mark]

b) For the equation in (a), deduce which substance is oxidised. [1 mark]

Q2 Describe the test you would carry out in order to distinguish between solid samples of sodium chloride and sodium bromide using silver nitrate solution and aqueous ammonia.
State your observations and write equations for the reactions which occur. [6 marks]

Q3 The halogen below iodine in Group 7 is astatine (At). Predict, giving an explanation:

a) the physical state of astatine at r.t.p., [3 marks]

b) whether or not silver astatide will dissolve in concentrated ammonia solution. [3 marks]

Don't skip this page — it could cost you £15 000...

Let me explain... the other night I was watching Who Wants to Be a Millionaire, and this question was on for £32 000:

Which of the these elements is a halogen?
A Argon B Nitrogen
C Fluorine D Sodium

Bet Mr Redmond from Wiltshire wishes he paid more attention in Chemistry now, eh. Ha sucker...

Disproportionation and Water Treatment

Here's comes another page jam-packed with golden nuggets of halogen fun. Oh yes, I kid you not. This page is the Alton Towers of AS Chemistry... white-knuckle excitement all the way...

Halogens undergo **Disproportionation** with Alkalis

The halogens will react with cold dilute alkali solutions.
In these reactions, the halogen is simultaneously oxidised and reduced (called **disproportionation**)...

$$X_2 + 2NaOH \rightarrow NaXO + NaX + H_2O$$

Ionic equation: $X_2 + 2OH^- \rightarrow XO^- + X^- + H_2O$

Oxidation number of X: 0 (in X_2), +1 (in XO^-), −1 (in X^-)

The halogens (except fluorine) can exist in a wide range of oxidation states. E.g. chlorine can exist as:

-1	0	+1
Cl^-	Cl_2	ClO^-
chloride	chlorine	chlorate(I)

Chlorine and **Sodium Hydroxide** make Bleach

If you mix chlorine gas with sodium hydroxide at **room temperature**, the above reaction takes place and you get **sodium chlorate(I) solution**, $NaClO_{(aq)}$, which just happens to be common household **bleach**.

$$2NaOH_{(aq)} + Cl_{2\,(aq)} \rightarrow NaClO_{(aq)} + NaCl_{(aq)} + H_2O_{(l)}$$

Oxidation number: **0** (in Cl_2), **+1** (in NaClO), **–1** (in NaCl)

The oxidation number of Cl goes up and down so, you guessed it, it's disproportionation. Hurray.

The sodium chlorate(I) solution (bleach) has loads of uses — it's used in **water treatment**, to bleach **paper** and **textiles**... and it's good for **cleaning toilets**, too. Handy...

Chlorine is used to Kill Bacteria in Water

When you mix chlorine with water, it undergoes disproportionation.
You end up with a mixture of hydrochloric acid and **chloric(I) acid** (also called hypochlorous acid).

$$Cl_{2(g)} + H_2O_{(l)} \rightleftharpoons HCl_{(aq)} + HClO_{(aq)}$$

Oxidation number of Cl: 0 (in Cl_2), −1 (in HCl, hydrochloric acid), +1 (in HClO, chloric(I) acid)

Aqueous chloric(I) acid **ionises** to make **chlorate(I) ions** (also called hypochlorite ions).

$$HClO_{(aq)} + H_2O_{(l)} \rightleftharpoons ClO^-_{(aq)} + H_3O^+_{(aq)}$$

Chlorate(I) ions **kill bacteria**.

So, **adding chlorine** (or a compound containing chlorate(I) ions) to water can make it safe to **drink** or **swim** in.

Disproportionation and Water Treatment

Chlorine in Water — There are **Benefits, Risks** and **Ethical Implications**

1) Clean drinking water is amazingly important — around the world almost **two million people die** every year from waterborne diseases like cholera, typhoid and dysentery because they have to drink dirty water.
2) In the UK now we're lucky, because our drinking water is **treated** to make it safe. **Chlorine** is an important part of water treatment:
   - It **kills disease-causing microorganisms** (see previous page).
   - Some chlorine remains in the water and **prevents reinfection** further down the supply.
   - It prevents the growth of **algae**, eliminating **bad tastes** and **smells**, and **removes discolouration** caused by organic compounds.

Brian gives Susie the water treatment

3) However, there are risks from using chlorine to treat water:
   - **Chlorine gas** is **very harmful** if it's breathed in — it irritates the **respiratory system**. **Liquid chlorine** on the skin or eyes causes severe **chemical burns**. Accidents involving chlorine could be really serious, or fatal.
   - Water contains a variety of organic compounds, e.g. from the decomposition of plants. Chlorine reacts with these compounds to form **chlorinated hydrocarbons**, e.g. chloromethane (CH_3Cl) — and many of these chlorinated hydrocarbons are carcinogenic (cancer-causing). However, this increased cancer risk is small compared to the risks from untreated water — a cholera epidemic, say, could kill thousands of people.
4) There are ethical considerations too. We don't get a **choice** about having our water chlorinated — some people object to this as forced 'mass medication'.

And Some Areas Have **Fluoridated** Water

In some areas of the UK **fluoride ions** are also added to drinking water. Health officials recommend this because it helps to prevent **tooth decay** — there's **loads** of good evidence for this.

There's a **small** amount of evidence linking fluoridated water to a slightly increased risk of some **bone cancers**. Most **toothpaste** is fluoridated, so some people think extra fluoride ions in water is unnecessary.

Practice Questions

Q1 Write the equation for the reaction of chlorine with water. State underneath the oxidation numbers of the chlorine.

Q2 How is common household bleach formed?

Q3 What are the benefits of adding chlorine to drinking water?

Exam Questions

Q1 If chlorine gas and sodium hydroxide are allowed to mix at room temperature, sodium chlorate(I) is formed.

a) This is a disproportionation reaction. Give the ionic equation for the reaction and use it to explain what is meant by disproportionation. [4 marks]

b) Give two uses of sodium chlorate(I). [2 marks]

Q2 Iodide ions react with chlorate(I) ions and water to form iodine, chloride ions and hydroxide ions.

a) Write a balanced equation for this reaction. [2 marks]

b) Show by use of oxidation states which substance has been oxidised and which has been reduced. [2 marks]

c) The reaction mixture is shaken with an organic solvent. What colour solution is formed with the organic solvent? [1 mark]

Remain seated until the page comes to a halt. Please exit to the right...

Oooh, what a lovely page, if I do say so myself. I bet the question of how bleach is made and how chlorine reacts with sodium hydroxide has plagued your mind since childhood. Well now you know. And remember... anything that chlorine can do, bromine and iodine can generally do as well. Eeee... it's just fun, fun, fun all the way.

Basic Stuff

This module's all about organic chemistry... carbon compounds, in other words. Read on...

There are Loads of Ways of Representing Organic Compounds

TYPE OF FORMULA	WHAT IT SHOWS YOU	FORMULA FOR BUTAN-1-OL
General formula	An algebraic formula that can describe **any member** of a family of compounds.	$C_nH_{2n+1}OH$ (for all alcohols)
Empirical formula	The **simplest ratio** of atoms of each element in a compound (cancel the numbers down if possible). (So ethane, C_2H_6, has the empirical formula CH_3.)	$C_4H_{10}O$
Molecular formula	The **actual** number of atoms of each element in a molecule.	$C_4H_{10}O$
Structural formula	Shows the atoms **carbon by carbon**, with the attached hydrogens and **functional groups**.	$CH_3CH_2CH_2CH_2OH$ or $CH_3(CH_2)_3OH$
Displayed formula	Shows how all the atoms are **arranged**, and all the bonds between them.	H–C(H)(H)–C(H)(H)–C(H)(H)–C(H)(H)–OH
Skeletal formula	Shows the **bonds** of the carbon skeleton **only**, with any functional groups. The hydrogen and carbon atoms aren't shown. This is handy for drawing large complicated structures, like cyclic hydrocarbons.	OH (zigzag chain)

A functional group is a reactive part of a molecule — it gives it many of its chemical properties.

The Alkanes are the Simplest Group of Organic Compounds

1) Organic chemistry is more about **groups** of similar chemicals than individual compounds.
2) These groups are called **homologous series**. A homologous series is a bunch of compounds that have the same **functional group** and **general formula**. Consecutive members of a homologous series differ by $-CH_2-$.
3) The simplest homologous series is the **alkanes**. They're **straight chain** molecules that contain only **carbon** and **hydrogen** atoms. There's a lot more about the alkanes on page 54.
4) The **general formula** for alkanes is $\mathbf{C_nH_{2n+2}}$. So the first alkane in the series is $C_1H_{(2\times1)+2} = CH_4$ (you don't need to write the 1 in C_1), the second is $C_2H_{(2\times2)+2} = C_2H_6$, the seventeenth is $C_{17}H_{(2\times17)+2} = C_{17}H_{36}$, and so on...
5) You need to know the names of the **first ten** alkanes.

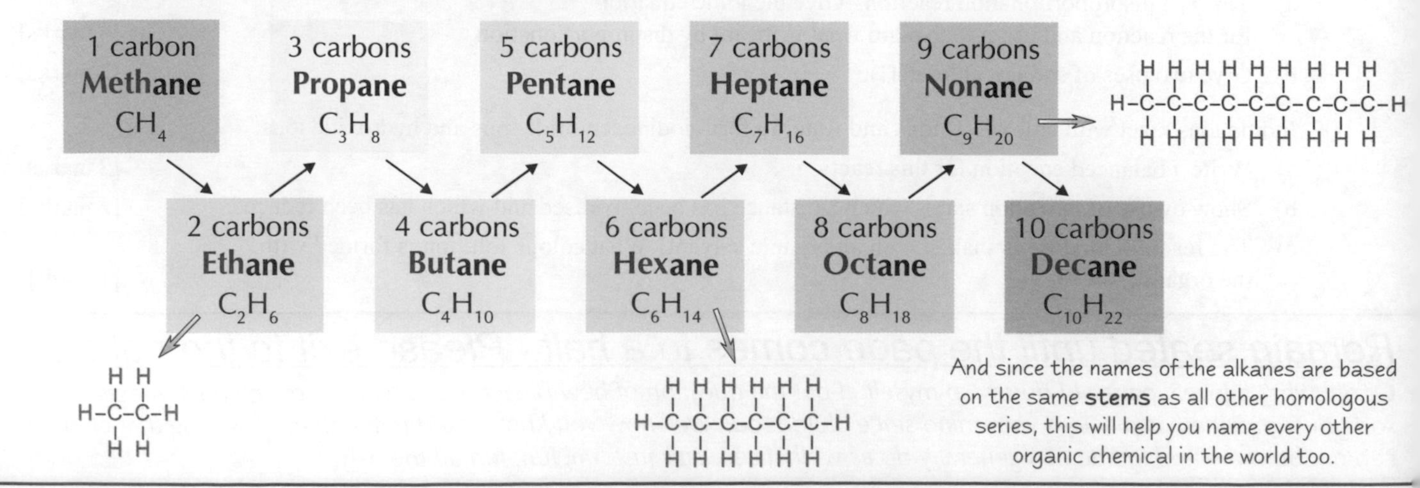

Basic Stuff

Nomenclature is a Fancy Word for the *Naming* of Organic Compounds

You can name any organic compound using these **rules** of nomenclature.

1) Count the carbon atoms in the **longest continuous chain** — this gives you the stem.
2) The **main functional group** of the molecule usually gives you the end of the name (the **suffix**) — see the table below.

No. of C	1	2	3	4	5	6
Stem	meth-	eth-	prop-	but-	pent-	hex-

Homologous series	Prefix or Suffix	Example
alkanes	-ane	propane $CH_3CH_2CH_3$
branched alkanes	alkyl- (-yl)	methylpropane $CH_3CH(CH_3)CH_3$
alkenes	-ene	propene $CH_3CH{=}CH_2$
halogenoalkanes	chloro- bromo- iodo-	chlorethane CH_3CH_2Cl
alcohols	-ol	ethanol CH_3CH_2OH
aldehydes	-al	ethanal CH_3CHO
ketones	-one	propanone CH_3COCH_3
cycloalkanes	cyclo- -ane	cyclohexane C_6H_{12}
arenes	benzene	ethylbenzene $C_6H_5C_2H_5$
esters	alkyl -oate	propyl ethanoate $CH_3COOCH_2CH_2CH_3$
carboxylic acids	-oic acid	ethanoic acid CH_3COOH

3) Number the **longest** carbon chain so that the main functional group has the lowest possible number. If there's more than one longest chain, pick the one with the **most side-chains**.
4) Any side-chains or less important functional groups are added as prefixes at the start of the name. Put them in **alphabetical** order, with the **number** of the carbon atom each is attached to.
5) If there's more than one **identical** side-chain or functional group, use **di-** (2), **tri-** (3) or **tetra-** (4) before that part of the name — but ignore this when working out the alphabetical order.

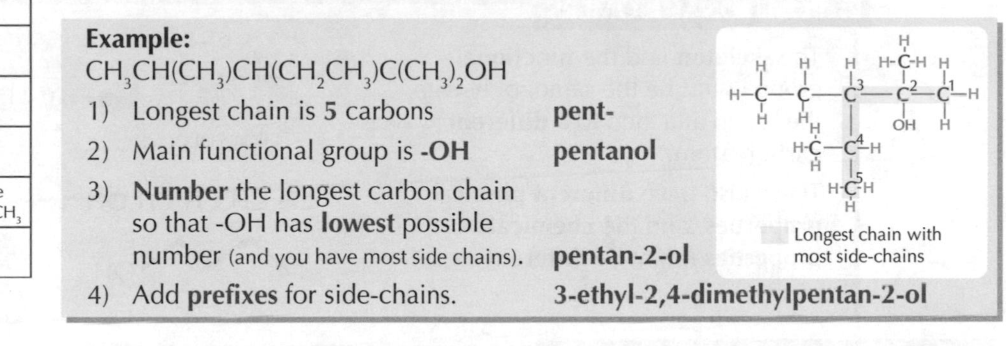

Example:
$CH_3CH(CH_3)CH(CH_2CH_3)C(CH_3)_2OH$

1) Longest chain is **5** carbons — **pent-**
2) Main functional group is **-OH** — **pentanol**
3) **Number** the longest carbon chain so that -OH has **lowest** possible number (and you have most side chains). — **pentan-2-ol**
4) Add **prefixes** for side-chains. — **3-ethyl-2,4-dimethylpentan-2-ol**

Longest chain with most side-chains

Practice Questions

Q1 Explain the difference between molecular formulas and structural formulas.

Q2 Draw the displayed formula for octane. Now write the structural formula.

Q3 In what order should prefixes be listed in the name of an organic compound?

Q4 Draw the displayed formula of 2,3,5-trimethylhexan-3-ol.

Q5 What is meant by the term 'homologous series'?

Q6 Write down the structural formula of the 8th compound in the homologous series with general formula $C_nH_{2n+1}OH$.

Exam Questions

Q1 1-bromobutane is prepared from butan-1-ol in this reaction: $C_4H_9OH + NaBr + H_2SO_4 \rightarrow C_4H_9Br + NaHSO_4 + H_2O$

a) Draw the displayed formulae for butan-1-ol and 1-bromobutane. [2 marks]

b) What is the functional group in butan-1-ol and why is it necessary to state its position on the carbon chain? [2 marks]

Q2 Give the systematic names of the following compounds.

A
```
   H  H  H  H  H
   |  |  |  |  |
H—C——C——C——C——C—H
   |  |  |  |  |
   H  H  H  H  H
```

B
```
   H CH3 H  H
   |  |  |  |
H—C——C——C——C—H
   |  |  |  |
   H  H  H  H
```

C
```
   H CH3 H
   |  |  |
H—C——C——C—H
   |  |  |
   H CH3 H
```
[6 marks]

It's as easy as 1,2,3-trimethylpentan-2-ol...

The best thing to do now is find some random organic compounds and work out their names using the rules. Then have a bash at it the other way around — read the name and draw the compound. It might seem a bit tedious now, but come the exam, you'll be thanking me. Talking of exams — read the questions carefully and check what type of formula they want.

Isomerism

Isomers have the same molecular formula, but different arrangements of atoms.
There are two main types of isomerism — structural isomerism and stereoisomerism.

Structural Isomers *have different* ***Structural Arrangements*** *of Atoms*

In structural isomers, the **molecular formula** is the same, but the **structural formula** is different.
There are **three** different types of structural isomer:

1. CHAIN ISOMERS

The **carbon skeleton** can be arranged differently — for example, as a **straight chain**, or **branched** in different ways.

These isomers have **similar chemical properties** — but their **physical properties**, like boiling point, will be **different** because of the change in shape of the molecule.

2. POSITIONAL ISOMERS

The **skeleton** and the **functional group** could be the same, only with the group attached to a **different carbon atom**.

These also have **different physical properties**, and the **chemical properties** might be **different** too.

3. FUNCTIONAL GROUP ISOMERS

The same atoms can be arranged into **different functional groups**.

These have very **different physical** and **chemical** properties.

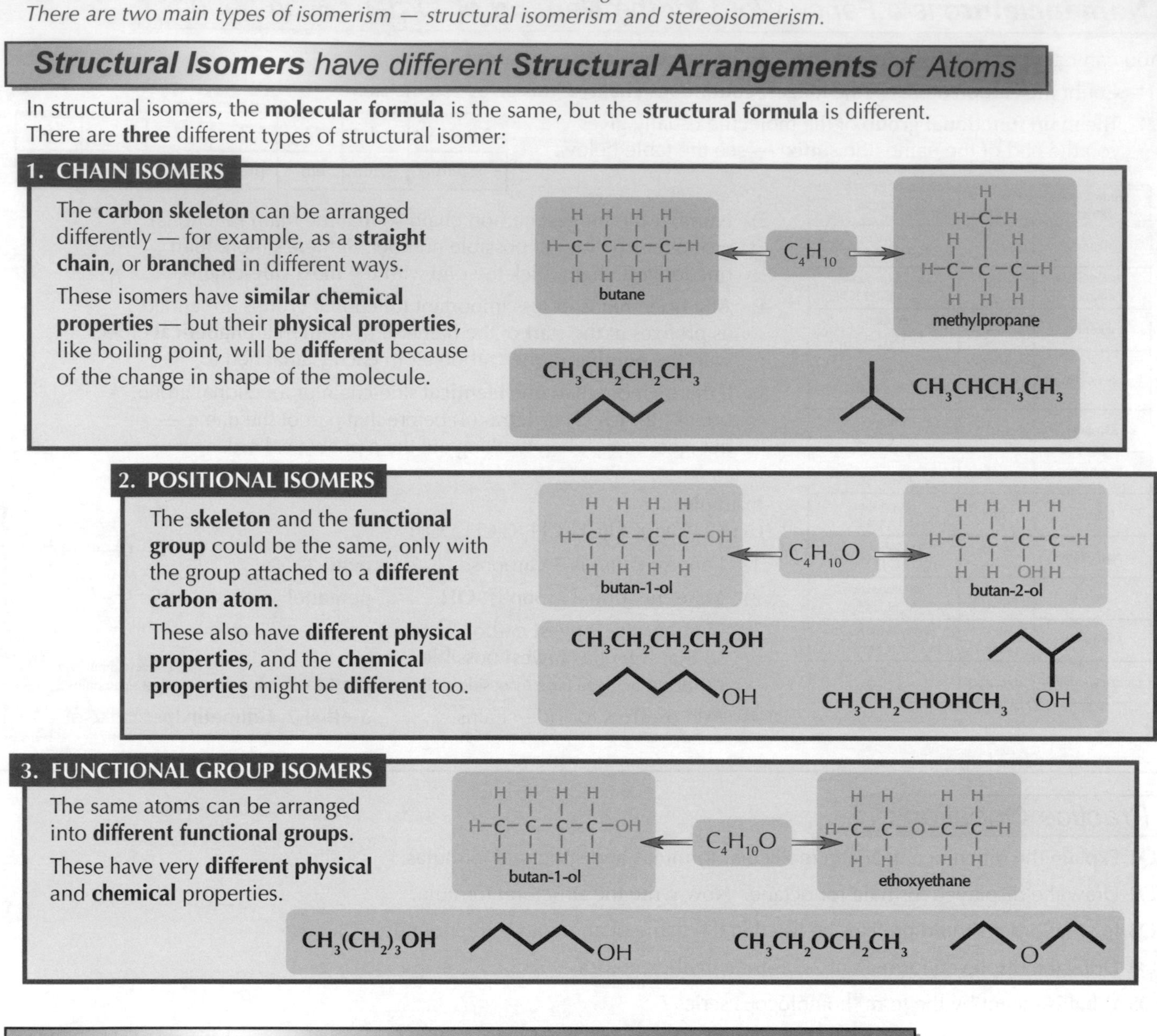

Don't be Fooled *— What Looks Like an Isomer Might* ***Not*** *Be*

Atoms can rotate as much as they like around single **C–C bonds**.
Remember this when you work out structural isomers — sometimes what looks like an isomer, isn't.

For example, **propanol** can only be put together in **two** different ways...

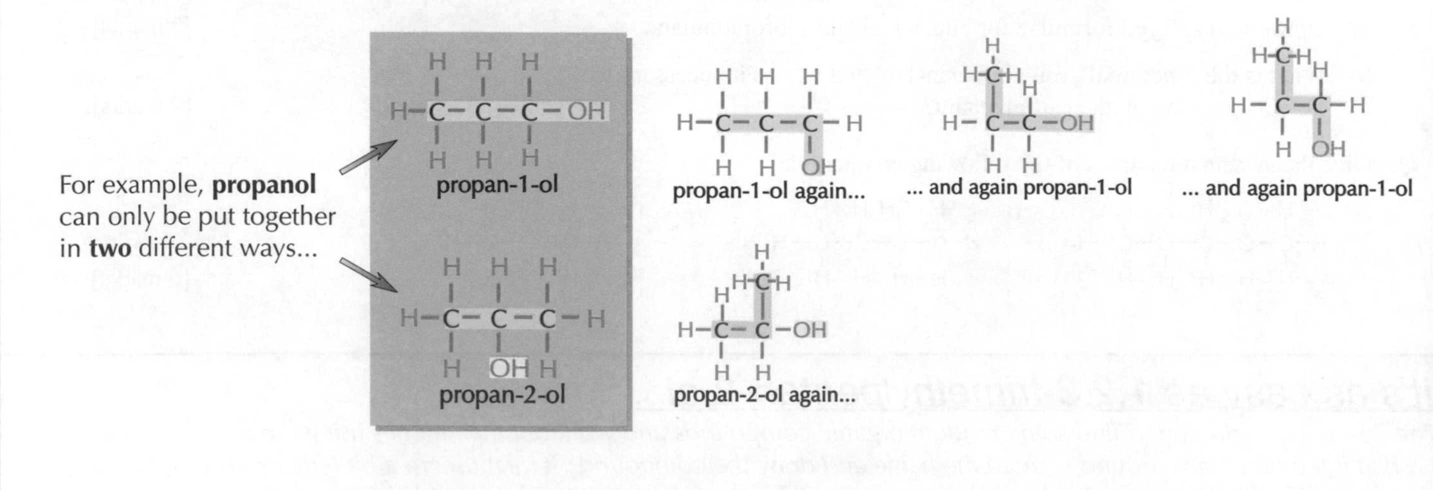

Isomerism

E/Z isomerism is a Type of Stereoisomerism

1) **Stereoisomers** have the same structural formula but a **different arrangement** in space.
(Just bear with me for a moment... that will become clearer, I promise.)
2) Some **alkenes** have stereoisomers — this is because there's a **lack of rotation** around the C=C double bond (see p62). When the double-bonded carbon atoms each have **two different atoms** or **groups** attached to them, you get an '**E-isomer**' and a '**Z-isomer**'.
For example, the double-bonded carbon atoms in but-2-ene each have an **H** and a **CH_3** group attached.

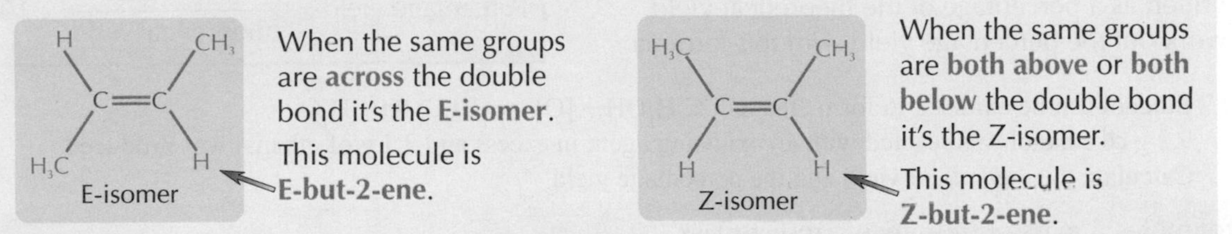

When the same groups are **across** the double bond it's the **E-isomer**. This molecule is **E-but-2-ene.**

When the same groups are **both above** or **both below** the double bond it's the Z-isomer. This molecule is **Z-but-2-ene.**

3) E/Z isomerism is sometimes called **cis-trans isomerism**, where '**cis**' means the **Z-isomer**, and '**trans**' means the **E-isomer**. So E-but-2-ene can be called trans-but-2-ene, and Z-but-2-ene can be called cis-but-2-ene.
4) **BUT** you can't use the cis-trans system if there are **more than two** different groups (other than hydrogen atoms) attached around the double bond.

This could be **trans-1-bromo-1-fluoropropene**, because the **Br** and **CH_3** are on **opposite** sides, or it could be **cis-1-bromo-1-fluoropropene**, because the **F** and **CH_3** are on the same side...

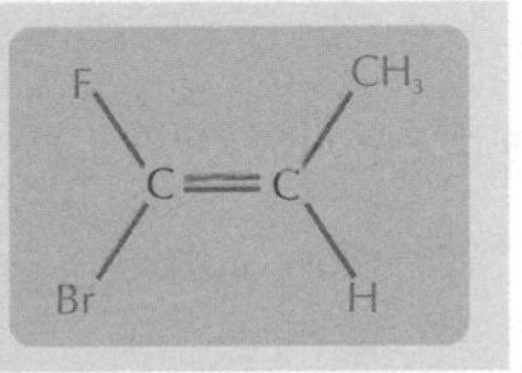

5) The E/Z system keeps on working though. Each of the groups linked to the double-bonded carbons is given a **priority**. If the two carbon atoms have their 'higher priority group' on **opposite** sides, then it's an **E isomer**. If the two carbon atoms have their 'higher priority group' on the **same** side, then it's a **Z isomer**. (You don't need to know the rules for deciding the order of these priorities.)
6) In the E/Z system, Br has a **higher priority** than F, so the names depend on where the Br atom is in relation to the CH_3 group (which has a higher priority than the H atom).

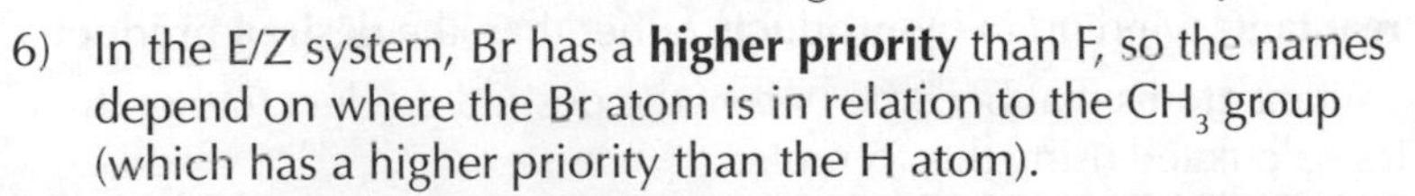

E-1-bromo-1-fluoropropene — Z-1-bromo-1-fluoropropene

Practice Questions

Q1 What are isomers?

Q2 Name the three types of structural isomerism.

Q3 What is a positional isomer?

Q4 What is stereoisomerism?

Q5 Why doesn't but-1-ene show E/Z isomerism?

Exam Question

Q1 a) There are five chain isomers of the alkane C_6H_{14}.
(i) Draw and name all five isomers of C_6H_{14}. [10 marks]
(ii) Explain what is meant by the term 'chain isomerism'. [2 marks]

b) There are four isomers of the alkene C_3H_5Cl.
(i) Draw and name the pair of stereoisomers. [4 marks]
(ii) Draw and name the two isomers which do not show stereoisomerism. [4 marks]

c) Alkanes and alkenes are both examples of a homologous series. What is a homologous series? [2 marks]

Human structural isomers...

Atom Economy and Percentage Yield

How to make a subject like chemistry even more exciting — introduce the word 'economy'...

*The **Theoretical Yield** of a Product is the **Maximum** you could get*

1) The **theoretical yield** is the **mass of product** that **should** be made in a reaction if **no** chemicals are '**lost**' in the process. You can use the **masses of reactants** and a **balanced equation** to calculate the theoretical yield for a reaction.
2) The **actual** mass of product (the **actual yield**) is always **less** than the theoretical yield. Some chemicals are always 'lost', e.g. some solution gets left on filter paper, or is lost during transfers between containers.
3) The **percentage yield** is the **actual** amount of product you collect, written as a percentage of the theoretical yield. You can work out the percentage yield with this formula:

$$\text{Percentage yield} = \frac{\text{actual yield}}{\text{theoretical yield}} \times 100\%$$

Example: Ethanol can be oxidised to form ethanal: $C_2H_5OH + [O] \rightarrow CH_3CHO + H_2O$
9.2 g of ethanol was reacted with an oxidising agent in excess and 2.1 g of ethanal was produced.
Calculate the theoretical yield and the percentage yield.

Number of moles = mass of substance ÷ molar mass

Moles of C_2H_5OH = 9.2 ÷ [(2 × 12) + (5 × 1) + 16 + 1] = 9.2 ÷ 46 = 0.2 moles

1 mole of C_2H_5OH produces 1 mole of CH_3CHO, so 0.2 moles of C_2H_5OH will produce 0.2 moles of CH_3CHO.

M of CH_3CHO = (2 × 12) + (4 × 1) + 16 = 44 g mol^{-1}

Theoretical yield (mass of CH_3CHO) = number of moles × M = 0.2 × 44 = **8.8 g**

So, if the actual yield was 2.1 g, the percentage yield = $\frac{\text{actual yield}}{\text{theoretical yield}} \times 100\% = \frac{2.1}{8.8} \times 100\% \approx$ **24%**

***Atom Economy** is a Measure of the **Efficiency** of a Reaction*

1) The **percentage yield** tells you how wasteful the **process** is — it's based on how much of the product is lost because of things like reactions not completing or losses during collection and purification.
2) But percentage yield doesn't measure how wasteful the **reaction** itself is. A reaction that has a 100% yield could still be very wasteful if a lot of the atoms from the **reactants** wind up in **by-products** rather than the **desired product**.
3) **Atom economy** is a measure of the proportion of reactant **atoms** that become part of the desired product (rather than by-products) in the **balanced** chemical equation. It's calculated using this formula:

$$\%\text{ atom economy} = \frac{\text{molecular mass of desired product}}{\text{sum of molecular masses of all products}} \times 100\%$$

4) In an **addition reaction**, the reactants **combine** to form a **single product**.
The atom economy for addition reactions is **always 100%** since no atoms are wasted.

For example, ethene (C_2H_4) and hydrogen react to form ethane (C_2H_6) in an addition reaction:

$$C_2H_4 + H_2 \rightarrow C_2H_6$$

The **only product** is ethane — the desired product. So no reactant atoms are wasted — the atom economy is **100%**.

5) A **substitution reaction** is one where some atoms from one reactant are **swapped** with atoms from another reactant. This type of reaction **always** results in **at least two products** — the desired product and at least one by-product.

An example is the reaction of bromomethane (CH_3Br) with sodium hydroxide (NaOH) to make methanol (CH_3OH):

$$CH_3Br + NaOH \rightarrow CH_3OH + NaBr$$

This is **more wasteful** than an addition reaction because the Na and Br atoms are not part of the desired product.

$$\%\text{ atom economy} = \frac{\text{molecular mass of desired product}}{\text{sum of molecular masses of all products}} \times 100\%$$

$$= \frac{M_r(CH_3OH)}{M_r(CH_3OH) + M_r(NaBr)} \times 100\%$$

$$= \frac{(12+(3\times1)+16+1)}{(12+(3\times1)+16+1)+(23+80)} \times 100\% = \frac{32}{32+103} \times 100\% = \mathbf{23.7\%}$$

Always make sure you're using a balanced equation.

Atom Economy and Percentage Yield

Reactions can Have **High Percentage Yields** and **Low Atom Economies**

Example: 0.475 g of CH_3Br reacts with an excess of NaOH in this reaction: $CH_3Br + NaOH \rightarrow CH_3OH + NaBr$
0.153 g of CH_3OH is produced. What is the percentage yield?

Number of moles = mass of substance ÷ molar mass

Moles of $CH_3Br = 0.475 \div (12 + 3 \times 1 + 80) = 0.475 \div 95 =$ **0.005 moles**

The reactant : product ratio is 1 : 1, so the maximum number of moles of CH_3OH is **0.005**.

Theoretical yield $= 0.005 \times M_r(CH_3OH) = 0.005 \times (12 + (3 \times 1) + 16 + 1) = 0.005 \times 32 =$ **0.160 g**

$$\text{percentage yield} = \frac{\text{actual yield}}{\text{theoretical yield}} \times 100\% = \frac{0.153}{0.160} \times 100\% = \mathbf{95.6\%}$$

So this reaction has a **very high percentage yield**, but, as you saw on the previous page, the **atom economy** is **low**.

It's Important to Develop Reactions with **High Atom Economies**

1) Companies in the chemical industry will often choose to use reactions with high atom economies. High atom economy has **environmental** and **economic benefits**.
2) A **low atom economy** means there's lots of **waste** produced. It costs money to **separate** the desired product from the waste products and more money to dispose of the waste products **safely** so they don't harm the environment.
3) Companies will usually have paid good money to buy the **reactant chemicals**. It's a **waste of money** if a high proportion of them end up as useless products.
4) Reactions with low atom economies are **less sustainable** (see p90). Many raw materials are in **limited supply**, so it makes sense to use them efficiently so they last as long as possible. Also, waste has to go somewhere — it's better for the environment if less is produced.

Finding uses for the by-products helps to solve some of the problems of low atom economy.

Practice Questions

Q1 How many products are there in an addition reaction?

Q2 Does the percentage yield for a reaction always have the same value as the percentage atom economy?

Q3 Why do reactions with high atom economy save chemical companies money and cause less environmental impact?

Exam Questions

Q1 Reactions 1 and 2 below show two possible ways of preparing the compound chloroethane (C_2H_5Cl):

1 $C_2H_5OH + PCl_5 \rightarrow C_2H_5Cl + POCl_3 + HCl$

2 $C_2H_4 + HCl \rightarrow C_2H_5Cl$

a) Which of these is an addition reaction? [1 mark]

b) Calculate the atom economy for reaction 1. [3 marks]

c) Reaction 2 has an atom economy of 100%. Explain why this is, in terms of the products of the reaction. [1 mark]

Q2 Phosphorus trichloride (PCl_3) reacts with chlorine to give phosphorus pentachloride (PCl_5):

$$PCl_3 + Cl_2 \rightleftharpoons PCl_5$$

a) If 0.275 g of PCl_3 reacts with 0.142 g of chlorine, what is the theoretical yield of PCl_5? [2 marks]

b) When this reaction is performed 0.198 g of PCl_5 is collected. Calculate the percentage yield. [1 mark]

c) Changing conditions such as temperature and pressure will alter the percentage yield of this reaction. Will changing these conditions affect the atom economy? Explain your answer. [2 marks]

I knew a Tommy Conomy once... strange bloke...

These pages shouldn't be too much trouble — you've survived worse already. Make sure that you get plenty of practice using the percentage yield and atom economy formulas. And whatever you do, don't get mixed up between percentage yield (which is to do with the process) and atom economy (which is to do with the reaction).

Alkanes

Alkanes are your basic hydrocarbons — like it says on the tin, they've got hydrogen and they've got carbon.

Alkanes are Saturated Hydrocarbons

1) Alkanes have the **general formula C_nH_{2n+2}.**
They've only got **carbon** and **hydrogen** atoms, so they're **hydrocarbons**.
2) Every carbon atom in an alkane has **four single bonds** with other atoms.
It's **impossible** for carbon to make more than four bonds, so alkanes are **saturated**.

Here are a few examples of alkanes —

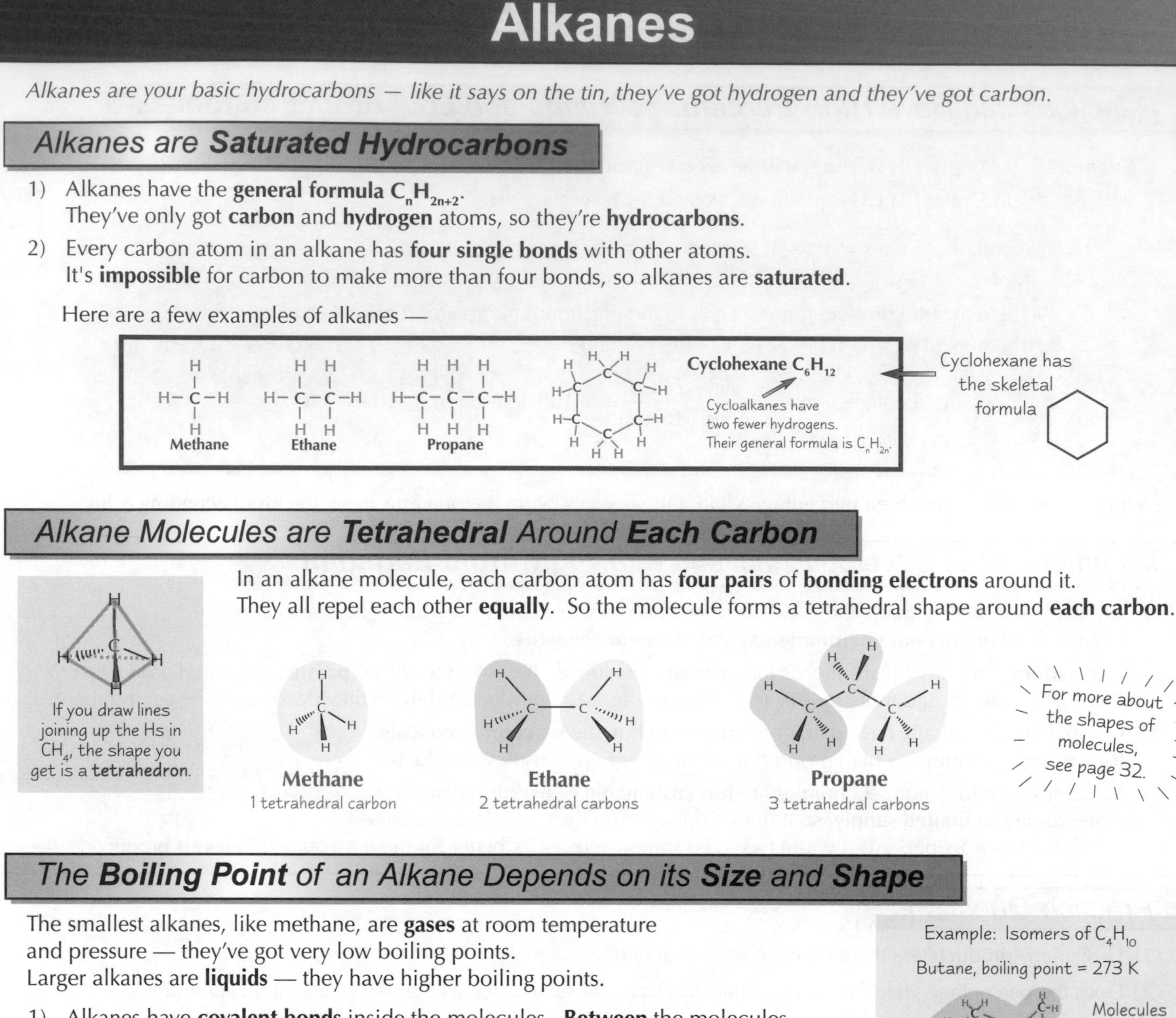

Alkane Molecules are Tetrahedral Around Each Carbon

In an alkane molecule, each carbon atom has **four pairs** of **bonding electrons** around it.
They all repel each other **equally**. So the molecule forms a tetrahedral shape around **each carbon**.

If you draw lines joining up the Hs in CH_4, the shape you get is a **tetrahedron**.

Methane 1 tetrahedral carbon

Ethane 2 tetrahedral carbons

Propane 3 tetrahedral carbons

For more about the shapes of molecules, see page 32.

The Boiling Point of an Alkane Depends on its Size and Shape

The smallest alkanes, like methane, are **gases** at room temperature and pressure — they've got very low boiling points.
Larger alkanes are **liquids** — they have higher boiling points.

1) Alkanes have **covalent bonds** inside the molecules. **Between** the molecules, there are **van der Waals** forces which hold them all together.
2) The **longer** the carbon chain, the **stronger** the van der Waals forces. This is because there's **more molecular surface area** and more electrons to interact.
3) So as the molecules get longer, it takes **more energy** to overcome the van der Waals forces and separate them, and the boiling point **rises**.
4) A **branched-chain** alkane has a **lower** boiling point than its straight-chain isomer. Branched-chain alkanes can't **pack closely** together and they have smaller **molecular surface areas** — so the van der Waals forces are reduced.

Example: Isomers of C_4H_{10}

Butane, boiling point = 273 K

Molecules can pack closely.

Methylpropane, boiling point = 261 K

Close packing isn't possible.

Alkanes Burn Completely in Oxygen

1) If you burn (**oxidise**) alkanes with enough **oxygen**, you get **carbon dioxide** and water — this is a **combustion reaction**.

Here's the equation for the combustion of propane — $C_3H_{8(g)} + 5O_{2(g)} \rightarrow 3CO_{2(g)} + 4H_2O_{(g)}$

2) Combustion reactions happen between **gases**, so liquid alkanes have to be **vaporised** first.
Smaller alkanes turn into **gases** more easily (they're more **volatile**), so they'll **burn** more easily too.
3) Larger alkanes release heaps more **energy** per mole because they have more bonds to react.
4) Because they release so much energy when they burn, alkanes make excellent fuels.
Propane is used as a **central heating** and **cooking** fuel. **Butane** is bottled and sold as **camping gas**.
Petrol and **diesel** are both made up of a mixture of alkanes too (and additives).

Alkanes

Burning **Alkanes** In **Limited Oxygen** Produces **Carbon Monoxide**

1) If there isn't much oxygen around, the alkane will still burn, but it will produce **carbon monoxide** and water.

For example, burning methane with not much O_2 —

$$2CH_{4(g)} + 3O_{2(g)} \rightarrow 2CO_{(g)} + 4H_2O_{(g)}$$

2) This is a problem because **carbon monoxide** is **poisonous**.

   1) The **oxygen** in your bloodstream is carried around by **haemoglobin**.
   2) **Carbon monoxide** is **better** at binding to haemoglobin than oxygen is. So if you breathe in air with a **high concentration** of carbon monoxide it will bind to the haemoglobin in your bloodstream **before** the oxygen can.
   3) This means that **less oxygen** will reach your cells. You will start to suffer from symptoms associated with **oxygen deprivation** — things like fatigue, headaches, and nausea. At very high concentrations of carbon monoxide it can even be fatal.

3) **Any** appliance that burns alkanes can produce carbon monoxide. This includes things like gas- or oil-fired boilers and heaters, gas stoves, and coal or wood fires. Cars also produce carbon monoxide.
4) All appliances that use an alkane-based fuel need to be **properly ventilated**. They should be checked and maintained regularly, and their sources of ventilation should **never** be blocked.
5) If you have any alkane burning appliances it's a good idea to have a **carbon monoxide detector** around.

Practice Questions

Q1 What's the general formula for alkanes?

Q2 What kind of intermolecular forces are there between alkane molecules?

Q3 Why do straight-chain alkanes have higher boiling points than branched-chain alkanes?

Q4 What are the combustion products of alkanes when there's plenty of oxygen around? And when oxygen is limited?

Exam Questions

Q1 The alkane ethane is a saturated hydrocarbon. It is mostly unreactive, but will react with oxygen in a combustion reaction.

a) What is a saturated hydrocarbon? [2 marks]

b) Write a balanced equation for the complete combustion of ethane. [2 marks]

Q2 Nonane is a hydrocarbon with the formula C_9H_{20}.

a) What homologous series does nonane belong to? [1 mark]

b) Which would you expect to have a higher boiling point, nonane or 2,2,3,3-tetramethylpentane? Explain your answer. [2 marks]

c) When nonane burns in a limited air supply the products are carbon monoxide and water.

(i) Write a balanced equation for the reaction. [1 mark]

(ii) Explain why carbon monoxide is such a dangerous gas. [2 marks]

d) Explain why burning 1 mole of nonane produces more energy than burning 1 mole of methane. [2 marks]

Tetrahedra — aren't they those monsters from Greek mythology...

Alkanes... they don't do much, so there's only so much the examiners can ask you. Which means:
(i) you need to understand exactly why two molecules containing the same atoms can have different boiling points,
(ii) you need to know why burning alkane-based fuels without enough oxygen is dangerous. No excuses now.

Petroleum

Petroleum is just a posh word for crude oil — the black, yukky stuff they get out of the ground from huge oil wells.

Crude Oil** is a Mixture of **Hydrocarbons

1) Petroleum or crude oil is mostly **alkanes**. They range from **smallish alkanes**, like propane, to **massive alkanes** with more than 50 carbons.
2) Crude oil isn't very useful as it is, but you can **separate** it into more useful bits (or **fractions**) by **fractional distillation**.

Here's how fractional distillation works — don't try this at home.

1) First, the crude oil is **vaporised** at about 350 °C.
2) The vaporised crude oil goes into the **fractionating column** and rises up through the trays. The largest hydrocarbons don't **vaporise** at all, because their boiling points are too high — they just run to the bottom and form a gooey **residue**.
3) As the crude oil vapour goes up the fractionating column, it gets **cooler**. Because of the different chain lengths, each fraction **condenses** at a different temperature. The fractions are **drawn off** at different levels in the column.
4) The hydrocarbons with the **lowest boiling points** don't condense. They're drawn off as **gases** at the top of the column.

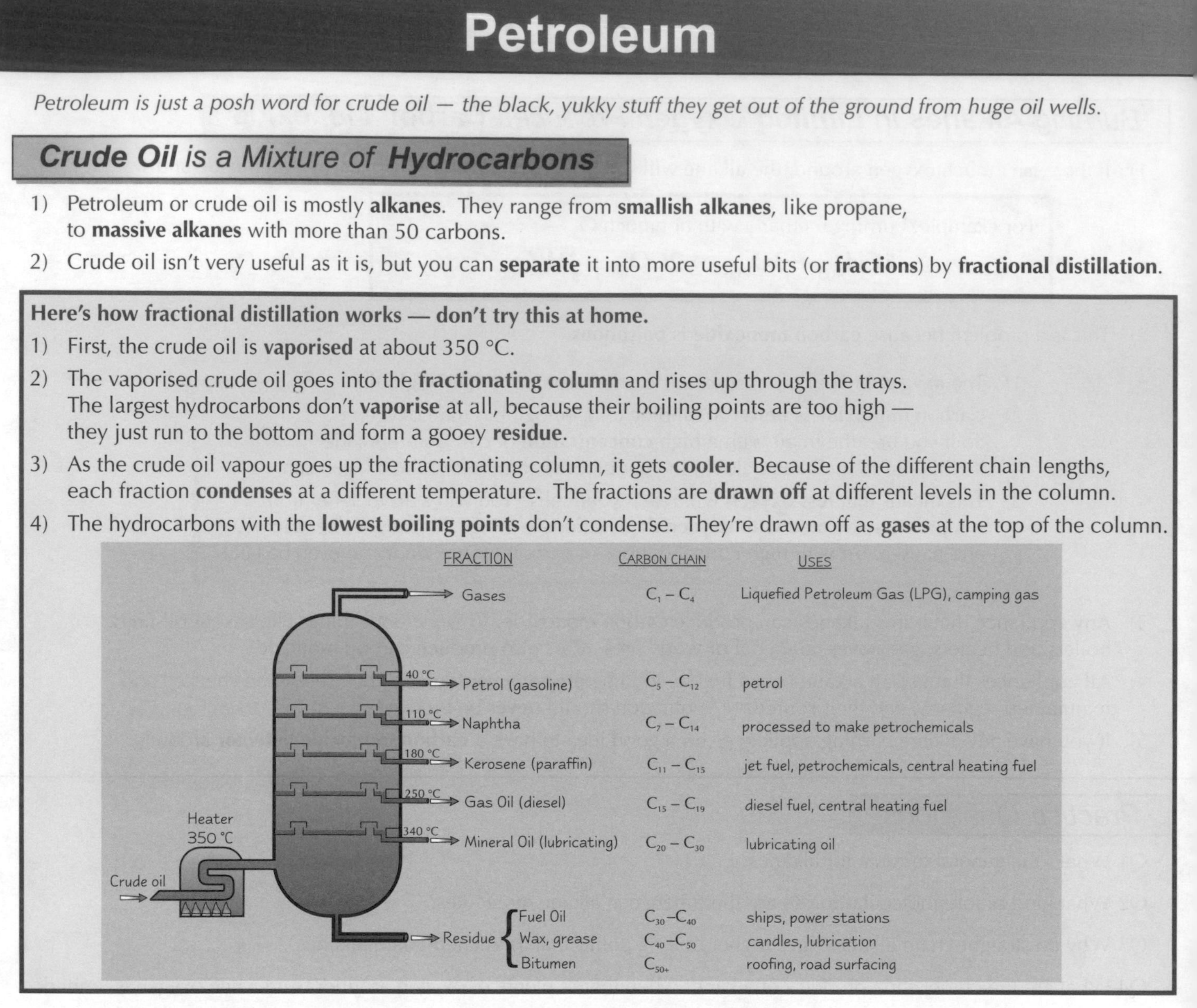

3) Most of the fractions are either used as **fuels** or processed to make **petrochemicals**. A **petrochemical** is any compound that is made from crude oil or any of its fractions and is not a fuel.

Heavy Fractions** can be **'Cracked'** to Make **Smaller Molecules

1) People want loads of the **light** fractions, like petrol and naphtha. They don't want so much of the **heavier** stuff like bitumen though.
2) To meet this demand, the less popular heavier fractions are **cracked**. Cracking is **breaking** long-chain alkanes into **smaller** hydrocarbons. It involves breaking the **C–C bonds**. You could crack **decane** like this —

$$\underset{\text{decane}}{C_{10}H_{22}} \rightarrow \underset{\text{ethene}}{C_2H_4} + \underset{\text{octane}}{C_8H_{18}}$$

3) The main way of doing this is **catalytic cracking**.
   - The heavier fractions are passed over a **catalyst** at a high temperature and a moderate pressure.
   - This breaks them up into **smaller molecules**.
   - Using a catalyst **cuts costs**, because the reaction can be done at a **lower** temperature and pressure. The catalyst also **speeds** up the reaction, and time is money and all that.

Aromatic hydrocarbons contain benzene rings.

4) This method of cracking gives a high percentage of **branched hydrocarbons** and **aromatic hydrocarbons** — these are particularly useful for making **petrol**.

Petroleum

*Hydrocarbons with a **High Octane Rating** Burn More **Smoothly***

1) Here's a super-quick whizz through how a **petrol engine** works:
   The **fuel/air** mixture is squashed by a **piston** and **ignited** with a spark, creating an **explosion**. This drives the piston up again, turning the **crankshaft**. Four pistons work **one after the other**, so that the engine runs smoothly.
2) The problem is, **straight-chain alkanes** in petrol tend to **auto-ignite** — when the fuel/air mixture is compressed they explode without being ignited by the spark. This extra explosion causes '**knocking**' in the engine.
3) To get rid of knocking and make combustion more efficient, **shorter branched-chain alkanes**, **cycloalkanes** and **arenes** are included in petrols, creating a **high octane rating**.

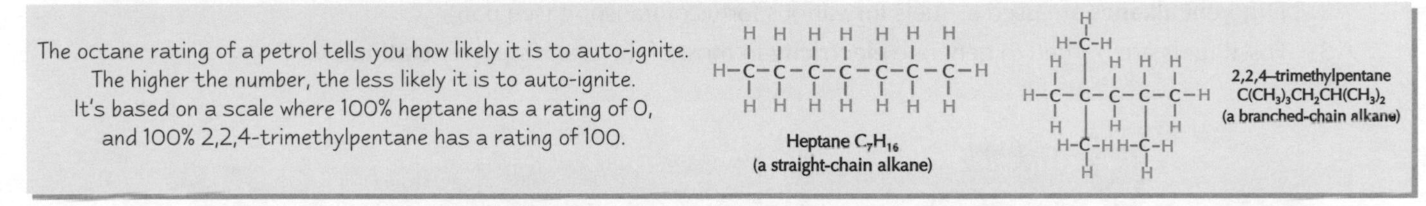

Straight-Chain Alkanes** are Made Into **Branched** or **Cyclic Hydrocarbons

Fuel manufacturers convert some of the **straight-chain alkanes** into **branched-chain alkanes** and **cyclic hydrocarbons** using isomerisation and reforming.

ISOMERISATION — STRAIGHT-CHAIN TO BRANCHED-CHAIN

Isomerisation occurs when you heat **straight-chain** alkanes with a **catalyst** stuck on inert aluminium oxide. The alkanes break up and join back together as **branched isomers**.

A **molecular sieve** (zeolite) is used to separate the isomers. **Straight-chain** molecules go through the sieve and are **recycled**.

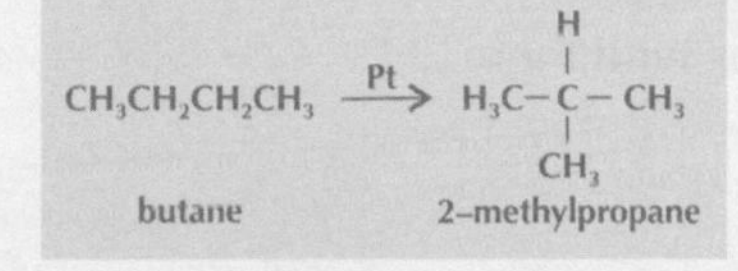

REFORMING — STRAIGHT-CHAIN TO CYCLIC

Reforming converts **alkanes** into **cyclic hydrocarbons**.

It uses a **catalyst** made of **platinum** and another metal. Again, you need to stick the catalyst on inert aluminium oxide.

$CH_3CH_2CH_2CH_2CH_2CH_3$ (hexane) $\xrightarrow{\text{Pt and metal}}$ cyclohexane $+ H_2$ $\longrightarrow$ benzene $+ 3H_2$

Practice Questions

Q1 What is the naphtha fraction of crude oil used for?

Q2 What is cracking?

Q3 Explain why isomerisation is carried out.

Exam Question

Q1 Crude oil is a source of fuels and petrochemicals. It's vaporised and separated into fractions using fractional distillation.

a) Some heavier fractions are processed using cracking.
   (i) Give one reason why cracking is carried out. [2 marks]
   (ii) Write an equation for the cracking of dodecane, $C_{12}H_{26}$. [1 mark]

b) Some hydrocarbons are processed using isomerisation or reforming, producing a petrol with a high octane rating. Petrols with a high octane rating burn more efficiently.
   (i) What kinds of compounds are desirable for a petrol that will burn efficiently? [3 marks]
   (ii) What effect do they have on the petrol's performance? [1 mark]
   (iii) Draw and name two isomers formed from pentane by isomerisation. [4 marks]

Crude oil — not the kind of oil you could take home to meet your mother...

This ain't the most exciting page in the history of the known universe. Although in a galaxy far, far away there may be lots of pages on even more boring topics. But, that's neither here nor there, cos you've got to learn the stuff anyway. Get fractional distillation and cracking straight in your brain and make sure you know why people bother to do them.

Fossil Fuels

Ah, fossils... so this is going to be a cool page about dinosaurs and stuff, I expect...

Fossil Fuels are Incredibly Useful — We Rely on Them for Loads of Things...

The three fossil fuels — coal, oil and natural gas — are major fuels. We use them to provide...

Energy

1) The combustion of fossil fuels is very **exothermic** — they give out large amounts of energy when they burn, which is why they make great **fuels**.
2) Different **alkanes** are used as fuels for various forms of transport (see p56).
3) Fossil fuels are burned to generate **electricity** in most of the world's **power stations**.

Raw Materials

1) Coal, oil and gas aren't just used as fuels, though. They're also important **raw materials** in the chemical industry. Hydrocarbons obtained from fossil fuels — especially oil — are used, either on their own or with other chemicals, for a whole range of purposes.
2) For example, almost all modern plastics are polymers (see p63) made with organic chemicals from fossil fuels.
3) Other products of the petrochemical industry include **solvents**, **detergents**, **adhesives** and **lubricants**.

In Fact... Maybe We Rely on Them Too Much

Fossil fuels are really useful — but there are a couple of major problems with them...

Burning Fuels Makes Greenhouse Gases

1) Right now we're burning more **carbon-based fossil fuels** (e.g. in transport, power stations etc.) than ever before. This is one factor that's helping to cause an increase in the amount of **carbon dioxide** in the atmosphere.
2) Carbon dioxide is a **greenhouse gas**. The extra carbon dioxide we're producing is contributing to **global warming** and **climate change** by enhancing the **greenhouse effect**. See pages 92-93.

Fossil Fuels are Non-Renewable

1) There's a finite amount of fossil fuels — and they're **running out**. Oil will be the first to go — and as it gets really scarce, it'll become more expensive. It's not **sustainable** to keep using fossil fuels willy-nilly (see page 90 for more on sustainability).
2) The developed world relies heavily on fossil fuels to produce **energy** for transport, heating and electricity generation, and to make **chemicals** like plastics and fibres.
3) Some estimates suggest that if we keep using them up at the rate we are doing, there could be just **45 years** worth of oil, **70 years** worth of gas and **250 years** worth of coal left in the ground. And we could run out even sooner, because countries like **China** and **India** are developing rapidly and increasing their energy needs. New supplies may be found, but eventually they'll run out too.
4) There are **alternative sources** of energy that can be used (see the next page), and most of the chemicals currently made from crude oil can be made from **coal** or **plants**. But while there are still reserves of fossil fuels most businesses aren't keen to spend money developing these alternatives.

A couple of overturned bins make a great alternative to motor transport

Fossil Fuels

There are Potential **Alternatives** to Fossil Fuels

So we need to do something about the fuel situation — and there are various options. **Plants** could be an important source of fuels for the future. They're great, because they're **renewable** — you can grow more if you need to.

Bioethanol

1) Ethanol can be used to fuel cars, either on its own or added to petrol. **Bioethanol** is ethanol that's produced from plants — it's made by the **fermentation of sugar** from crops such as maize.
2) Bioethanol's thought of as being **carbon neutral** — in other words, it has **no overall carbon emission** into the atmosphere. That's because all the CO_2 **released** when the fuel is burned was **removed** by the crop as it grew.
3) **BUT** — there are still carbon emissions if you consider the **whole** process. Making the fertilisers and powering agricultural machinery will probably involve burning fossil fuels. It's still better than petrol though, and it does conserve crude oil supplies.

Biodiesel

1) **Biodiesel** is another fuel that can come from plants. As the name suggests, it can be used in **diesel engines** — you can use 100% biodiesel or a mixture of biodiesel and conventional diesel.
2) It's made by refining renewable **fats and oils**, such as vegetable oils (biodiesel can even be made from used restaurant fryer oil).
3) Like bioethanol, biodiesel can be a **carbon neutral** fuel (but the same big **BUT** from point 3 above applies here, too).

There are some potential problems with using crops to make fuels:

- It's possible that **developed countries** (like us) will create a huge demand as they try and find fossil fuel alternatives. Poorer **developing countries** (in South America, say) will use this as a way of **earning money** and rush to convert their farming land to produce these 'crops for fuels', which may mean they won't grow enough **food** to eat.
- There are also worries that in some places **forests** are being **cleared** to make room for biofuel crops. The crops absorb **far less CO_2** than the forest did — so this defeats one of the main objects of growing biofuels.

Practice Questions

Q1 Describe two uses of fossil fuels.

Q2 Describe two disadvantages of using fossil fuels.

Q3 What is the raw material for the production of bioethanol?

Exam Questions

Q1 Various alternative fuels for transport have been proposed. One of these is 'bioethanol' made from sugar.

a) Name the process used to produce ethanol from sugar. [1 mark]

b) Explain why ethanol produced this way is considered to be carbon neutral. [2 marks]

c) Describe the possible negative effect on developing countries of growing crops for conversion to fuel. [2 marks]

Q2 Coal, oil and natural gas are described as fossil fuels.

a) Why do coal, oil and gas make good fuels? [1 mark]

b) Why does fossil fuel use contribute to global warming? [2 marks]

c) Why are fossil fuels described as non-renewable? [2 marks]

d) Describe one other use of fossil fuels, other than being burned as fuels. [1 mark]

[Insert predictable 'fossil' joke about the age of your teacher here]

The question of what we're going to do to replace fossil fuels is a pretty vital one — because they are going to run out and, in the case of oil at least, that could well happen within your lifetime. Biofuels look like they could be an answer but they're hardly problem-free either. And as if the world's fuel problems weren't enough, you've got AS exams coming up too...

Alkanes — Substitution Reactions

Oooh, eh... mechanisms. You might like them. You might not. But you've gotta learn 'em. Reactions don't happen instantaneously — there are often a few steps. And mechanisms show you what they are.

There are Two Types of Bond Fission — Homolytic and Heterolytic

Breaking a covalent bond is called **bond fission**. A single covalent bond is a shared pair of electrons between two atoms. It can break in two ways:

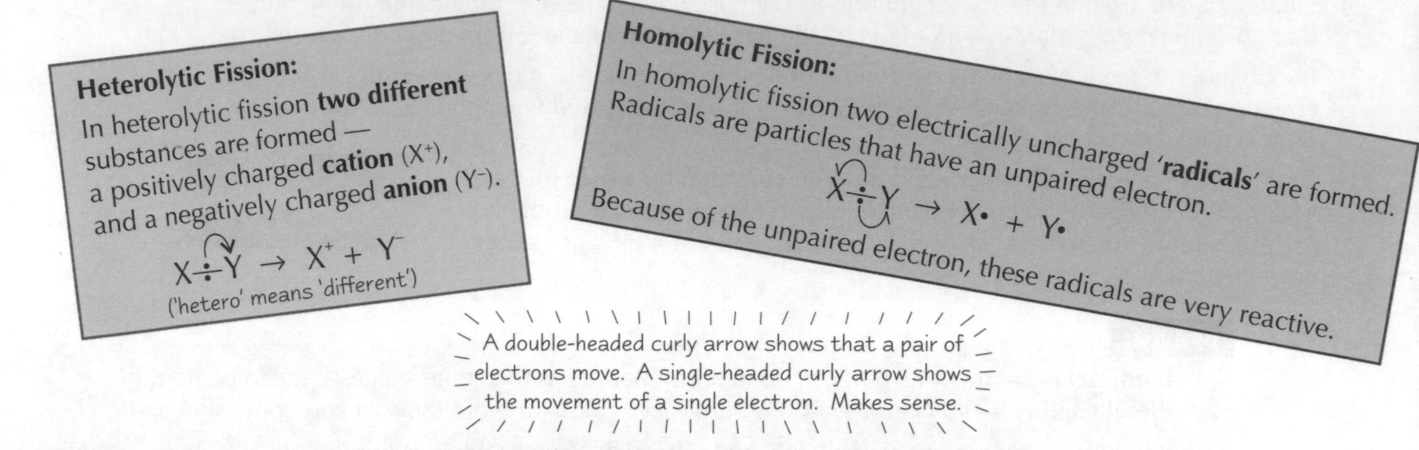

Halogens React with Alkanes, Forming Halogenoalkanes

1) Halogens react with alkanes in **photochemical** reactions. Photochemical reactions are started by **light** — this reaction requires **ultraviolet light** in particular to get going.
2) A hydrogen atom is **substituted** (replaced) by chlorine or bromine. This is a **free-radical substitution reaction**.

Chlorine and **methane** react with a bit of a bang to form **chloromethane**:

$$CH_4 + Cl_2 \xrightarrow{UV} CH_3Cl + HCl$$

Hallo Jen.
Go away Nigel.

The **reaction mechanism** has three stages:

Initiation reactions — free radicals are produced.

1) Sunlight provides enough energy to break the Cl-Cl bond — this is **photodissociation**.

$$Cl_2 \xrightarrow{UV} 2Cl\cdot$$

2) The bond splits **equally** and each atom gets to keep one electron — **homolytic fission**. The atom becomes a highly reactive **free radical**, Cl·, because of its **unpaired electron**.

Propagation reactions — free radicals are used up and created in a chain reaction.

1) Cl· attacks a **methane** molecule: $Cl\cdot + CH_4 \rightarrow \cdot CH_3 + HCl$
2) The new **methyl free radical**, $\cdot CH_3$, can attack another Cl_2 molecule: $\cdot CH_3 + Cl_2 \rightarrow CH_3Cl + Cl\cdot$
3) The new Cl· can attack **another** CH_4 molecule, and so on, until all the Cl_2 or CH_4 molecules are wiped out.

Termination reactions — free radicals are mopped up.

1) If two free radicals join together, they make a **stable molecule**.
2) There are **heaps** of possible termination reactions. Here are a couple of them to give you the idea: $Cl\cdot + \cdot CH_3 \rightarrow CH_3Cl$
$\cdot CH_3 + \cdot CH_3 \rightarrow C_2H_6$

Some products formed will be trace impurities in the final sample.

The reaction between bromine and methane works in exactly the same way.

$$CH_4 + Br_2 \xrightarrow{UV} CH_3Br + HBr$$

Alkanes — Substitution Reactions

The Problem is — You End Up With a **Mixture of Products**

1) The big problem with free-radical substitution is that you **don't only get chloromethane**, but a **mixture of products**.
2) If there's **too much chlorine** in the reaction mixture, some of the remaining **hydrogen atoms** on the **chloromethane molecule** will be swapped for chlorine atoms.
   The propagation reactions happen again, this time to make **dichloromethane**.

$$Cl\cdot + CH_3Cl \rightarrow CH_2Cl\cdot + HCl$$

$$CH_2Cl\cdot + Cl_2 \rightarrow \mathbf{CH_2Cl_2} + Cl\cdot$$

dichloromethane

3) It doesn't stop there. Another substitution reaction can take place to form **trichloromethane**.

$$Cl\cdot + CH_2Cl_2 \rightarrow CHCl_2\cdot + HCl$$

$$CHCl_2\cdot + Cl_2 \rightarrow \mathbf{CHCl_3} + Cl\cdot$$

trichloromethane

4) **Tetrachloromethane** (CCl_4) is formed in the last possible substitution. There are no more hydrogens attached to the carbon atom, so the substitution process has to stop.
5) So the end product is a mixture of CH_3Cl, CH_2Cl_2, $CHCl_3$ and CCl_4. This is a nuisance, because you have to separate the **chloromethane** from the other three unwanted by-products.
6) The best way of reducing the chance of these by-products forming is to have an **excess of methane**. This means there's a greater chance of a chlorine radical colliding only with a **methane molecule** and not a **chloromethane molecule**.

Practice Questions

Q1 What's a free radical?

Q2 What's homolytic fission?

Q3 What's photodissociation?

Q4 Complete this equation: $CH_4 + Cl_2 \xrightarrow{UV}$

Q5 Write down three possible products, other than chloromethane, from the photochemical reaction between CH_4 and Cl_2.

Exam Questions

Q1 When irradiated with UV light, methane gas will react with bromine to form a mixture of several organic compounds.

(a) Name the type of mechanism involved in this reaction. [1 mark]

(b) Write an overall equation to show the formation of bromomethane from methane and bromine. [1 mark]

(c) Write down the two equations in the propagation step for the formation of CH_3Br. [2 marks]

(d) (i) Explain why a tiny amount of ethane is found in the product mixture. [1 mark]

(ii) Name the mechanistic step that leads to the formation of ethane. [1 mark]

(iii) Write the equation for the formation of ethane in this reaction. [1 mark]

(e) Name the major product formed when a large excess of bromine reacts with methane in the presence of UV light. [1 mark]

Q2 The alkane ethane is a saturated hydrocarbon. It is mostly unreactive, but will react with bromine in a photochemical reaction.

Write an equation and outline the mechanism for the photochemical reaction of bromine with ethane. You should assume ethane is in excess. [6 marks]

This page is like... totally radical, man...

Mechanisms can be an absolute pain in the bum to learn, but unfortunately reactions are what Chemistry's all about. If you don't like it, you should have taken art — no mechanisms in that, just pretty pictures. Ah well, there's no going back now. You've just got to sit down and learn the stuff. Keep hacking away at it, till you know it all off by heart.

Alkenes and Polymers

Alkenes are short. But join lots of them together and you get polymers, which are very long. That's these pages in a nutshell.

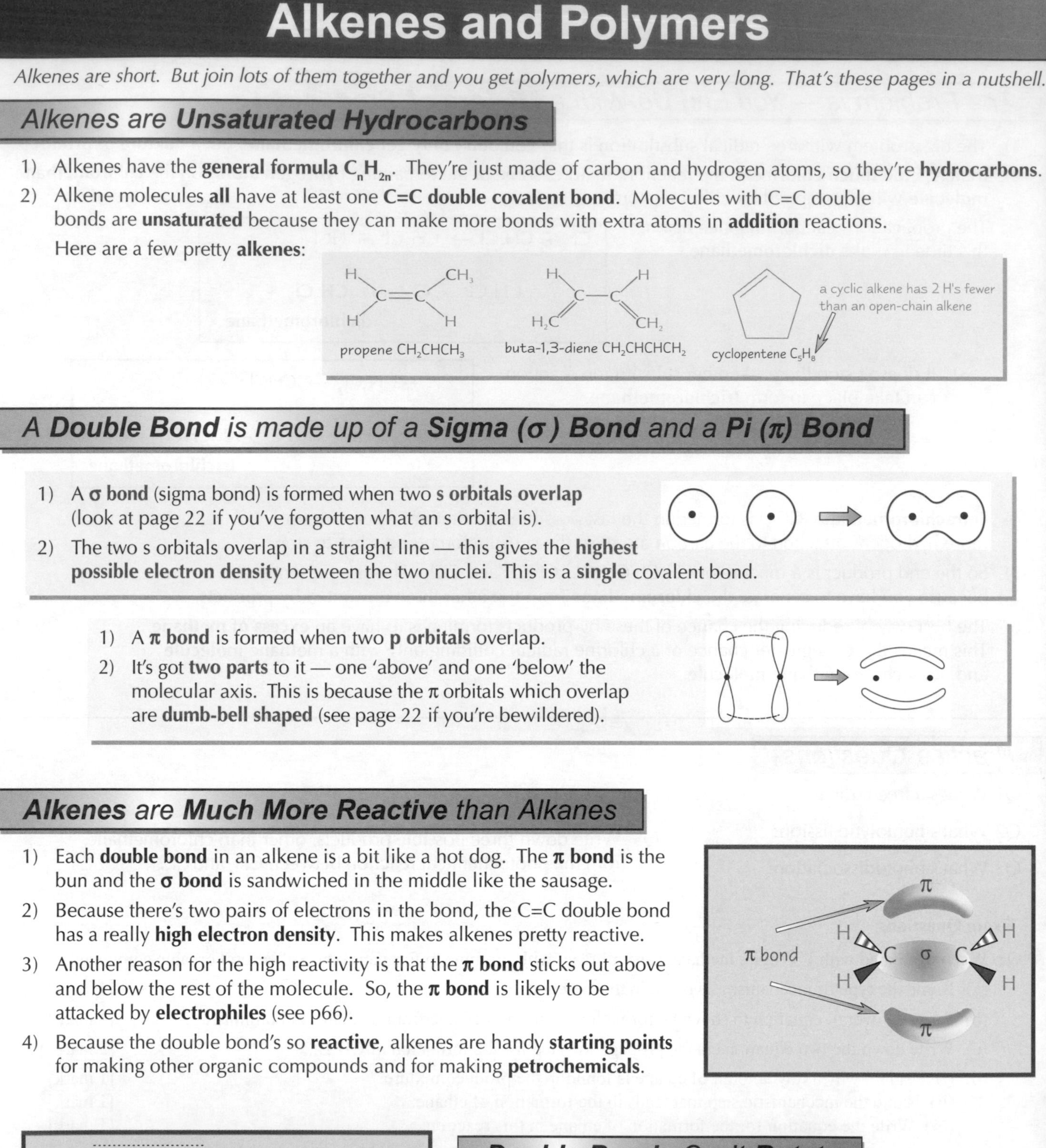

Alkenes are Unsaturated Hydrocarbons

1) Alkenes have the **general formula C_nH_{2n}**. They're just made of carbon and hydrogen atoms, so they're **hydrocarbons**.
2) Alkene molecules **all** have at least one **C=C double covalent bond**. Molecules with C=C double bonds are **unsaturated** because they can make more bonds with extra atoms in **addition** reactions.

Here are a few pretty **alkenes**:

A Double Bond is made up of a Sigma (σ) Bond and a Pi (π) Bond

1) A **σ bond** (sigma bond) is formed when two **s orbitals overlap** (look at page 22 if you've forgotten what an s orbital is).
2) The two s orbitals overlap in a straight line — this gives the **highest possible electron density** between the two nuclei. This is a **single** covalent bond.

1) A **π bond** is formed when two **p orbitals** overlap.
2) It's got **two parts** to it — one 'above' and one 'below' the molecular axis. This is because the π orbitals which overlap are **dumb-bell shaped** (see page 22 if you're bewildered).

Alkenes are Much More Reactive than Alkanes

1) Each **double bond** in an alkene is a bit like a hot dog. The **π bond** is the bun and the **σ bond** is sandwiched in the middle like the sausage.
2) Because there's two pairs of electrons in the bond, the C=C double bond has a really **high electron density**. This makes alkenes pretty reactive.
3) Another reason for the high reactivity is that the **π bond** sticks out above and below the rest of the molecule. So, the **π bond** is likely to be attacked by **electrophiles** (see p66).
4) Because the double bond's so **reactive**, alkenes are handy **starting points** for making other organic compounds and for making **petrochemicals**.

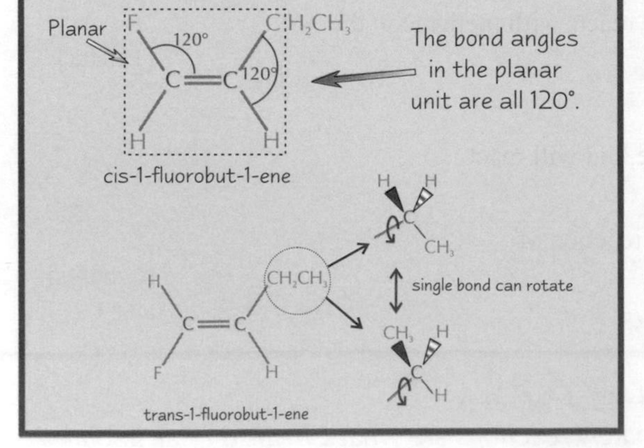

Double Bonds Can't Rotate

1) The carbon atoms in the C=C double bond can't **rotate**. This is because the p orbitals have to **overlap** to form a **π bond**. The C=C double bond and the atoms bonded to these carbons are **planar** (flat) and **rigid** (they can't bend or twist much).
2) Ethene, C_2H_4, is completely planar, but in larger alkenes, only the >C=C< unit is planar — atoms can still rotate around other **single bonds** within the molecule.
3) The **restricted rotation** around the C=C double bond is what causes **cis-trans** or **E/Z isomerism** (see p51).

Alkenes and Polymers

Alkenes **Join Up** to form **Addition Polymers**

1) The **double bonds** in alkenes can open up and join together to make long chains called **polymers**. It's kind of like they're holding hands in a big line. The individual, small alkenes are called **monomers**.
2) This is called **addition polymerisation**. For example, **poly(ethene)** is made by the **addition polymerisation** of **ethene**.

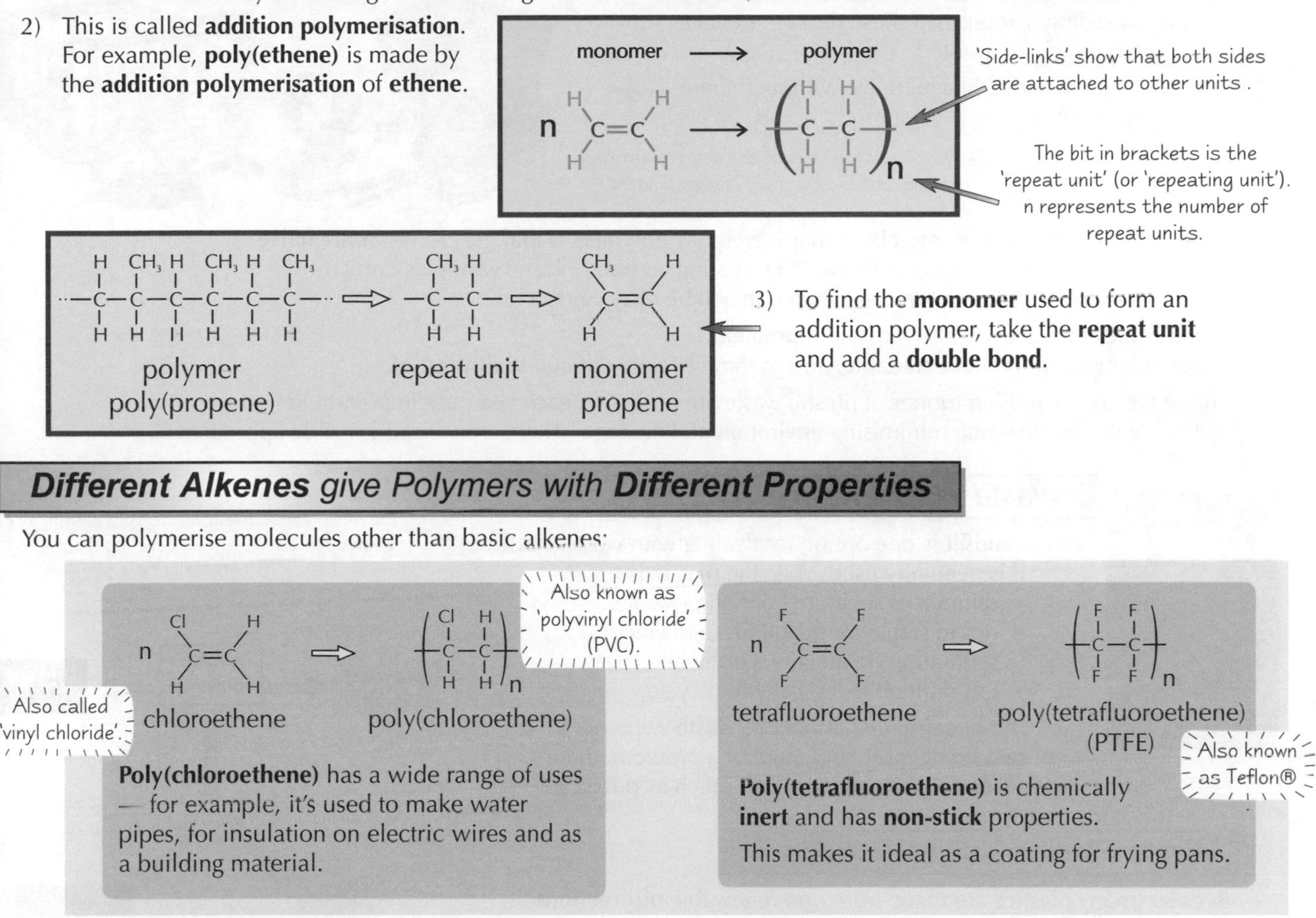

3) To find the **monomer** used to form an addition polymer, take the **repeat unit** and add a **double bond**.

Different Alkenes give Polymers with **Different Properties**

You can polymerise molecules other than basic alkenes:

Poly(chloroethene) has a wide range of uses — for example, it's used to make water pipes, for insulation on electric wires and as a building material.

Poly(tetrafluoroethene) is chemically **inert** and has **non-stick** properties. This makes it ideal as a coating for frying pans.

Practice Questions

Q1 What is an alkene?

Q2 Describe the arrangement of electrons in a single bond and in a double bond.

Q3 What is addition polymerisation?

Exam Question

Q1 One of the most important products made from crude oil is ethene.

a) Draw diagrams to show how the orbitals interact in the bonding between the carbon atoms in ethene. [3 marks]

b) Describe the shape of an ethene molecule and explain, in terms of bonding, why it has this shape. [3 marks]

c) Explain why ethene is particularly useful to the petrochemical industry. [2 marks]

d) Chloroethene CH_2=CHCl forms the polymer poly(chloroethene), commonly known as PVC. Write an equation for the polymerisation of chloroethene, including a full structural formula showing the repeating unit in poly(chloroethene). [2 marks]

Alkenes — join up today, your polymer needs YOU...

There's a lot here, make no mistake. Some of it's kinda tricky too... that σ and π bond stuff, for example. The rest should be more straightforward — but just make <u>very sure</u> that if they give you the structure of a polymer in the exam, you could draw the monomer that it's made from (including that vital double bond). And vice versa, of course. Okay... cup of tea time.

Polymers and the Environment

Polymers are amazingly useful. But they have one big drawback...

Polymers — **Useful** but Difficult to **Get Rid Of**

1) Synthetic polymers have loads of **advantages**, so they're incredibly widespread these days — we take them pretty much for granted.
Just imagine what you'd have to live without if there were no polymers...

(Okay... I could live without the polystyrene head, but the rest of this stuff is pretty useful.)

2) One of the really useful things about many everyday polymers is that they're very **unreactive**. This means food doesn't react with the PTFE coating on pans, plastic windows don't rot, plastic crates can be left out in the rain and they'll be okay, and so on.
3) But this **lack** of reactivity also leads to a **problem**. Most polymers aren't **biodegradable**, and so they're really difficult to **dispose of**.
4) In the UK over **2 million** tonnes of plastic waste are produced each year. It's important to find ways to get rid of this waste while minimising **environmental damage**. There are various possible approaches...

Waste Plastics can be **Buried**

1) **Landfill** is one option for dealing with waste plastics. It is generally used when the plastic is:
   - difficult to separate from other waste,
   - not in sufficient quantities to make separation financially worthwhile,
   - too difficult technically to recycle.
2) But because the **amount of waste** we generate is becoming more and more of a problem, there's a need to **reduce** landfill as much as possible.

Landfill means taking waste to a landfill site, compacting it, and then covering it with soil.

Waste plastics → burying in landfill; Waste plastics → burning as fuel (see below); Waste plastics → sorting → cracking → processing → other chemicals / new plastics; sorting → remoulding → new objects

Waste Plastics can be **Recycled**

Because many plastics are made from non-renewable **oil-fractions**, it makes sense to recycle plastics as much as possible.

There's more than one way to recycle plastics.
After **sorting** into different types:

- some plastics (poly(propene), for example) can be **melted** and **remoulded**,
- some plastics can be **cracked** into **monomers**, and these can be use to make more plastics or other chemicals.

Plastic products are usually marked to make sorting easier. The different numbers show different polymers
e.g. ♲3 = PVC, and ♲5 = poly(propene)

Waste Plastics can be **Burned**

1) If recycling isn't possible for whatever reason, waste plastics can be burned — and the heat can be used to generate **electricity**.
2) This process needs to be carefully **controlled** to reduce **toxic** gases. For example, polymers that contain **chlorine** (such as **PVC**) produce **HCl** when they're burned — this has to be removed.
3) Waste gases from the combustion are passed through **scrubbers** which can **neutralise** gases such as HCl by allowing them to react with a **base**.

Rex and Dirk enjoy some waist plastic.

Polymers and the Environment

Biodegradable Polymers **Decompose** *in the* **Right Conditions**

Scientists can now make **biodegradable** polymers — ones that naturally **decompose**.

1) **Biodegradable polymers** decompose pretty quickly in certain conditions — because organisms can digest them. (You might get asked about 'compostable' polymers as well as 'biodegradable' ones. These two terms mean more or less the same thing — 'compostable' just means it has to decay fairly quickly, "at the speed of compost".)
2) Biodegradable polymers can be made from materials such as **starch** (from maize and other plants) and from the hydrocarbon **isoprene** (2-methyl-1,3-butadiene). So, biodegradable polymers can be produced from **renewable** raw materials or from **oil fractions**:

> Using **renewable** raw material has several **advantages**.
> (i) Raw materials aren't going to **run out** like oil will.
> (ii) When polymers biodegrade, **carbon dioxide** (a greenhouse gas — see p92) is produced. If your polymer is **plant-based**, then the CO_2 released as it decomposes is the same CO_2 absorbed by the plant when it grew. But with an **oil-based** biodegradable polymer, you're effectively transferring carbon from the oil to the atmosphere.
> (iii) Over their 'lifetime' some plant-based polymers **save energy** compared to oil-based plastics.

But whatever raw material you use, at the moment the energy for making polymers usually comes from fossil fuels.

3) Even though they're biodegradable, these polymers still need the right conditions before they'll decompose. You **couldn't** necessarily just put them in a landfill and expect them to perish away — because there's a lack of moisture and oxygen under all that compressed soil. You need to chuck them on a big compost heap. This means that you still need to **collect** and **separate** the biodegradable polymers from non-biodegradable plastics. At the moment, they're also **more expensive** than oil-based equivalents.

4) There are various potential uses — e.g. plastic sheeting used to protect plants from the frost can be made from poly(ethene) with **starch grains** embedded in it. In time the starch is broken down by **microorganisms** and the remaining poly(ethene) crumbles into dust. There's no need to collect and dispose of the old sheeting.

Practice Questions

Q1 Many plastics are unreactive. Describe one benefit and one disadvantage of this.

Q2 Which harmful gas is produced during the combustion of PVC?

Q3 Describe three ways in which used polymers such as poly(propene) can be handled.

Q4 What is a compostable polymer? Name two things compostable polymers can be made from.

Exam Questions

Q1 Waste plastics can be disposed of by burning.

a) Describe one advantage of disposing of waste plastics by burning. [1 mark]

b) Describe a disadvantage of burning waste plastic that contains chlorine, and explain how the impact of this disadvantage could be reduced. [2 marks]

Q2 Describe one way in which waste poly(propene) could be recycled into new plastic objects. [2 marks]

Q3 Apart from being biodegradable, describe TWO benefits of using starch- or maize-based polymers instead of oil-based polymers. [2 marks]

Phil's my recycled plastic plane — but I don't know where to land Phil...

You might have noticed that all this recycling business is a hot topic these days. And not just in the usual places, such as Chemistry books. No, no, no... recycling even makes it regularly onto the news as well. This suits examiners just fine — they like you to know how useful and important chemistry is. So learn this stuff, pass your exam, and do some recycling.

Reactions of Alkenes

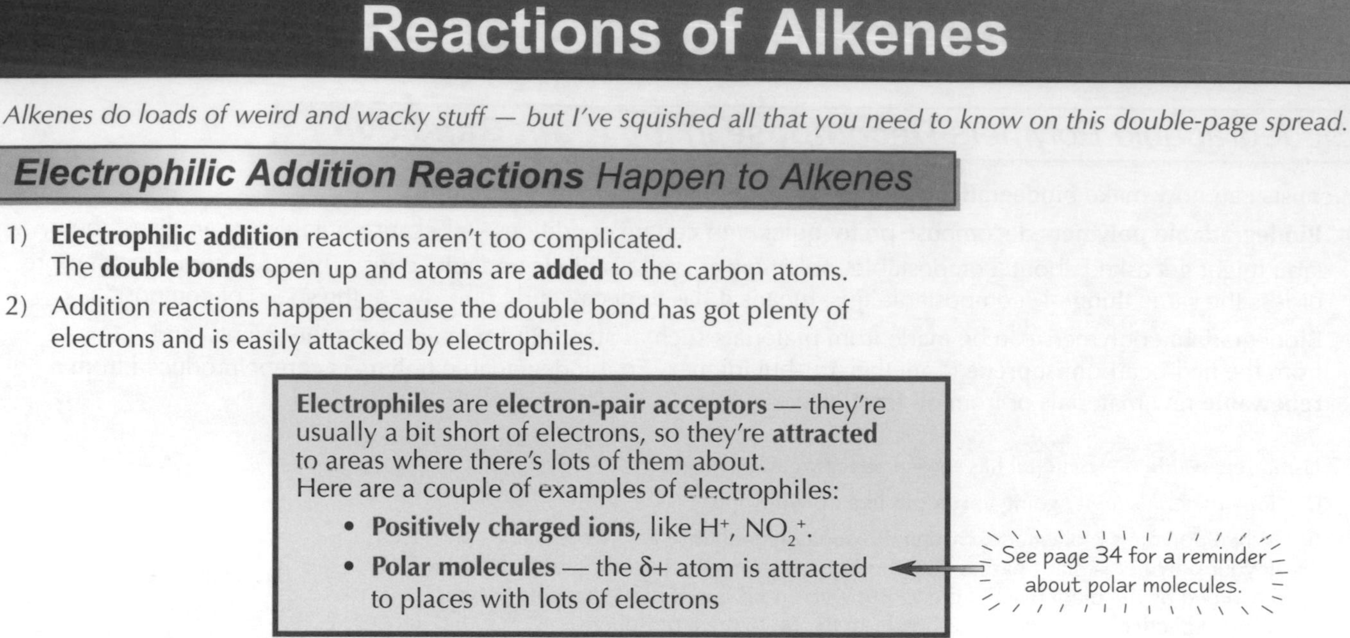

Alkenes do loads of weird and wacky stuff — but I've squished all that you need to know on this double-page spread.

Electrophilic Addition Reactions Happen to Alkenes

1) **Electrophilic addition** reactions aren't too complicated. The **double bonds** open up and atoms are **added** to the carbon atoms.
2) Addition reactions happen because the double bond has got plenty of electrons and is easily attacked by electrophiles.

> **Electrophiles** are **electron-pair acceptors** — they're usually a bit short of electrons, so they're **attracted** to areas where there's lots of them about.
> Here are a couple of examples of electrophiles:
> - **Positively charged ions**, like H^+, NO_2^+.
> - **Polar molecules** — the δ+ atom is attracted to places with lots of electrons

See page 34 for a reminder about polar molecules.

3) The double bond is also **nucleophilic** — it's attracted to places that don't have enough **electrons**.

Adding **Hydrogen** to C=C Bonds Produces **Alkanes**

1) Ethene will react with **hydrogen** gas to produce ethane. It needs a **nickel catalyst** and a temperature of **150 °C** though.

$$H_2C{=}CH_2 + H_2 \xrightarrow[150\ ^\circ C]{Ni} CH_3CH_3$$

2) **Margarine's** made by **'hydrogenating' unsaturated vegetable oils**. By removing some **double bonds**, you raise the **melting point** of the oil so that it becomes **solid** at room temperature.

Use **Bromine Water** to Test for C=C Double Bonds

When you shake an alkene with **orange bromine water**, the solution quickly **decolourises**. Bromine is added across the double bond to form a colourless **dibromoalkane** — this happens by **electrophilic addition**.

Here's the mechanism...

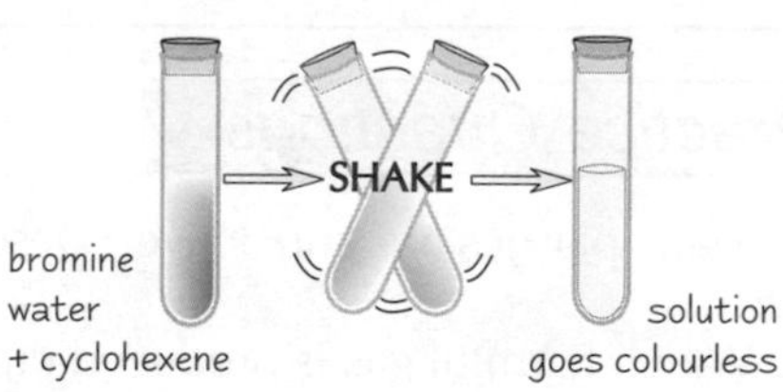

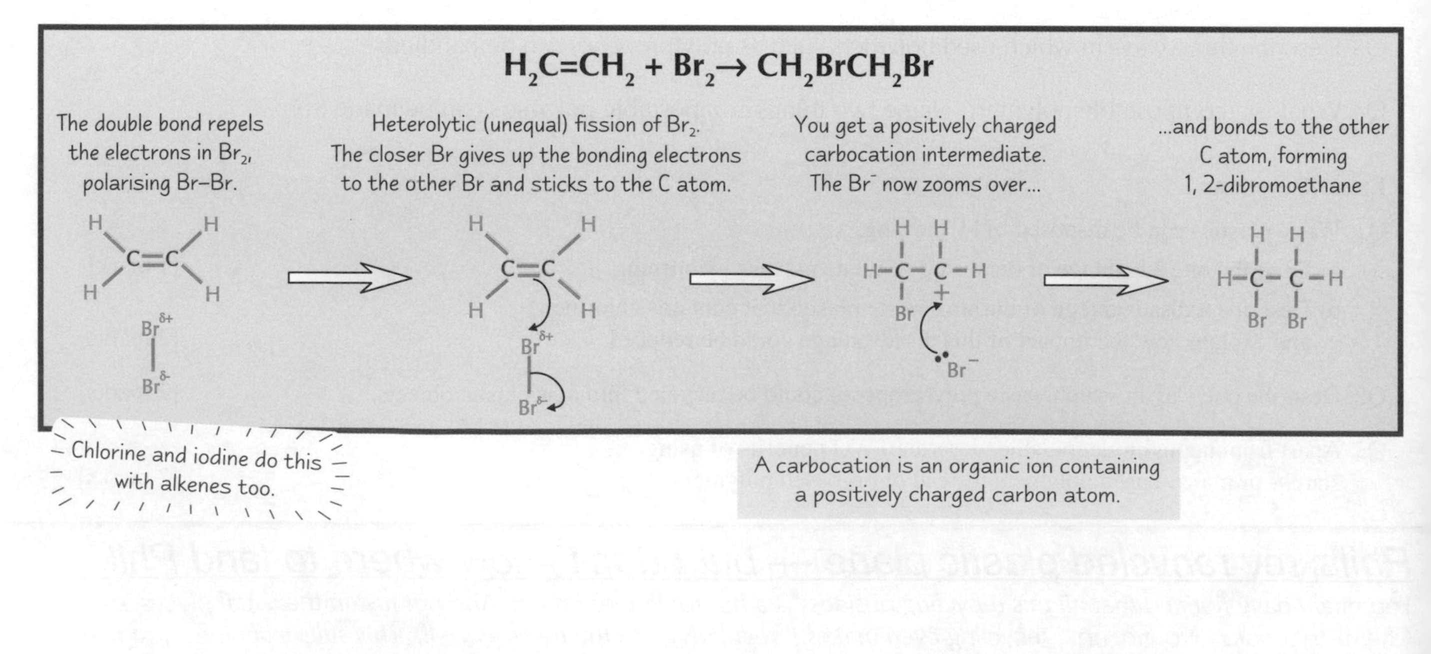

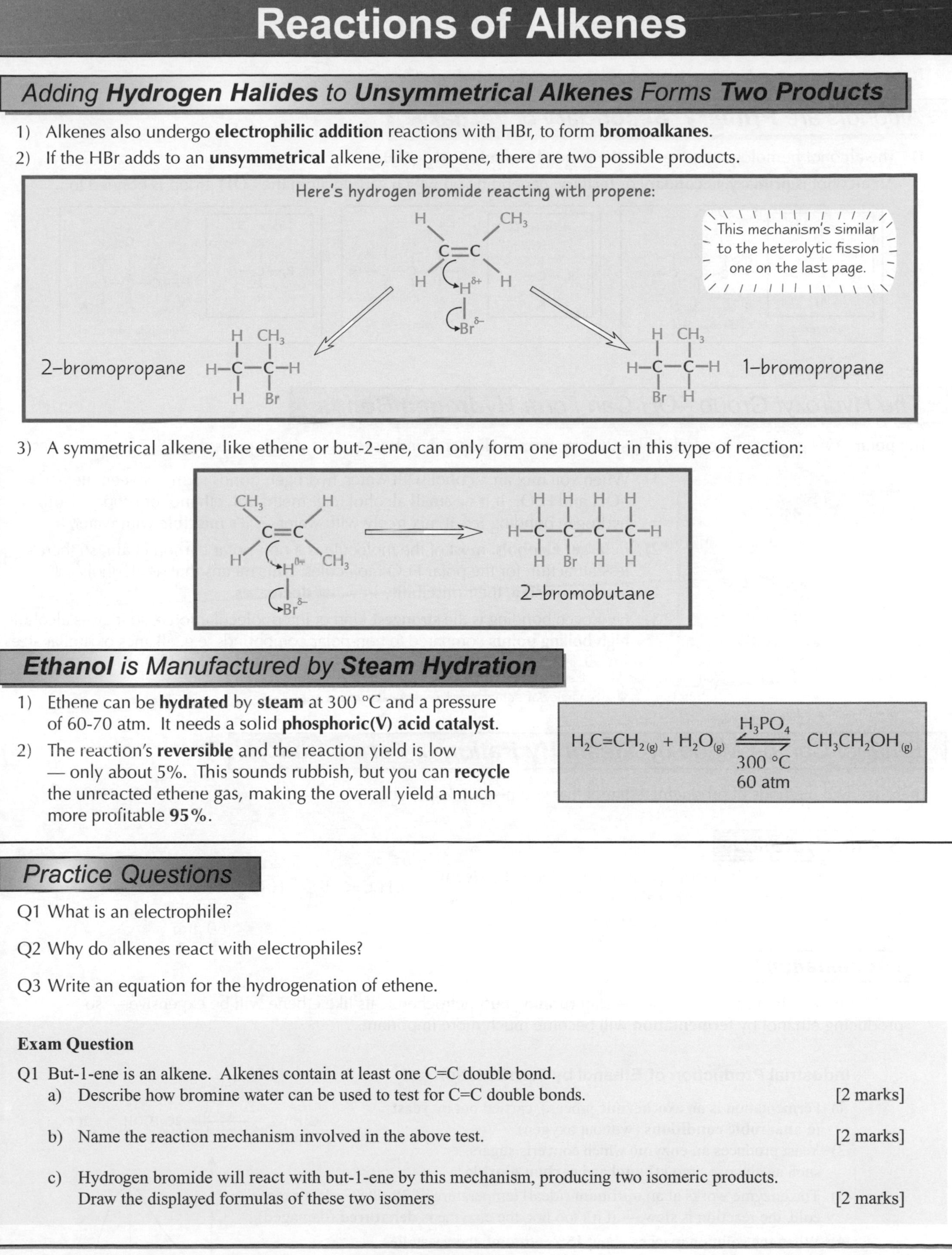

Reactions of Alkenes

*Adding **Hydrogen Halides** to **Unsymmetrical Alkenes** Forms **Two Products***

1) Alkenes also undergo **electrophilic addition** reactions with HBr, to form **bromoalkanes**.
2) If the HBr adds to an **unsymmetrical** alkene, like propene, there are two possible products.

3) A symmetrical alkene, like ethene or but-2-ene, can only form one product in this type of reaction:

*Ethanol is Manufactured by **Steam Hydration***

1) Ethene can be **hydrated** by **steam** at 300 °C and a pressure of 60-70 atm. It needs a solid **phosphoric(V) acid catalyst**.
2) The reaction's **reversible** and the reaction yield is low — only about 5%. This sounds rubbish, but you can **recycle** the unreacted ethene gas, making the overall yield a much more profitable **95%**.

$$H_2C{=}CH_{2(g)} + H_2O_{(g)} \underset{300\ °C,\ 60\ atm}{\overset{H_3PO_4}{\rightleftharpoons}} CH_3CH_2OH_{(g)}$$

Practice Questions

Q1 What is an electrophile?

Q2 Why do alkenes react with electrophiles?

Q3 Write an equation for the hydrogenation of ethene.

Exam Question

Q1 But-1-ene is an alkene. Alkenes contain at least one C=C double bond.

a) Describe how bromine water can be used to test for C=C double bonds. [2 marks]

b) Name the reaction mechanism involved in the above test. [2 marks]

c) Hydrogen bromide will react with but-1-ene by this mechanism, producing two isomeric products. Draw the displayed formulas of these two isomers [2 marks]

This section is free from all GM ingredients...

Mechanisms are another of those classics that examiners just love. You need to know the electrophilic addition examples on these pages, so shut the book and scribble them out. Make sure you know the tests for double bonds too. They aren't as handy in real life as, say, a tin opener, but you won't need a tin opener in the exam. Unless your exam paper comes in a tin.

Alcohols

These two pages could well be enough to put you off alcohols for life...

Alcohols are **Primary**, **Secondary** or **Tertiary**

1) The alcohol homologous series has the **general formula $C_nH_{2n+1}OH$**.
2) An alcohol is **primary**, **secondary** or **tertiary**, depending on which carbon atom the **–OH** group is bonded to...

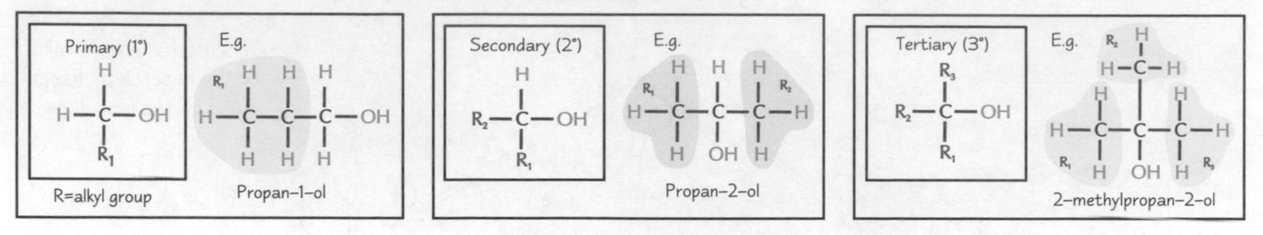

The Hydroxyl Group –OH Can Form **Hydrogen Bonds**

The **polar** –OH group on alcohols helps them to form **hydrogen bonds** (see p35), which gives them certain properties...

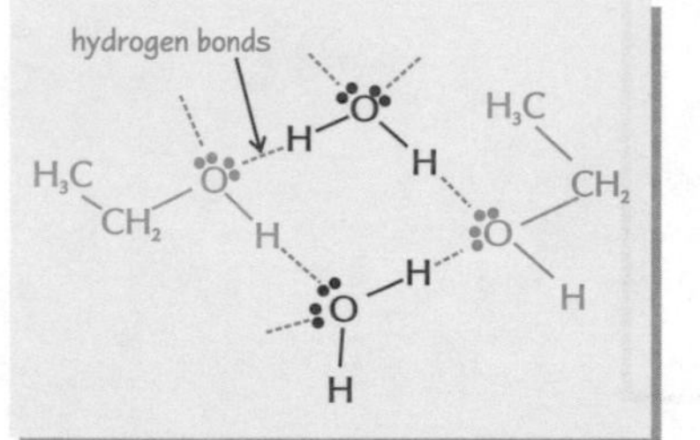

1) When you mix an alcohol with water, hydrogen bonds form between the **–OH** and **H_2O**. If it's a **small** alcohol (e.g. methanol, ethanol or propan-1-ol), hydrogen bonding lets it mix freely with water — it's **miscible** with water.
2) In **larger alcohols**, most of the molecule is a non-polar carbon chain, so there's less attraction for the polar H_2O molecules. This means that as alcohols **increase in size**, their miscibility in water **decreases.**
3) Hydrogen bonding is the **strongest** kind of intermolecular force, so it gives alcohols **high boiling points** compared to non-polar compounds, e.g. alkanes of similar sizes.

You might also hear it said that alcohols have relatively low volatility. Volatility is the tendency of something to evaporate into a gas.

Ethanol Can be Made by **Steam Hydration** or **Fermentation**

There are two methods of producing ethanol that you need to know about:

Steam Hydration

At the moment most industrial ethanol is produced by **steam hydration of ethene** with a **phosphoric acid catalyst** (see p67). The ethene comes from cracking heavy fractions of crude oil.

$$H_2C{=}CH_{2(g)} + H_2O_{(g)} \underset{300\ ^\circ C,\ 60\ atm}{\overset{H_3PO_4}{\rightleftharpoons}} CH_3CH_2OH_{(g)}$$

Fermentation

In the future, when crude oil supplies start **running out**, petrochemicals like ethene will be expensive — so producing ethanol by **fermentation** will become much more important...

Industrial Production of Ethanol by Fermentation

1) Fermentation is an **exothermic** process, carried out by **yeast** in **anaerobic conditions** (without oxygen).
2) Yeast produces an **enzyme** which converts sugars, such as glucose, into **ethanol** and **carbon dioxide**.
3) The enzyme works at an **optimum** (ideal) temperature of **30-40 °C**. If it's too cold, the reaction is **slow** — if it's too hot, the enzyme is **denatured** (damaged).
4) When the solution reaches about **15% ethanol**, the yeast dies. **Fractional distillation** is used to increase the concentration of ethanol.
5) Fermentation is **low-tech** — it uses cheap equipment and **renewable resources**. The ethanol produced by this method has to be **purified** though.

$$\underset{\text{glucose}}{C_6H_{12}O_{6(aq)}} \xrightarrow[\text{yeast}]{\text{warm}} 2C_2H_5OH_{(aq)} + 2CO_{2(g)}$$

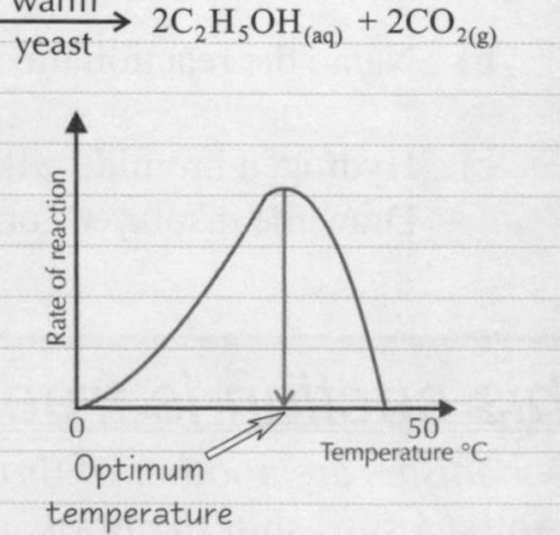

Alcohols

Alcohols Have a **Wide Variety** of Uses

1) **Ethanol** is the alcohol found in **alcoholic drinks.**
2) **Methylated spirits** is an industrial **solvent**. It's basically ethanol, with some methanol and purple dye added to make it **undrinkable** and tax-exempt. Ethanol will dissolve **polar**, **non-polar** and some **ionic compounds.**
3) Ethanol is also being used increasingly as a **fuel**, particularly in countries with few oil reserves.
4) **Unleaded petrol** contains 5% methanol and 15% MTBE (an ether made from methanol) to improve combustion.
5) Methanol is important as a **feedstock** (starting point) for manufacturing organic chemicals, e.g. plastics and dyes.

Alcohols can be **Dehydrated** to Form **Alkenes**

1) You can make ethene by **eliminating** water from **ethanol** in a **dehydration reaction.**

$$C_2H_5OH \rightarrow CH_2{=}CH_2 + H_2O$$

2) The ethanol is mixed with an **acid catalyst** and heated to **170 °C**.
3) The ethene that is produced is collected over water.
4) The acid catalyst used is either **concentrated sulfuric acid** (H_2SO_4) or **concentrated phosphoric acid** (H_3PO_4)

Reacting a **Carboxylic Acid** With **Ethanol** Produces an **Ester**

1) If you warm **ethanol** with a **carboxylic acid** (like ethanoic acid) and a **strong acid catalyst** (concentrated sulfuric acid will do), it forms an ester (**ethyl ethanoate** in this case).
2) The **O–H** bond in ethanol is broken in the **esterification** reaction.

$$C_2H_5OH + CH_3COOH \rightleftharpoons CH_3C(=O){-}O{-}CH_2CH_3 + H_2O$$

This stuff smells of pear drops — esters generally smell fruity.

Practice Questions

Q1 What is the general formula for an alcohol?

Q2 How do the boiling points of alcohols compare with the boiling points of similarly sized alkanes?

Q3 Give three uses of alcohols.

Exam Questions

Q1 Butanol C_4H_9OH has four chain and positional isomers. Name each isomer and class it as primary, secondary or tertiary. [8 marks]

a) $CH_3CH_2CH_2CH_2OH$ (H–C–C–C–C–OH, all other positions H)

b) $CH_3CH(OH)CH_3$ with a CH_3 on the central... (H–C–C–C–H with CH₃ above and OH below the central carbon)

c) $CH_3CH_2CH(OH)CH_3$ (H–C–C–C–C–H with OH on the third carbon)

d) $HOCH_2CH(CH_3)CH_3$ (H–C–C–C–H with CH₂OH above the central carbon)

[6 marks]

Q2 Ethanol is a useful alcohol.

a) State whether ethanol is a primary, secondary or tertiary alcohol, and explain why. [2 marks]

b) Industrially, ethanol can be produced by fermentation of glucose, $C_6H_{12}O_6$.
(i) Write a balanced equation for this reaction. [1 mark]
(ii) State two optimum conditions for fermentation. [2 marks]

c) At present most ethanol is produced by the acid-catalysed hydration of ethene. Why is this? Why might this change in the future? [3 marks]

Euuurghh, what a page... I think I need a drink...

Not much to learn here — a few basic definitions, some fiddly explanations of properties in terms of bonding, 4 or 5 uses, a couple of industrial processes, a dehydration reaction, an esterification reaction... Like I said, not much here at all. Think I'm going to faint. *[THWACK]*

Oxidation of Alcohols

Another page of alcohol reactions. Probably not what you wanted for Christmas...

*The Simple way to Oxidise Alcohols is to **Burn Them***

It doesn't take much to set ethanol alight and it burns with a **pale blue flame**. The C–C and C–H bonds are broken as the ethanol is **completely oxidised** to make carbon dioxide and water. This is a **combustion** reaction.

$$C_2H_5OH_{(l)} + 3O_{2(g)} \rightarrow 2CO_{2(g)} + 3H_2O_{(g)}$$

If you burn any alcohol along with plenty of oxygen you get carbon dioxide and water as products.
But if you want to end up with something more interesting, you need a more sophisticated way of oxidising...

*How Much an Alcohol can be **Oxidised** Depends on its **Structure***

You can use the **oxidising agent acidified potassium dichromate(VI)** ($K_2Cr_2O_7/H_2SO_4$) to **mildly** oxidise alcohols.

- **Primary** alcohols are oxidised to **aldehydes** and then to **carboxylic acids**.
- **Secondary** alcohols are oxidised to **ketones** only.
- **Tertiary** alcohols won't be oxidised.

The orange dichromate(VI) ion is reduced to the green chromium(III) ion, Cr^{3+}.

Aldehydes and **ketones** are **carbonyl** compounds — they have the functional group C=O.
Their general formula is $\mathbf{C_nH_{2n}O}$.

1) **Aldehydes** have a **hydrogen** and **one alkyl group** attached to the carbonyl carbon atom.
E.g.

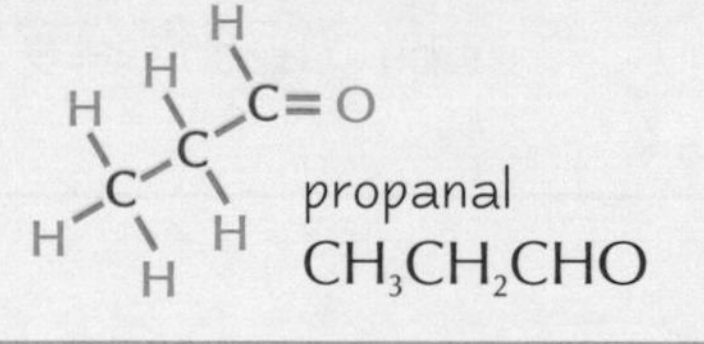

propanal
CH_3CH_2CHO

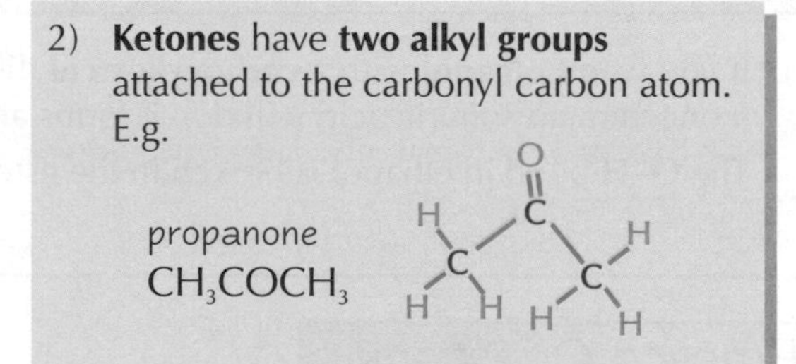

2) **Ketones** have **two alkyl groups** attached to the carbonyl carbon atom.
E.g.

propanone
CH_3COCH_3

Primary** Alcohols will Oxidise to **Aldehydes** and **Carboxylic Acids

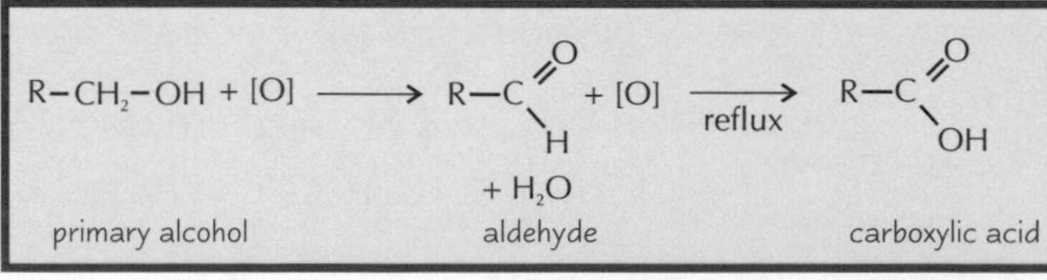

[O] = oxidising agent

You can control how **far** the alcohol is oxidised by controlling the **reaction conditions** (see next page).

Secondary** Alcohols will Oxidise to **Ketones

$$R_1-\overset{\displaystyle H}{\underset{\displaystyle R_2}{C}}-OH + [O] \xrightarrow[\text{reflux}]{} \begin{matrix} R_1 \\ R_2 \end{matrix}\!\!>C=O + H_2O$$

1) Refluxing a secondary alcohol, e.g. propan-2-ol, with acidified dichromate(VI) will produce a **ketone**.
2) Ketones can't be oxidised easily, so even prolonged refluxing won't produce anything more.

***Tertiary** Alcohols can't be Oxidised Easily*

Tertiary alcohols don't react with potassium dichromate(VI) at all — the solution stays orange. The only way to oxidise tertiary alcohols is by **burning** them.

Oxidation of Alcohols

Distil for an **Aldehyde**, and **Reflux** for a **Carboxylic Acid**

You can control how **far** a primary alcohol is oxidised by controlling the **reaction conditions**:

Oxidising Primary Alcohols

1) Gently heating ethanol with potassium dichromate(VI) solution and sulfuric acid in a test tube should produce "apple" smelling **ethanal** (an aldehyde). However, it's **really tricky** to control the amount of heat and the aldehyde is usually oxidised to form "vinegar" smelling **ethanoic acid**.

Reflux Apparatus

water out
Liebig condenser
water in
round bottomed flask
anti-bumping granules (added to make boiling smoother)
heat

2) To get just the **aldehyde**, you need to get it out of the oxidising solution **as soon** as it's formed. You can do this by gently heating excess alcohol with a **controlled** amount of oxidising agent in **distillation apparatus**, so the aldehyde (which boils at a lower temperature than the alcohol) is distilled off **immediately**.

3) To produce the **carboxylic acid**, the alcohol has to be **vigorously oxidised**. The alcohol is mixed with excess oxidising agent and heated under **reflux**. Heating under reflux means you can increase the **temperature** of an organic reaction to boiling without losing **volatile** solvents, reactants or products. Any vaporised compounds are cooled, condense and drip back into the reaction mixture. Handy, hey?

Practice Questions

Q1 What's the difference between an aldehyde and a ketone?

Q2 What will acidified potassium dichromate(VI) oxidise secondary alcohols to?

Q3 What is the colour change when potassium dichromate(VI) is reduced?

Q4 Why are anti-bumping granules used in distillation and reflux?

Exam Question

Q1 A student wanted to produce the aldehyde propanal from propanol, and set up reflux apparatus using acidified potassium dichromate(VI) as the oxidising agent.

a) Draw a labelled diagram of reflux apparatus. Explain why reflux apparatus is arranged in this way. [2 marks]

b) The student tested his product and found that he had not produced propanal.
(i) What is the student's product? [1 mark]
(ii) Write equations to show the two-stage reaction. You may use [O] to represent the oxidising agent. [2 marks]
(iii) What technique should the student have used and why? [2 marks]

c) The student also tried to oxidise 2-methylpropan-2-ol, unsuccessfully.
(i) Draw the full structural formula for 2-methylpropan-2-ol. [1 mark]
(ii) Why is it not possible to oxidise 2-methylpropan-2-ol with an oxidising agent? [1 mark]

I.... I .... just can't do it, R2...

Don't give up now. Only as a fully-trained Chemistry Jedi, with the force as your ally, can you take on the Examiner. If you quit now, if you choose the easy path as Wader did, all the marks you've fought for will be lost. Be strong. Don't give in to hate — that leads to the dark side... *(Only a few more pages to go before you're done with this section...)*

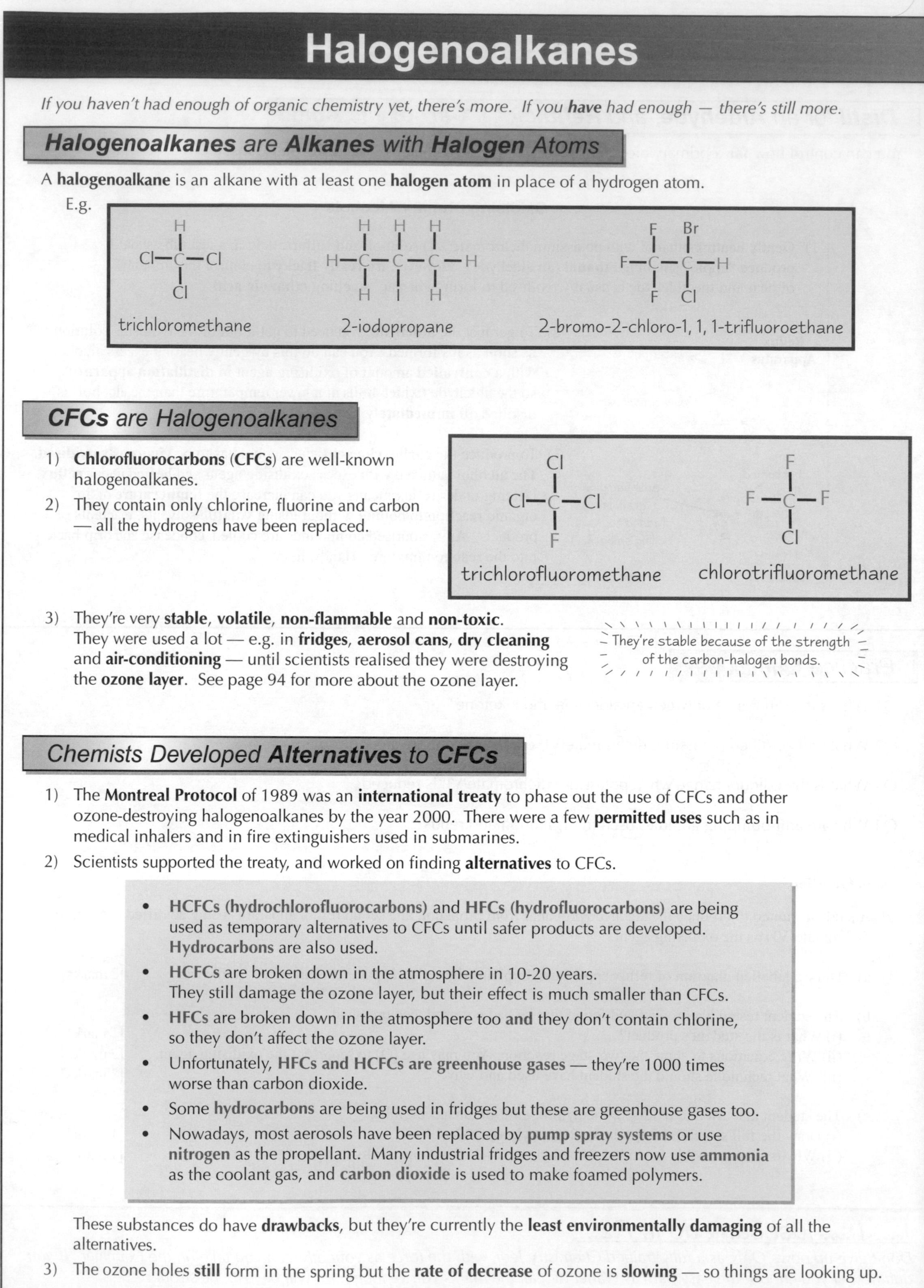

Halogenoalkanes

*If you haven't had enough of organic chemistry yet, there's more. If you **have** had enough — there's still more.*

Halogenoalkanes are Alkanes with Halogen Atoms

A **halogenoalkane** is an alkane with at least one **halogen atom** in place of a hydrogen atom.

E.g.

trichloromethane — 2-iodopropane — 2-bromo-2-chloro-1, 1, 1-trifluoroethane

CFCs are Halogenoalkanes

1) **Chlorofluorocarbons** (**CFCs**) are well-known halogenoalkanes.
2) They contain only chlorine, fluorine and carbon — all the hydrogens have been replaced.

trichlorofluoromethane — chlorotrifluoromethane

3) They're very **stable**, **volatile**, **non-flammable** and **non-toxic**. They were used a lot — e.g. in **fridges**, **aerosol cans**, **dry cleaning** and **air-conditioning** — until scientists realised they were destroying the **ozone layer**. See page 94 for more about the ozone layer.

They're stable because of the strength of the carbon-halogen bonds.

Chemists Developed Alternatives to CFCs

1) The **Montreal Protocol** of 1989 was an **international treaty** to phase out the use of CFCs and other ozone-destroying halogenoalkanes by the year 2000. There were a few **permitted uses** such as in medical inhalers and in fire extinguishers used in submarines.
2) Scientists supported the treaty, and worked on finding **alternatives** to CFCs.

- **HCFCs (hydrochlorofluorocarbons)** and **HFCs (hydrofluorocarbons)** are being used as temporary alternatives to CFCs until safer products are developed. **Hydrocarbons** are also used.
- **HCFCs** are broken down in the atmosphere in 10-20 years. They still damage the ozone layer, but their effect is much smaller than CFCs.
- **HFCs** are broken down in the atmosphere too **and** they don't contain chlorine, so they don't affect the ozone layer.
- Unfortunately, **HFCs and HCFCs are greenhouse gases** — they're 1000 times worse than carbon dioxide.
- Some **hydrocarbons** are being used in fridges but these are greenhouse gases too.
- Nowadays, most aerosols have been replaced by **pump spray systems** or use **nitrogen** as the propellant. Many industrial fridges and freezers now use **ammonia** as the coolant gas, and **carbon dioxide** is used to make foamed polymers.

These substances do have **drawbacks**, but they're currently the **least environmentally damaging** of all the alternatives.

3) The ozone holes **still** form in the spring but the **rate of decrease** of ozone is **slowing** — so things are looking up.

Halogenoalkanes

PVC and PTFE are (Polymer) Halogenoalkanes

The plastics PVC and PTFE are halogenoalkanes.
Look back at page 63 and make sure you know how they're made, and some of their uses.

The Carbon–Halogen Bond in Halogenoalkanes is Polar

1) Halogens are much more **electronegative** than carbon. So, the **carbon–halogen bond** is **polar**.
2) The **δ+ carbon** doesn't have enough electrons. This means it can be attacked by a **nucleophile**. A nucleophile's an **electron-rich** ion or molecule. It donates an **electron pair** to somewhere without enough electrons.
3) **OH⁻**, **CN⁻** and **NH_3** are all **nucleophiles** which react with halogenoalkanes. **Water's** a nucleophile too, but it reacts slowly.

$-C^{\delta+}-Br^{\delta-}$

Halogenoalkanes can be Hydrolysed to make Alcohols

Bromoethane can be **hydrolysed** to **ethanol**.
You have to use **warm aqueous sodium** or **potassium hydroxide** or it won't work.

Hydrolysis is when water breaks bonds.

$$CH_3CH_2Br + OH^- \xrightarrow[\text{reflux}]{OH^- / H_2O} C_2H_5OH + Br^-$$

If you don't know what 'reflux' is check out page 71.

Here's what happens. It's a nice simple **one-step mechanism**.

This is a nucleophilic substitution reaction.

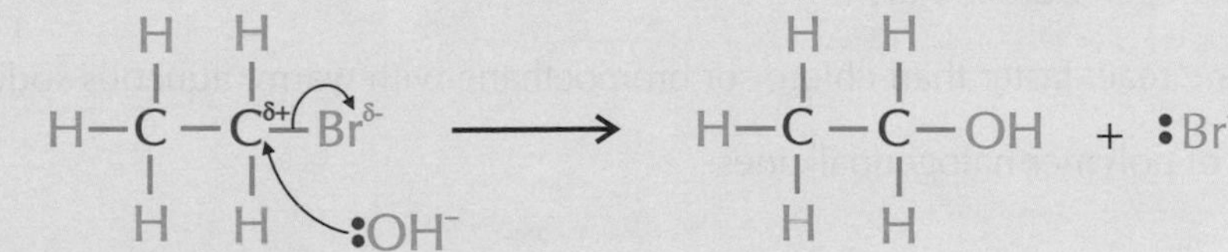

1) OH⁻ is the **nucleophile** which provides a **pair of electrons** for the $C^{\delta+}$.
2) The C-Br bond breaks **heterolytically** — **both** electrons from the bond are taken by **Br⁻**.
3) **Br⁻** falls off as **OH⁻** bonds to the carbon.

Iodoalkanes are Hydrolysed the Fastest

1) How quickly different halogenoalkanes are hydrolysed depends on **bond enthalpy** — see p80 for more on this.
2) **Weaker** carbon-halogen bonds **break** more easily — so they react **faster**.
3) **Iodoalkanes** have the **weakest bonds**, so they hydrolyse the **fastest**.
4) **Fluoroalkanes** have the **strongest bonds**, so they're the **slowest** at hydrolysing.

bond	bond enthalpy kJ mol⁻¹
C–F	467
C–Cl	346
C–Br	290
C–I	228

Faster hydrolysis as bond enthalpy decreases (the bonds are getting weaker).

Halogenoalkanes

Use **Silver Nitrate** to Compare **Reaction Rates** of Halogenoalkanes

1) When you mix a **halogenoalkane** with water, it reacts to form an **alcohol**.

$$R\text{–}X + 2H_2O \rightarrow R\text{–}OH + H_3O^+ + X^-$$

H_2O is the nucleophile.

R–X represents the halogenoalkane — the R stands for the alkyl group, the X is the halogen.

2) If you put **silver nitrate solution** in the mixture too, the silver ions react with the **halide ions** as soon as they form, giving a **silver halide precipitate** (see page 45).

$$Ag^+_{(aq)} + X^-_{(aq)} \rightarrow AgX_{(s)}$$

3) To compare the reactivities, set up four flasks each containing a different halogenoalkane, ethanol (as a solvent) and dilute silver nitrate solution.

You need to use an iodoalkane, a bromoalkane, a chloroalkane and a fluoroalkane. They should be the same in all other respects to make it a fair test.

4) You can 'measure' the rates of the reactions by timing how quickly each silver halide is **precipitated**, using the good old 'timing how long it takes the cross to disappear method' (not its official name). To do this, stick a piece of paper with a **cross** on it under each flask and measure how long it takes until you can't see the cross any more.

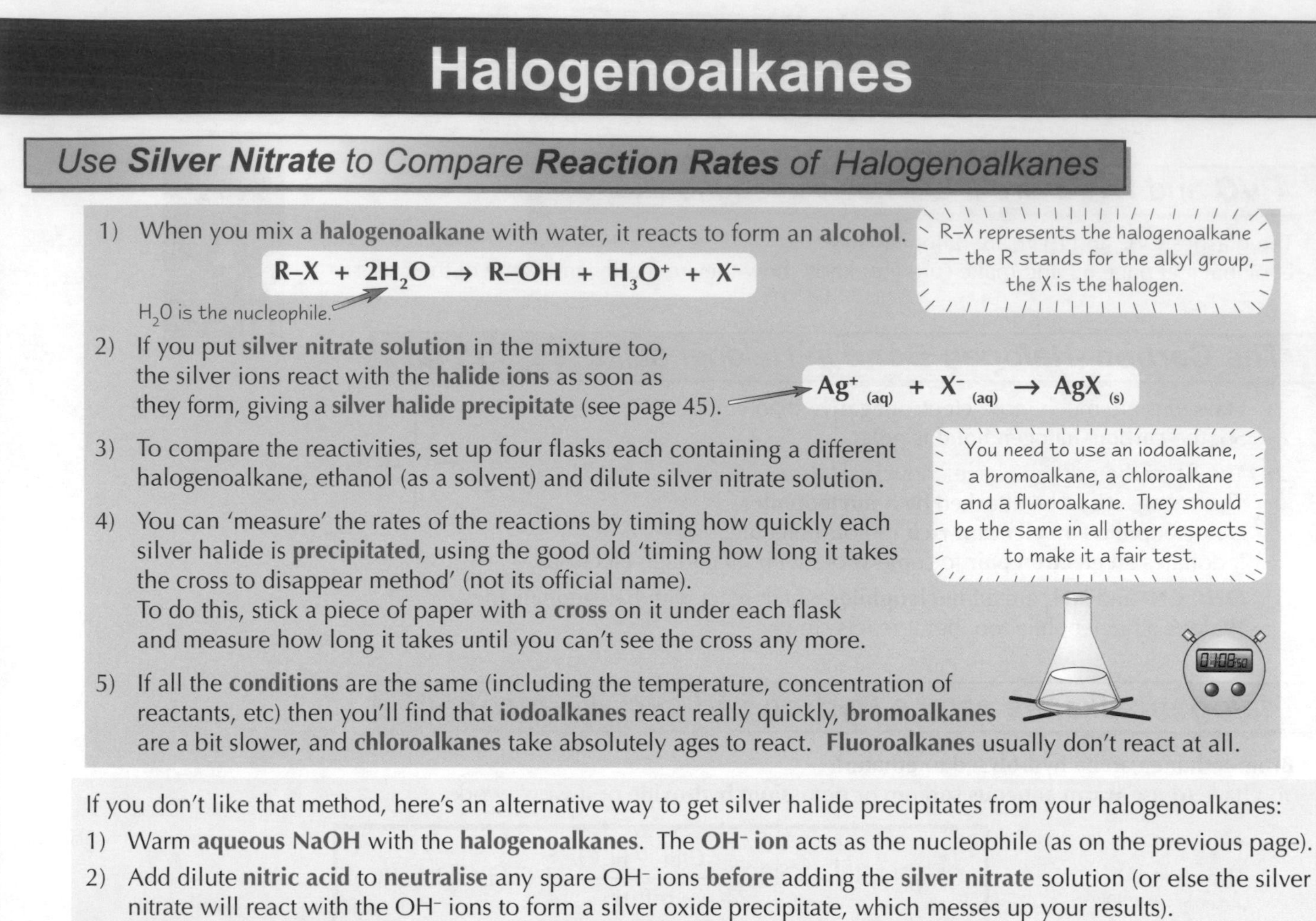

5) If all the **conditions** are the same (including the temperature, concentration of reactants, etc) then you'll find that **iodoalkanes** react really quickly, **bromoalkanes** are a bit slower, and **chloroalkanes** take absolutely ages to react. **Fluoroalkanes** usually don't react at all.

If you don't like that method, here's an alternative way to get silver halide precipitates from your halogenoalkanes:

1) Warm **aqueous NaOH** with the **halogenoalkanes**. The **OH^- ion** acts as the nucleophile (as on the previous page).
2) Add dilute **nitric acid** to **neutralise** any spare OH^- ions **before** adding the **silver nitrate** solution (or else the silver nitrate will react with the OH^- ions to form a silver oxide precipitate, which messes up your results).

Practice Questions

Q1 What is a nucleophile?

Q2 Why is the carbon-halogen bond polar?

Q3 Why does iodoethane react faster than chloro- or bromoethane with warm, aqueous sodium hydroxide?

Q4 Give two examples of polymer halogenoalkanes.

Exam Questions

Q1 Freon-11 (trichlorofluoromethane) is a compound that was used for many years in fridges. Its use is now banned along with other similar compounds.

a) What name is given to this type of halogenoalkane? [1 mark]

b) Give three properties that these compounds have that makes them very useful. [3 marks]

Q2 The halogenoalkane chloromethane is a substance that was formerly used as a refrigerant.

a) Draw the structure of this molecule. [1 mark]

b) Give the mechanism for the hydrolysis of this molecule by warm sodium hydroxide solution. [3 marks]

c) What would be observed if silver nitrate solution was added to the products of the reaction in part b)? [2 marks]

d) What difference would you expect in the rate of hydrolysis in part b) if you used iodomethane instead of chloromethane? [1 mark]

Polar bonds are like premium bonds, but are only bought by penguins...

Polar bonds get in just about every area of Chemistry. If you still think they're something to do with either bears or mints, you need to flick back to Unit 1: Section 2 and have a good read of page 34. Make sure you learn the stuff about CFCs and that hole in the ozone layer — it's always coming up in exams. Ruin the examiner's day and get it right.

Analytical Techniques

If you've got some stuff and don't know what it is, don't taste it. Stick it in an infrared spectrometer or a mass spectrometer instead. You'll wind up with some scary looking graphs. But just learn the basics, and you'll be fine.

Infrared Spectroscopy Helps You *Identify Organic Molecules*

1) In infrared (IR) spectroscopy, a beam of **IR radiation** is passed through a sample of a chemical.
2) The IR radiation is absorbed by the **covalent bonds** in the molecules, increasing their **vibrational** energy.
3) **Bonds between different atoms** absorb **different frequencies** of IR radiation. Bonds in different **places** in a molecule absorb different frequencies too — so the O–H group in an **alcohol** and the O–H in a **carboxylic acid** absorb different frequencies.

   This table shows what **frequencies** different bonds absorb:

This tells you what the peak on the graph will look like.

Functional group	Where it's found	Frequency/ Wavenumber (cm^{-1})	Type of absorption
C–H	most organic molecules	2800 - 3100	strong, sharp
O–H	alcohols	3200 - 3550	strong, broad
O–H	carboxylic acids	2500 - 3300	medium, broad
C=O	aldehydes, ketones, carboxylic acids	1680 - 1750	strong, sharp

You don't need to learn this data, but you do need to understand how to use it.

4) An infrared spectrometer produces a **graph** that shows you what frequencies of radiation the molecules are absorbing. You can use it to identify the **functional groups** in a molecule:

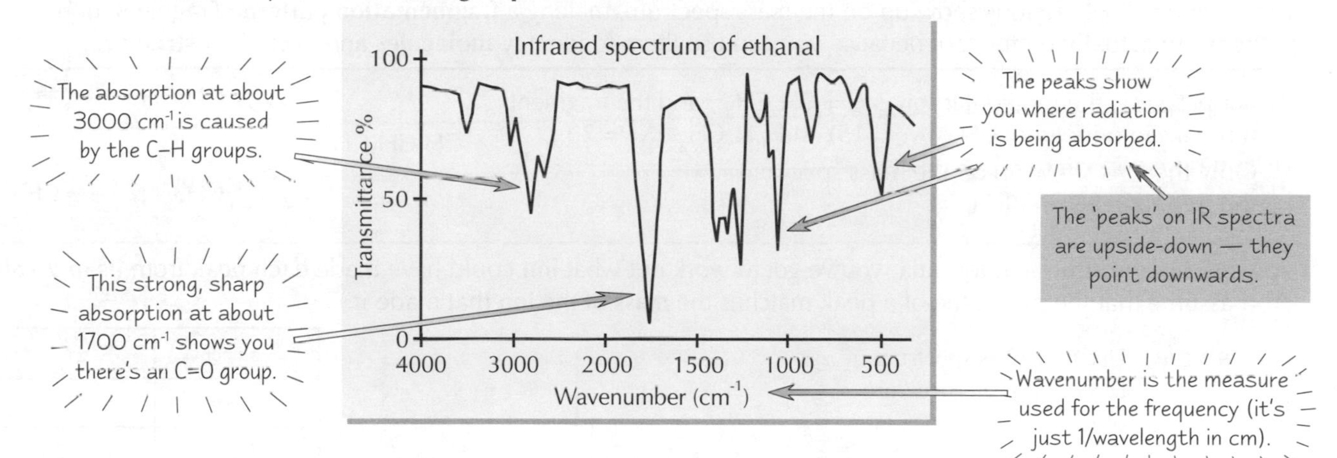

This also means that you can tell if a functional group has **changed** during a reaction. For example, if you **oxidise** an **alcohol** to an **aldehyde** you'll see the O–H absorption **disappear** from the spectrum, and a C=O absorption **appear**.

Infrared Spectroscopy Helps Catch Drunk Drivers

1) If a person's suspected of drink driving, they're **breathalysed**.
2) First a very quick test is done by the roadside — if it says that the driver's over the limit, they're taken into a police station for a more **accurate test** using **infrared spectroscopy**.
3) The **amount** of **ethanol vapour** in the driver's breath is found by measuring the **intensity** of the peak corresponding to the **C–H bond** in the IR spectrum. It's chosen because it's **not affected** by any **water vapour** in the breath.

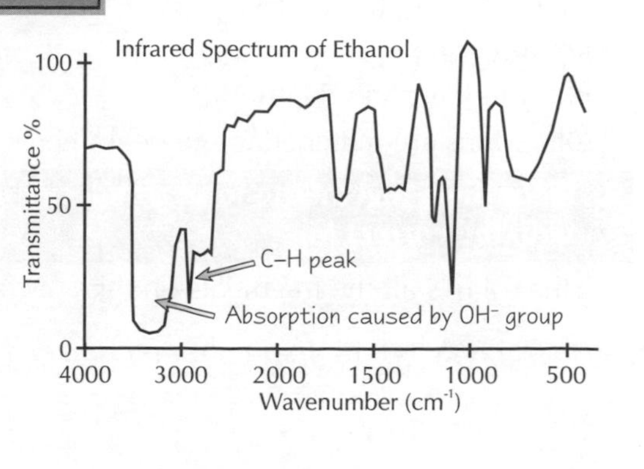

Analytical Techniques

Mass Spectrometry Can Help to Identify Compounds

1) You saw on page 8 how you can use a mass spectrum showing the relative isotopic abundances of an element to work out its relative atomic mass. You need to make sure you can remember how to do this. You can also get mass spectra for **molecular samples**.
2) A mass spectrum is produced by a mass spectrometer. The molecules in the sample are bombarded with electrons and a **molecular ion**, $M^+_{(g)}$, is formed when the bombarding electrons remove an electron from the molecule.
3) To find the relative molecular mass of a compound you look at the **molecular ion peak** (the **M peak**). The mass/charge value of the molecular ion peak is the **molecular mass**.

Assuming the ion has a 1+ charge, which it normally will have.

The **y-axis** gives the **abundance of ions**, often as a percentage.

The **x-axis** units are given as a '**mass/charge**' ratio.

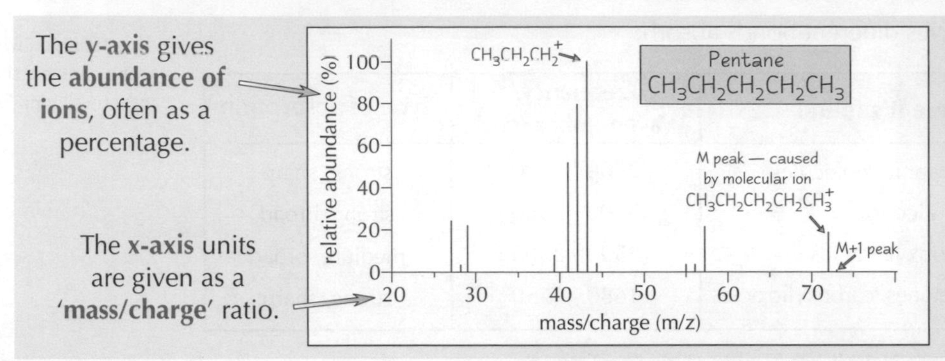

Here's the mass spectrum of pentane. Its M peak is at 72 — so the compound's M_r is 72.

For most organic compounds the M peak is the one with the second highest mass/charge ratio.

The smaller peak to the right of the M peak is called the M+1 peak — it's caused by the presence of the carbon isotope ^{13}C (you don't need to worry about this at AS).

The *Molecular Ion* can be *Broken* into *Smaller Fragments*

The bombarding electrons make some of the molecular ions break up into **fragments**. The fragments that are ions show up on the mass spectrum, making a **fragmentation pattern**. Fragmentation patterns are actually pretty cool because you can use them to identify **molecules** and even their **structure**.

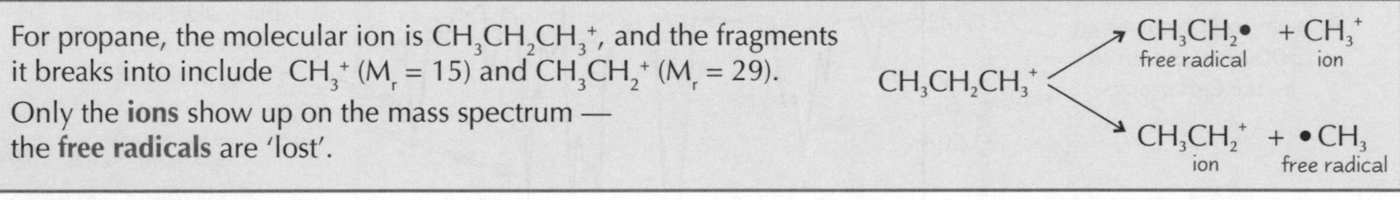

For propane, the molecular ion is $CH_3CH_2CH_3^+$, and the fragments it breaks into include CH_3^+ (M_r = 15) and $CH_3CH_2^+$ (M_r = 29). Only the **ions** show up on the mass spectrum — the **free radicals** are 'lost'.

To work out the structural formula, you've got to work out what **ion** could have made each peak from its **m/z value**. (You assume that the m/z value of a peak matches the **mass** of the ion that made it.)

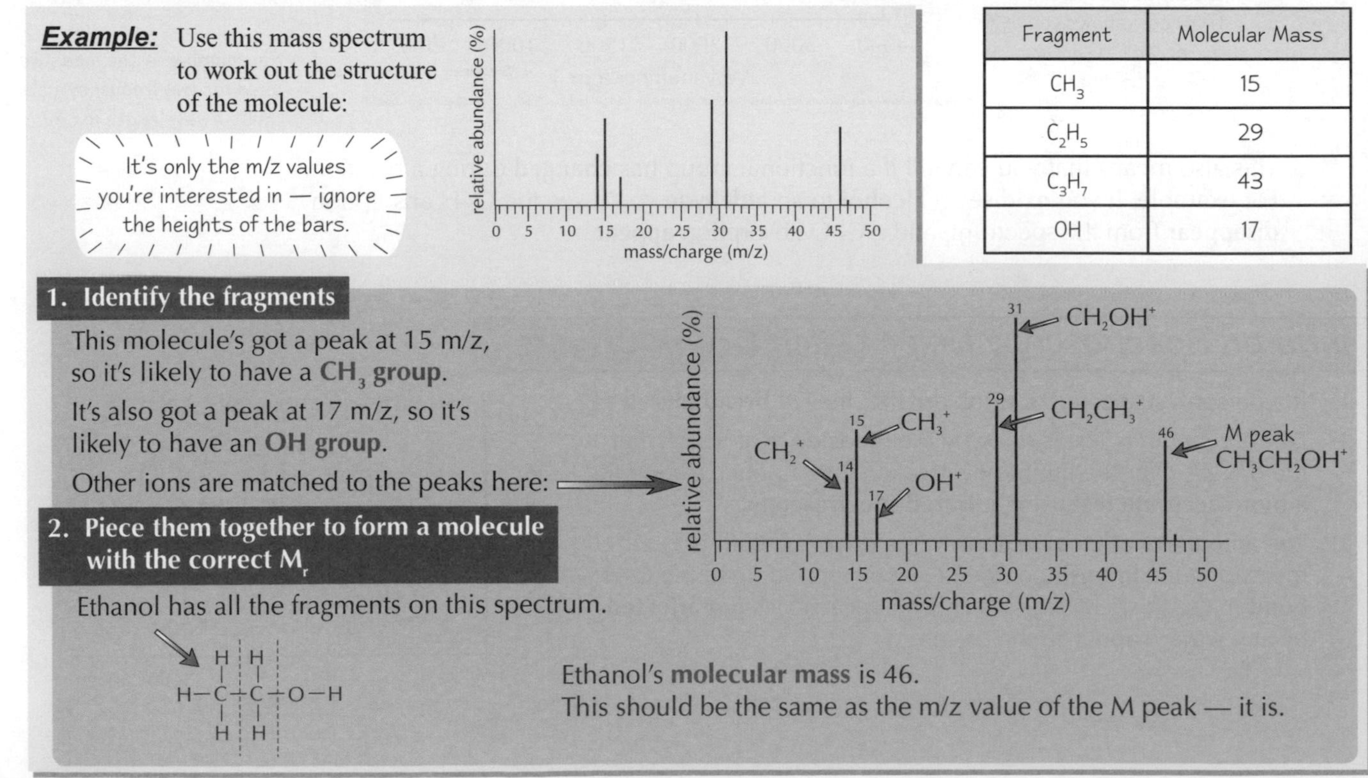

Example: Use this mass spectrum to work out the structure of the molecule:

It's only the m/z values you're interested in — ignore the heights of the bars.

Fragment	Molecular Mass
CH_3	15
C_2H_5	29
C_3H_7	43
OH	17

1. Identify the fragments

This molecule's got a peak at 15 m/z, so it's likely to have a **CH_3 group**.

It's also got a peak at 17 m/z, so it's likely to have an **OH group**.

Other ions are matched to the peaks here:

2. Piece them together to form a molecule with the correct M_r

Ethanol has all the fragments on this spectrum.

Ethanol's **molecular mass** is 46. This should be the same as the m/z value of the M peak — it is.

Analytical Techniques

Mass Spectrometry is Used to Differentiate Between Similar Molecules

1) Even if two **different compounds** contain **the same atoms**, you can still tell them apart with mass spectrometry because they won't produce exactly the same set of fragments.
2) The formulas of **propanal** and **propanone** are shown on the right. They've got the same M_r, but different structures, so they produce some **different fragments**. For example, propanal will have a C_2H_5 fragment but propanone won't.

propanal — propanone

3) Every compound produces a different mass spectrum — so the spectrum's like a **fingerprint** for the compound. Large computer **databases** of mass spectra can be used to identify a compound from its spectrum.

Mass Spectrometry Has Many Uses

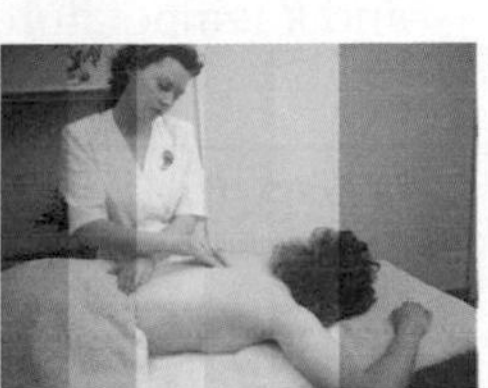
A massage spectrum

Mass spectrometry is actually used by scientists out there in the real world. Here are a couple of examples.

1) **Probes to Mars** have carried small mass spectrometers to study the composition of the surface of Mars and to look for molecules that might suggest that life existed on the planet.
2) Mass spectrometry can also be used to measure the **levels of pollutants** present in the environment, e.g. the amount of lead or pesticides entering the food chain via vegetables.

Practice Questions

Q1 Which parts of a molecule absorb infrared energy?

Q2 Why do most infrared spectra of organic molecules have a strong, sharp peak at around 3000 cm^{-1}?

Q3 What is meant by the molecular ion?

Q4 What is the M peak?

Q5 Give two uses of mass spectrometry.

Exam Questions

Q1 A molecule with a molecular mass of 74 produces the IR spectrum shown on the right.

a) Which functional groups are responsible for peaks A and B? [2 marks]

b) Give the molecular formula and name of this molecule. Explain your answer. [3 marks]

Use the infrared absorption data on p75.

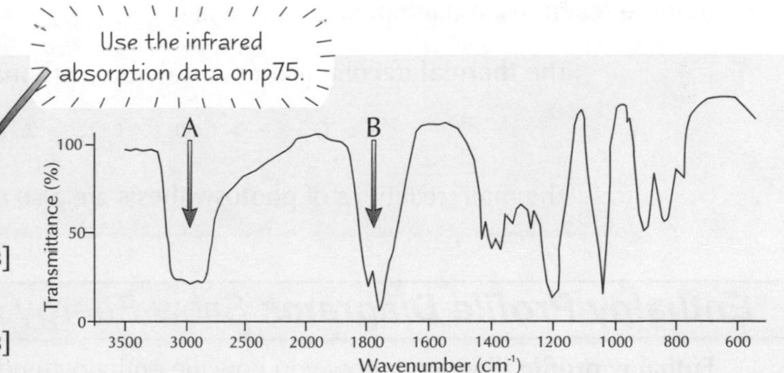

Q2 Below is the mass spectrum of an organic compound, Q.

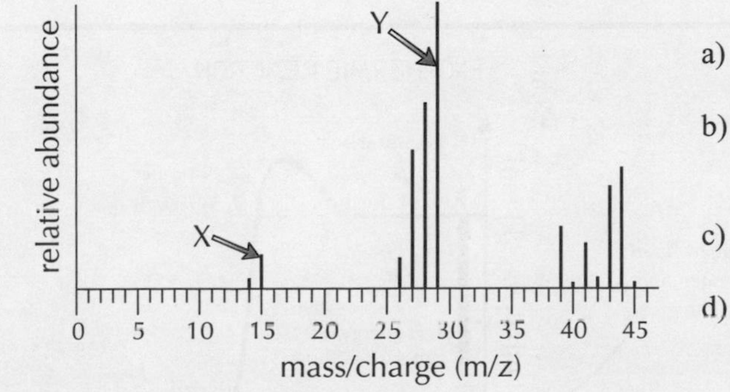

a) What is the M_r of compound Q? [1 mark]

b) What fragments are the peaks marked X and Y most likely to correspond to? [2 marks]

c) Suggest a structure for this compound. [1 mark]

d) Why is it unlikely that this compound is an alcohol? [2 marks]

Use the clues, identify a molecule — mass spectrometry my dear Watson...

Luckily you don't have to remember what any of the infrared spectrum graphs look like. But you need to be able to interpret them — they're bound to turn up in the exam. It's handy if you can learn the molecular masses of the common mass spec fragments, but if you forget them, you can work them out from the relative atomic masses of the atoms in each fragment.

Enthalpy Changes

A whole new module to enjoy — but don't forget, Big Brother is watching...

Chemical Reactions Often Have Enthalpy Changes

When chemical reactions happen, some bonds are **broken** and some bonds are **made**. More often than not, this'll cause a **change in energy**. The souped-up chemistry term for this is **enthalpy change** —

> **Enthalpy change**, ΔH (delta H), is the heat energy transferred in a reaction at **constant pressure**. The units of ΔH are **kJ mol⁻¹**.

You write ΔH° to show that the elements were in their **standard states** (i.e. their states at a pressure of 100 kPa), and that the measurements were made under **standard conditions**. Standard conditions are **100 kPa (about 1 atm) pressure** and a temperature of **298 K** (25 °C). The next page explains why this is necessary.

Reactions can be either Exothermic or Endothermic

> **Exothermic** reactions **give out** energy. ΔH is **negative**.

In exothermic reactions, the temperature often goes **up**.

Oxidation is exothermic. Here are 2 examples:

- The **combustion** of a fuel like methane → $CH_{4(g)} + 2O_{2(g)} \longrightarrow CO_{2(g)} + 2H_2O_{(l)}$ $\Delta H_c^\circ = -890 \text{ kJ mol}^{-1}$ **exothermic**
- The oxidation of **carbohydrates**, such as glucose, $C_6H_{12}O_6$, in respiration.

The symbols ΔH_c° and ΔH_r° (below) are explained on the next page.

> **Endothermic** reactions **absorb** energy. ΔH is **positive**.

In these reactions, the temperature often **falls**.

The **thermal decomposition** of calcium carbonate is endothermic.

$$CaCO_{3(s)} \longrightarrow CaO_{(s)} + CO_{2(g)} \quad \Delta H_r^\circ = +178 \text{ kJ mol}^{-1} \textbf{ endothermic}$$

The main reactions of **photosynthesis** are also endothermic — sunlight supplies the energy.

Enthalpy Profile Diagrams Show Energy Change in Reactions

1) **Enthalpy profile diagrams** show you how the enthalpy (energy) changes during reactions.
2) The **activation energy**, E_a, is the minimum amount of energy needed to begin breaking reactant bonds and start a chemical reaction.

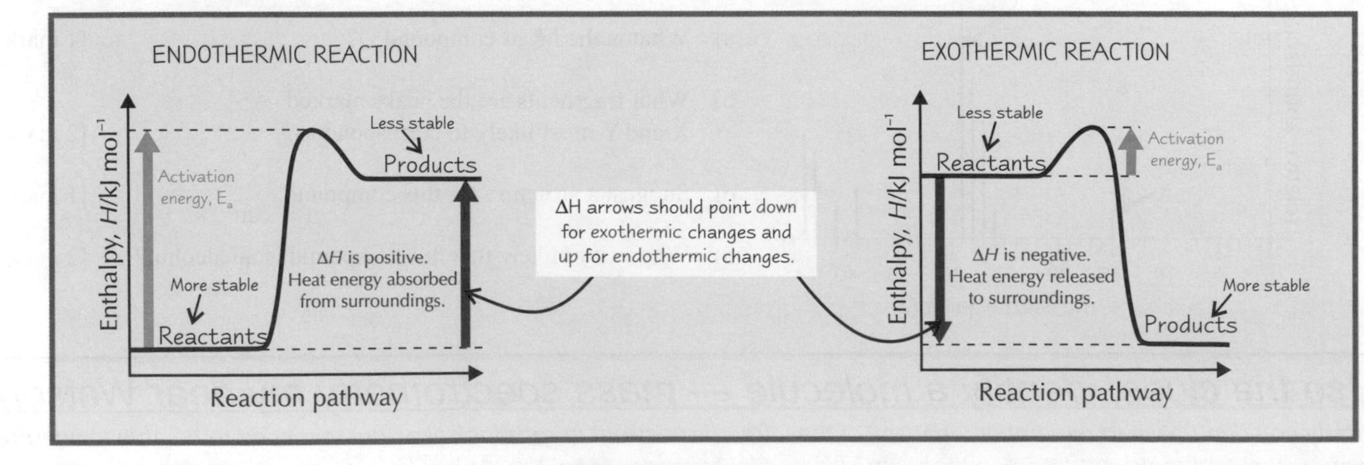

3) The **less enthalpy** a substance has, the **more stable** it is.

Enthalpy Changes

You Need to Specify the **Conditions** for **Enthalpy Changes**

1) You can't directly measure the **actual** enthalpy of a system. In practice, that doesn't matter, because it's only ever **enthalpy change** that matters. You can find enthalpy changes either by **experiment** or in **textbooks**.
2) Enthalpy changes you find in textbooks are usually **standard** enthalpy changes — enthalpy changes under **standard conditions** (**298 K** and **100 kPa**).
3) This is important because changes in enthalpy are affected by **temperature** and **pressure** — using standard conditions means that everyone can know **exactly** what the enthalpy change is describing.

There are Different Types of ΔH Depending On the **Reaction**

1) **Standard enthalpy change of reaction**, ΔH_r°, is the enthalpy change when the reaction occurs in the **molar quantities** shown in the **chemical equation**, under standard conditions in their standard states.

2) **Standard enthalpy change of formation**, ΔH_f°, is the enthalpy change when **1 mole** of a **compound** is formed from its **elements** in their standard states under standard conditions, e.g. $2C_{(s)} + 3H_{2(g)} + \frac{1}{2}O_{2(g)} \longrightarrow C_2H_5OH_{(l)}$

3) **Standard enthalpy change of combustion**, ΔH_c°, is the enthalpy change when **1 mole** of a substance is completely **burned in oxygen** under standard conditions.

Practice Questions

Q1 Explain the terms exothermic and endothermic, giving an example reaction in each case.

Q2 Draw and label enthalpy profile diagrams for an exothermic and an endothermic reaction.

Q3 Define standard enthalpy of formation and standard enthalpy of combustion.

Exam Questions

Q1 Hydrogen peroxide, H_2O_2, can decompose into water and oxygen. $2H_2O_{2(l)} \longrightarrow 2H_2O_{(l)} + O_{2(g)}$ $\Delta H_r^\circ = -98$ kJ mol^{-1}

Draw an enthalpy profile diagram for this reaction. Mark on the activation energy and ΔH. [3 marks]

Q2 Methanol, CH_3OH, when blended with petrol, can be used as a fuel. $\Delta H_c^\circ[CH_3OH] = -726$ kJ mol^{-1}.

a) Write an equation, including state symbols, for the standard enthalpy change of combustion of methanol. [2 marks]

b) Write an equation, including state symbols, for the standard enthalpy change of formation of methanol. [2 marks]

c) Liquid petroleum gas is a fuel that contains propane, C_3H_8. Give two reasons why the following equation does not represent a standard enthalpy change of combustion. [2 marks]

$$2C_3H_{8(g)} + 10O_{2(g)} \rightarrow 8H_2O_{(g)} + 6CO_{2(g)} \qquad \Delta H_r = -4113 \text{ kJ mol}^{-1}$$

Q3 Coal is mainly carbon. It is burned as a fuel. $\Delta H_c^\circ = -393.5$ kJ mol^{-1}

a) Write an equation, including state symbols, for the standard enthalpy change of combustion of carbon. [2 marks]

b) Explain why the standard enthalpy change of formation of carbon dioxide will also be -393.5 kJ mol^{-1} [1 mark]

c) How much energy would be released when 1 tonne of carbon is burned? (1 tonne = 1000 kg) [2 marks]

It's getting hot in here, so take off all your bonds...

Quite a few definitions here. And you need to know them all. If you're going to bother learning them, you might as well do it properly and learn all the pernickety details. They probably seem about as useful as a dead fly in your custard right now, but all will be revealed over the next few pages. Learn them now, so you've got a bit of a head start.

More on Enthalpy Changes

I bonded with my friend straight away. Now we're on the waiting list to be surgically separated.

Reactions are all about Breaking and Making Bonds

You can only break bonds if you've got enough energy.

When reactions happen, **reactant bonds** are **broken** and **product bonds** are **formed**.

1) You **need** energy to break bonds, so bond breaking is **endothermic** (ΔH is **positive**).
2) Energy is **released** when bonds are formed, so this is **exothermic** (ΔH is **negative**).
3) The **enthalpy change** for a reaction is the **overall effect** of these two changes. If you need **more** energy to **break** bonds than is released when bonds are made, ΔH is **positive**. If it's less, ΔH is negative.

You need Energy to Break the Attraction between Atoms and Ions

1) In ionic bonding, **positive** and **negative ions** are attracted to each other. In covalent molecules, the **positive nuclei** are attracted to the **negative** charge of the shared electrons in a covalent bond.
2) You need energy to **break** this attraction — **stronger** bonds take more energy to break. The **amount of energy** you need per mole is called the **bond dissociation enthalpy**. (Of course it's got a fancy name — this is chemistry.)
3) Bond dissociation enthalpies always involve bond breaking in **gaseous compounds**. This makes comparisons fair.

Average Bond Enthalpies are not Exact

1) Water (H_2O) has got **two O–H bonds**. You'd think it'd take the same amount of energy to break them both... but it **doesn't**.

> The **first** bond, $H–OH_{(g)}$: $E(H–OH) = +492$ kJ mol^{-1}
>
> The **second** bond, $H–O_{(g)}$: $E(H–O) = +428$ kJ mol^{-1}
>
> (OH^- is a bit easier to break apart because of the extra electron repulsion.)
>
> So, the **average** bond enthalpy is $\frac{492 + 428}{2}$ = **+460 kJ mol^{-1}**.

2) The **data book** says the bond enthalpy for O–H is +463 kJ mol^{-1}. It's a bit different because it's the average for a **much bigger range** of molecules, not just water. For example, it includes the O–H bonds in alcohols and carboxylic acids too.
3) So when you look up an **average bond enthalpy**, what you get is:

the energy needed to break one mole of bonds in the gas phase, averaged over many different compounds

You can find out Enthalpy Changes in the Lab

1) To measure the **enthalpy change** for a reaction you only need to know **two things** —
   - the **number of moles** of the stuff that's reacting,
   - the change in **temperature**.
2) How you go about doing the experiment depends on what type of reaction it is.
3)
   - To find the enthalpy of **combustion** of a **flammable liquid**, you burn it — using apparatus like this...
   - As the fuel burns, it heats the water. You can work out the **heat absorbed** by the water if you know the **mass of water**, the **temperature change of the water** (ΔT), and the **specific heat capacity of water** (= 4.18 J g^{-1} K^{-1}). See the next page for all the details.
   - Ideally all the heat given out by the fuel as it burns would be **absorbed** by the water — allowing you to work out the enthalpy change of combustion (see the next page). In practice though, you **always** lose some heat (as you heat the apparatus and the surroundings).

Stirrer
Thermometer
Water
Combustion chamber
Air
Fuel (reactant)

The **specific heat capacity** of a substance is the amount of heat energy it takes to raise the temperature of 1 g of that substance by 1 K.

4) Calorimetry can also be used to calculate an enthalpy change for a reaction that happens **in solution**, such as **neutralisation** or **displacement**. For a neutralisation reaction, combine known quantities of acid and alkali in an insulated container, and measure the temperature change. The **heat given out** can be calculated using the formula on the next page.

More on Enthalpy Changes

Calculate **Enthalpy Changes** Using the **Equation q = mcΔT**

It seems there's a snazzy equation for everything these days, and enthalpy change is no exception:

$q = mc\Delta T$ where, q = heat lost or gained (in joules). This is the same as the enthalpy change if the pressure is constant.
m = mass of water in the calorimeter, or solution in the insulated container (in grams)
c = specific heat capacity of water (4.18 J $g^{-1}K^{-1}$)
ΔT = the change in temperature of the water or solution

Example: In a laboratory experiment, 1.16 g of an organic liquid fuel was completely burned in oxygen. The heat formed during this combustion raised the temperature of 100 g of water from 295.3 K to 357.8 K. Calculate the standard enthalpy of combustion, $\Delta H_c^{\ominus}$, of the fuel. Its M_r is 58.

Remember — m is the mass of water, NOT the mass of fuel.

1. First off, you need to calculate the **amount of heat** given out by the fuel using $\boldsymbol{q = mc\Delta T}$.
$q = mc\Delta T$
$q = 100 \times 4.18 \times (357.8 - 295.3) = 26\,125 \text{ J} = 26.125 \text{ kJ}$ ⟸ Change the amount of heat from J to kJ.

2. The standard enthalpy of combustion involves 1 mole of fuel. So next you need to find out **how many moles** of fuel produced this heat. It's back to the old $n = \frac{\text{mass}}{\text{M}}$ equation.
$$n = \frac{1.16}{58} = 0.02 \text{ moles of fuel}$$

3. So, the heat produced by 1 mole of fuel = $\frac{-26.125}{0.02}$ ⟸ It's negative because combustion is an exothermic reaction.
$\approx$ **-1306 kJ mol⁻¹**. This is the standard enthalpy change of combustion.

The actual $\Delta H_c^{\ominus}$ of this compound is -1615 kJ mol^{-1} — lots of heat has been **lost** and not measured. For example it's likely a bit would escape through the **calorimeter** and also the fuel might not **combust completely**.

Practice Questions

Q1 Briefly describe an experiment that could be carried out to find the enthalpy change of a reaction.

Q2 Why is the enthalpy change determined in a laboratory likely to be lower than the value shown in a data book?

Q3 What equation is used to calculate the heat change in a chemical reaction?

Exam Questions

Q1 A 50 cm^3 sample of 0.200 M copper(II) sulfate solution placed in a polystyrene beaker gave a temperature increase of 2.6 K when excess zinc powder was added and stirred. Calculate the enthalpy change when 1 mole of zinc reacts. Assume that the specific heat capacity for the solution is 4.18 J $g^{-1}K^{-1}$. Ignore the increase in volume due to the zinc.
The equation for the reaction is: $Zn_{(s)} + CuSO_{4(aq)} \rightarrow Cu_{(s)} + ZnSO_{4(aq)}$ [8 marks]

Q2 a) Explain why bond enthalpies determine whether a reaction is exothermic or endothermic. [3 marks]

b) Calculate the temperature change that should be produced when 1 kg of water is heated by burning 6 g of coal. Assume the coal is pure carbon.
[The specific heat capacity of water is 4.18 J $g^{-1}K^{-1}$. For carbon, $\Delta H_c^{\ominus} = -393.5$ kJ mol^{-1}] [4 marks]

If you can't stand the enthalpy, get out of the chemistry class...

Reactions are like pulling your Lego spaceship apart and building something new. Sometimes the bits get stuck together and you need to use loads of energy to pull 'em apart. Okay, so energy's not really released when you stick them together, but you can't have everything — and it wasn't that bad an analogy up till now. Ah, well... you'd best get on and learn this stuff.

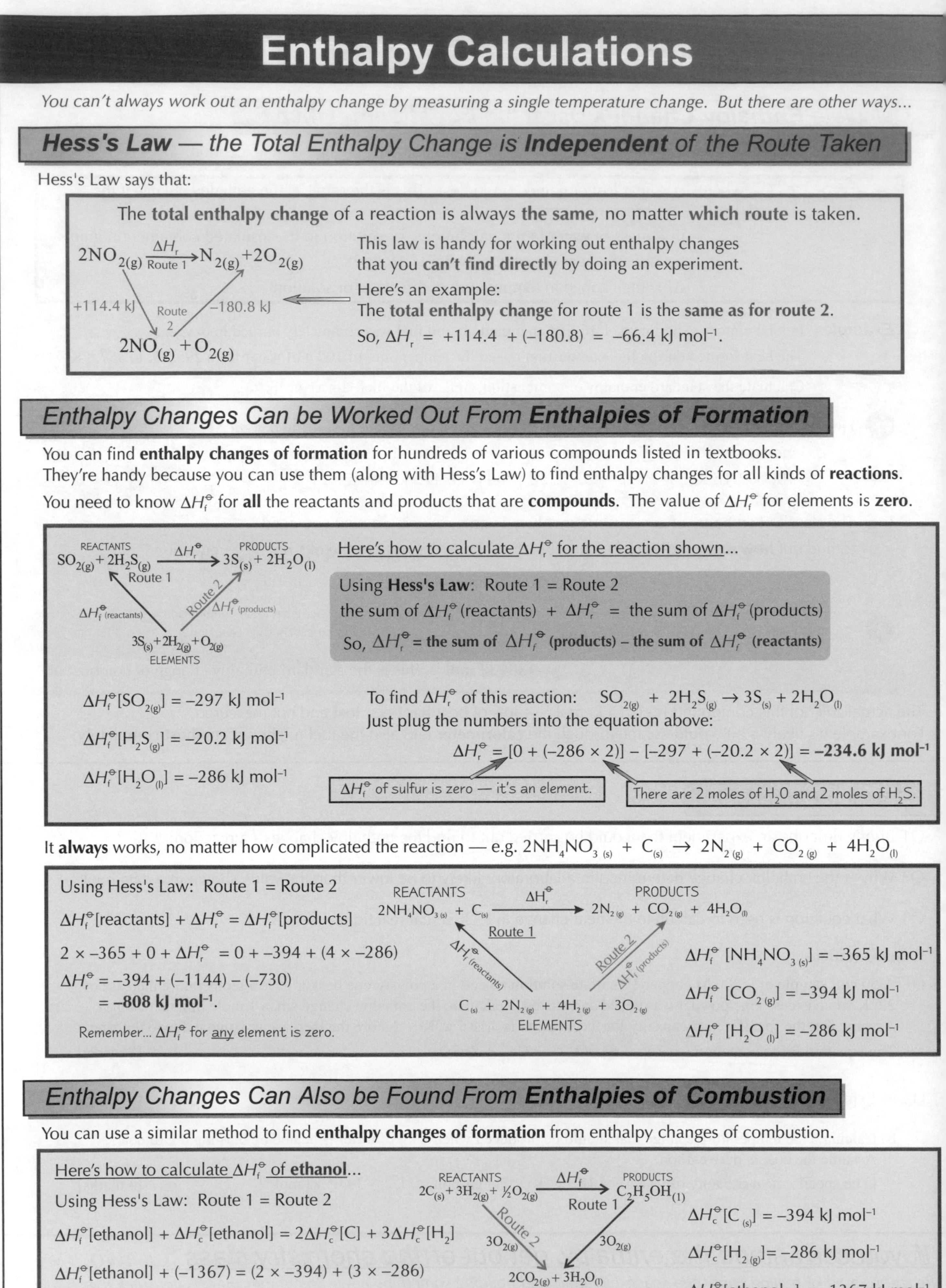

Enthalpy Calculations

You can't always work out an enthalpy change by measuring a single temperature change. But there are other ways...

Hess's Law — the Total Enthalpy Change is Independent of the Route Taken

Hess's Law says that:

The **total enthalpy change** of a reaction is always **the same**, no matter **which route** is taken.

This law is handy for working out enthalpy changes that you **can't find directly** by doing an experiment.

Here's an example:

The **total enthalpy change** for route 1 is the **same as for route 2**.

So, ΔH_r = +114.4 + (−180.8) = −66.4 kJ mol⁻¹.

Enthalpy Changes Can be Worked Out From Enthalpies of Formation

You can find **enthalpy changes of formation** for hundreds of various compounds listed in textbooks.
They're handy because you can use them (along with Hess's Law) to find enthalpy changes for all kinds of **reactions**.
You need to know ΔH_f° for **all** the reactants and products that are **compounds**. The value of ΔH_f° for elements is **zero**.

Here's how to calculate ΔH_r° for the reaction shown...

Using **Hess's Law**: Route 1 = Route 2

the sum of ΔH_f° (reactants) + ΔH_r° = the sum of ΔH_f° (products)

So, ΔH_r° = **the sum of ΔH_f° (products) − the sum of ΔH_f° (reactants)**

$\Delta H_f^\circ[SO_{2(g)}]$ = −297 kJ mol⁻¹

$\Delta H_f^\circ[H_2S_{(g)}]$ = −20.2 kJ mol⁻¹

$\Delta H_f^\circ[H_2O_{(l)}]$ = −286 kJ mol⁻¹

To find ΔH_r° of this reaction: $SO_{2(g)} + 2H_2S_{(g)} \rightarrow 3S_{(s)} + 2H_2O_{(l)}$

Just plug the numbers into the equation above:

ΔH_r° = [0 + (−286 × 2)] − [−297 + (−20.2 × 2)] = **−234.6 kJ mol⁻¹**

ΔH_f° of sulfur is zero — it's an element.

There are 2 moles of H_2O and 2 moles of H_2S.

It **always** works, no matter how complicated the reaction — e.g. $2NH_4NO_{3\,(s)} + C_{(s)} \rightarrow 2N_{2\,(g)} + CO_{2\,(g)} + 4H_2O_{(l)}$

Using Hess's Law: Route 1 = Route 2

ΔH_f°[reactants] + ΔH_r° = ΔH_f°[products]

2 × −365 + 0 + ΔH_r° = 0 + −394 + (4 × −286)

ΔH_r° = −394 + (−1144) − (−730)
= **−808 kJ mol⁻¹**.

Remember... ΔH_f° for <u>any</u> element is zero.

$\Delta H_f^\circ[NH_4NO_{3\,(s)}]$ = −365 kJ mol⁻¹

$\Delta H_f^\circ[CO_{2\,(g)}]$ = −394 kJ mol⁻¹

$\Delta H_f^\circ[H_2O_{(l)}]$ = −286 kJ mol⁻¹

Enthalpy Changes Can Also be Found From Enthalpies of Combustion

You can use a similar method to find **enthalpy changes of formation** from enthalpy changes of combustion.

<u>Here's how to calculate ΔH_f° of **ethanol**...</u>

Using Hess's Law: Route 1 = Route 2

ΔH_f°[ethanol] + ΔH_c°[ethanol] = $2\Delta H_c^\circ[C] + 3\Delta H_c^\circ[H_2]$

ΔH_f°[ethanol] + (−1367) = (2 × −394) + (3 × −286)

ΔH_f°[ethanol] = −788 + −858 − (−1367) = **−279 kJ mol⁻¹**.

$\Delta H_c^\circ[C_{(s)}]$ = −394 kJ mol⁻¹

$\Delta H_c^\circ[H_{2\,(g)}]$ = −286 kJ mol⁻¹

ΔH_c°[ethanol $_{(l)}$] = −1367 kJ mol⁻¹

Enthalpy Calculations

Enthalpy Changes Can Be Calculated using Average Bond Enthalpies

1) You **need** energy to break bonds, so bond breaking is an **endothermic** process (ΔH is **positive**).
2) Energy is **released** when bonds are formed, so this is an **exothermic** process (ΔH is **negative**).
3) The **enthalpy change** for a reaction is the **overall effect** of these two changes. If you need **more** energy to **break** bonds than is released when bonds are made, ΔH is **positive**. If it's **less**, ΔH is **negative**.

Enthalpy Change of Reaction = Total Energy Absorbed to Break Bonds − Total Energy Released in Making Bonds

4) **Average bond enthalpies** (the **energy needed** to **break** a bond, or the **energy given out** when a bond **forms**) are published in textbooks to help calculate **enthalpy changes** of **reactions**. They're pretty straightforward to use...

Example: Calculate the overall enthalpy change for this reaction:
$N_2 + 3H_2 \rightarrow 2NH_3$
Use the average bond enthalpy values in the table.

Bond	Average Bond Enthalpy
N≡N	945 kJ mol^{-1}
H–H	436 kJ mol^{-1}
N–H	391 kJ mol^{-1}

Bonds broken: 1 × N≡N bond broken = 1 × 945 = 945 kJ mol^{-1}
3 × H–H bonds broken = 3 × 436 = 1308 kJ mol^{-1}
Total Energy Absorbed = 945 + 1308 = **2253 kJ mol^{-1}**

Bonds formed: 6 × N–H bonds formed = 6 × 391 = 2346 kJ mol^{-1}
Total Energy Released = **2346 kJ mol^{-1}**

Enthalpy Change of Reaction = 2253 – 2346 = **–93 kJ mol^{-1}**

If you can't remember which value to subtract from which, just take the smaller number from the bigger one then add the sign at the end — positive if 'bonds broken' was the bigger number (endothermic), negative if 'bonds formed' was bigger (exothermic).

Practice Questions

Q1 What is Hess's Law?

Q2 What is the standard enthalpy change of formation of any element?

Q3 Describe how you can make a "Hess's Law triangle" to find the standard enthalpy change of a reaction using standard enthalpy changes of formation.

Exam Questions

Q1 Using the facts that the standard enthalpy change of formation of $Al_2O_{3(s)}$ is –1676 kJ mol^{-1} and the standard enthalpy change of formation of $MgO_{(s)}$ is –602 kJ mol^{-1}, calculate the enthalpy change of the following reaction.

$$Al_2O_{3(s)} + 3Mg_{(s)} \rightarrow 2Al_{(s)} + 3MgO_{(s)}$$

[3 marks]

Q2 Calculate the enthalpy change for the reaction below (the fermentation of glucose).

$$C_6H_{12}O_{6\,(s)} \rightarrow 2C_2H_5OH_{(l)} + 2CO_{2\,(g)}$$

Use the following standard enthalpies of combustion in your calculations:
$\Delta H_c^{\ominus}$(glucose) = –2820 kJ mol^{-1} $\Delta H_c^{\ominus}$(ethanol) = –1367 kJ mol^{-1} [3 marks]

Q3 Calculate the standard enthalpy of formation of propane from carbon and hydrogen.

$$3C_{(s)} + 4H_{2\,(g)} \rightarrow C_3H_{8\,(g)}$$

Using the following data:
$\Delta H_c^{\ominus}$(propane) = –2220 kJ mol^{-1} $\Delta H_c^{\ominus}$ (carbon) = –394 kJ mol^{-1} $\Delta H_c^{\ominus}$ (hydrogen) = –286 kJ mol^{-1} [3 marks]

Q4 The table on the right shows some average bond enthalpy values.

Bond	C–H	C=O	O=O	O–H
Average Bond Enthalpy (kJ/mol)	435	805	498	464

The complete combustion of methane can be represented by the following equation:

$$CH_{4\,(g)} + 2O_{2\,(g)} \rightarrow CO_{2\,(g)} + 2H_2O_{(l)}$$

Use the table of bond enthalpies above to calculate the enthalpy change for the reaction. [4 marks]

To understand this lot, you're gonna need a bar of chocolate. Or two...

To get your head around those Hess diagrams, you're going to have to do more than skim them. It'll also help if you know the definitions for those standard enthalpy thingumabobs. If I were you, you know what I'd do... I'd read those Hess Cycle examples again and make sure you understand how the elements/compounds at each corner were chosen to be there.

Reaction Rates

The rate of a reaction is just how quickly it happens. Lots of things can make it go faster or slower.

*Particles **Must** Collide to **React***

1) Particles in liquids and gases are **always moving** and **colliding** with **each other**. They **don't** react every time though — only when the **conditions** are right. A reaction **won't** take place between two particles **unless** —

   - They collide in the **right direction**. They need to be **facing** each other the right way.
   - They collide with at least a certain **minimum** amount of kinetic (movement) **energy**.

   This stuff's called **Collision Theory**.
2) The **minimum amount of kinetic energy** particles need to react is called the **activation energy**. The particles need this much energy to **break the bonds** to start the reaction.
3) Reactions with **low activation energies** often happen **pretty easily**. But reactions with **high activation energies** don't. You need to give the particles extra energy by **heating** them.

To make this a bit clearer, here's another **enthalpy profile diagram**.

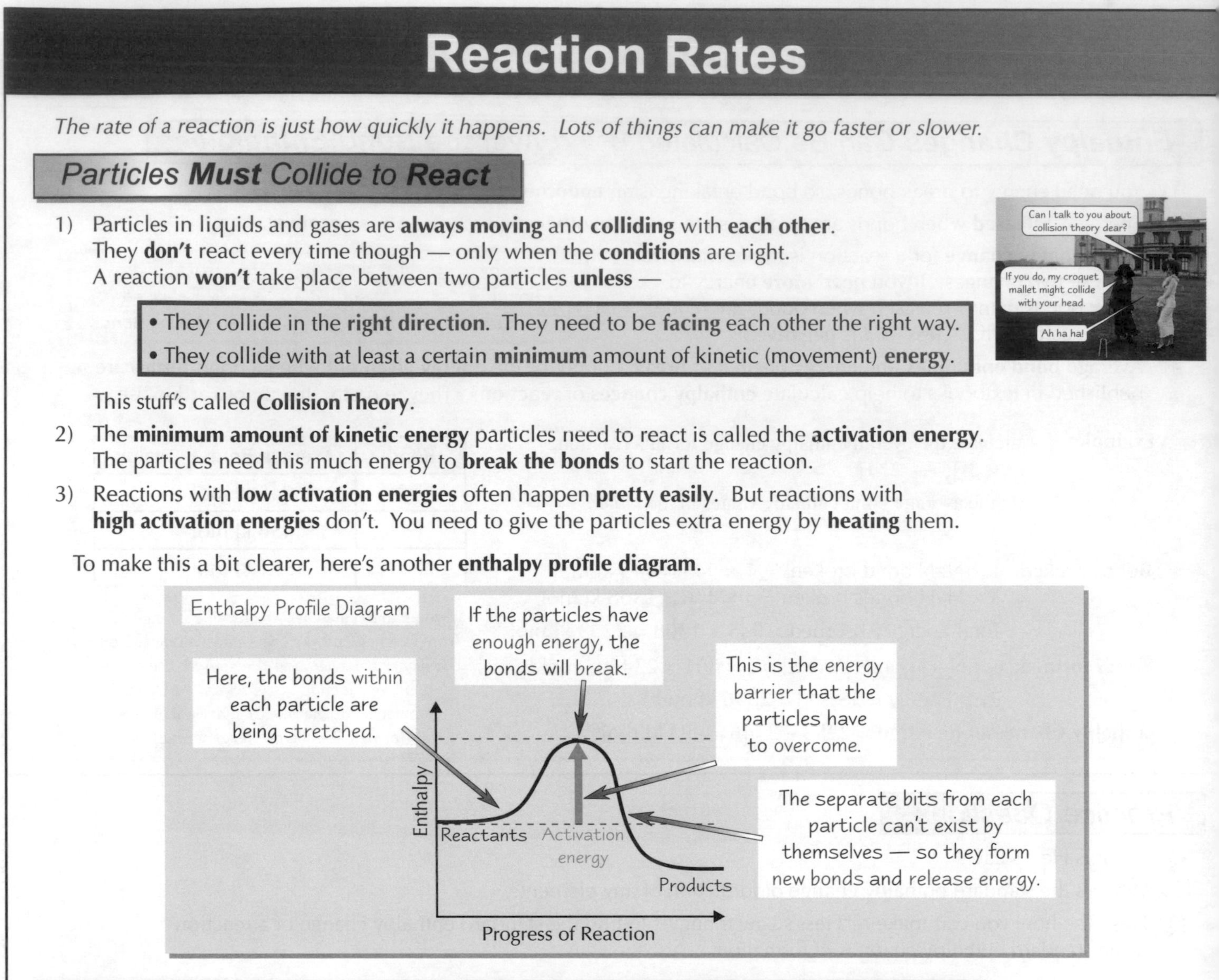

*Molecules in a Gas **Don't** all have the **Same Amount of Energy***

Imagine looking down on Oxford Street when it's teeming with people. You'll see some people ambling along **slowly**, some hurrying **quickly**, but most of them will be walking with a **moderate speed**. It's the same with the **molecules** in a **gas**. Some **don't have much kinetic energy** and move **slowly**. Others have **loads of kinetic energy** and **whizz** along. But most molecules are somewhere **in between**.

If you plot a **graph** of the **numbers of molecules** in a **gas** with different **kinetic energies** you get a **Maxwell-Boltzmann distribution**. It looks like this —

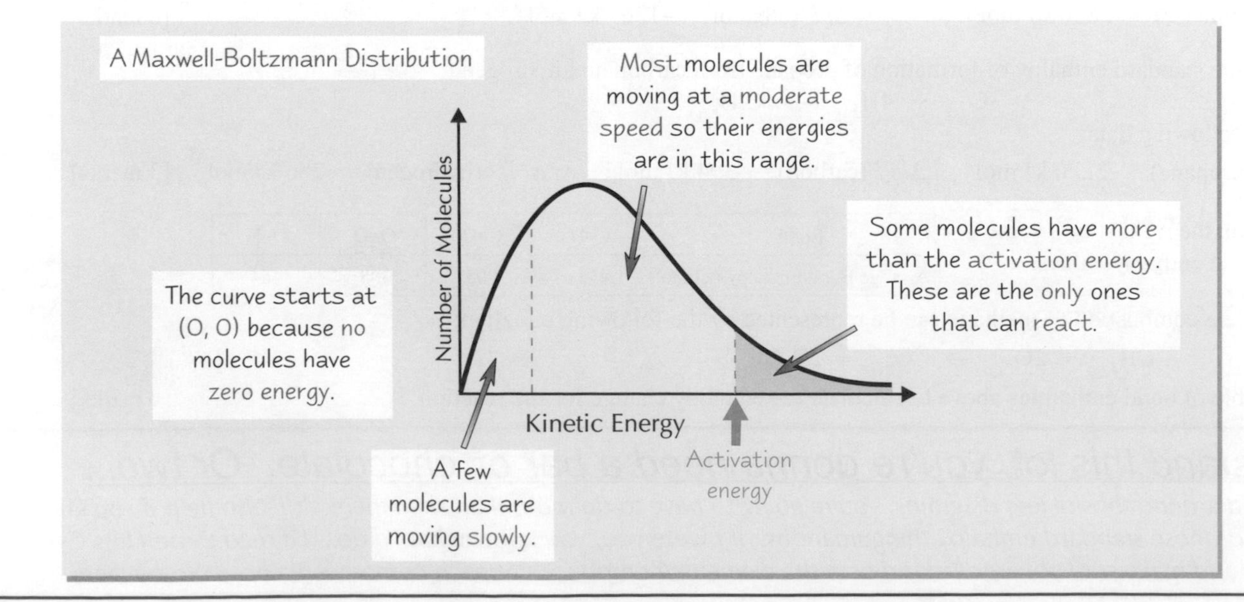

Reaction Rates

Increasing the Temperature makes Reactions Faster

1) If you increase the **temperature**, the particles will on average have more **kinetic energy** and will move **faster**.
2) So, a **greater proportion** of molecules will have more energy than the **activation energy** and be able to **react**. This changes the **shape** of the **Maxwell-Boltzmann distribution curve** — it pushes it over to the **right**.

The total number of molecules is still the same, which means the area under each curve must be the same.

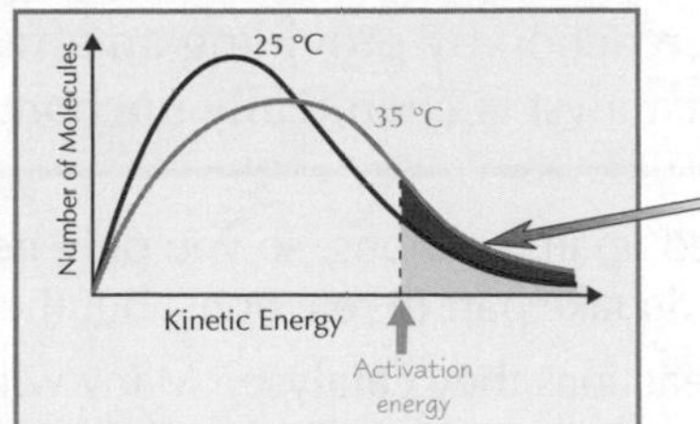

At higher temperatures, more molecules have more energy than the activation energy.

3) Because the molecules are flying about **faster**, they'll **collide more often**. This is **another reason** why increasing the temperature makes a reaction faster.

Concentration, Pressure and Catalysts also Affect the Reaction Rate

Increasing Concentration Speeds Up Reactions

If you increase the **concentration** of reactants in a **solution**, the particles will be **closer together** on average. If they're closer, they'll **collide more often**. If there are **more collisions**, they'll have **more chances** to react.

Increasing Pressure Speeds Up Reactions

If any of your reactants are **gases**, increasing the **pressure** will increase the rate of reaction. It's pretty much the same as increasing the **concentration** of a solution — at higher pressures, the particles will be **closer together**, increasing the chance of **successful collisions**.

Catalysts Can Speed Up Reactions

If one of the reactants is a solid, increasing its surface area makes the reaction faster too.

Catalysts are really useful. They **lower the activation energy** by providing a **different way** for the bonds to be broken and remade. If the activation energy's **lower**, more particles will have **enough energy** to react. There's heaps of information about catalysts on the next two pages.

Practice Questions

Q1 Explain the term 'activation energy'.

Q2 What is a Maxwell-Boltzmann distribution?

Q3 Name the four factors that affect the rate of a reaction.

Exam Questions

Q1 Nitrogen oxide (NO) and ozone (O_3) sometimes react to produce nitrogen dioxide (NO_2) and oxygen (O_2). How would increasing the pressure affect the rate of this reaction? Explain your answer. [2 marks]

Q2 On the right is a Maxwell-Boltzmann distribution curve for a sample of a substance at 25 °C.

a) Which of the curves X or Y shows the Maxwell-Boltzman distribution curve for the same sample at 15 °C ? [1 mark]

b) Explain how this curve shows that the reaction rate will be lower at 15 °C than at 25 °C. [2 marks]

X
Y
25 °C
Number of Molecules
Kinetic Energy
Activation Energy

Reaction Rates — cheaper than water rates

*This page isn't too hard to learn — no equations, no formulas... what more could you ask for. The only tricky thing might be the Maxwell-Boltzmann thingymajiggle. Remember, increasing concentration and pressure do exactly the same thing. The only difference is you increase the concentration of a **solution** and the pressure of a **gas**. Don't get them muddled.*

Catalysts

Catalysts were tantalisingly mentioned on the last page — here's the full story...

Catalysts Increase the Rate of Reactions

You can use **catalysts** to make chemical reactions happen **faster**. Learn this definition:

> A **catalyst** increases the **rate** of a reaction by providing an **alternative reaction pathway** with a **lower activation energy**. The catalyst is **chemically unchanged** at the end of the reaction.

1) Catalysts are **great**. They **don't** get used up in reactions, so you only need a **tiny bit** of catalyst to catalyse a **huge** amount of stuff. They **do** take part in reactions, but they're **remade** at the end.
2) Catalysts are **very fussy** about which reactions they catalyse. Many will usually **only** work on a single reaction.

An example of a catalyst is **iron**. It's used in the **Haber process** to make ammonia.

$$N_{2(g)} + 3H_{2(g)} \underset{}{\overset{Fe_{(s)}}{\rightleftharpoons}} 2NH_{3(g)}$$

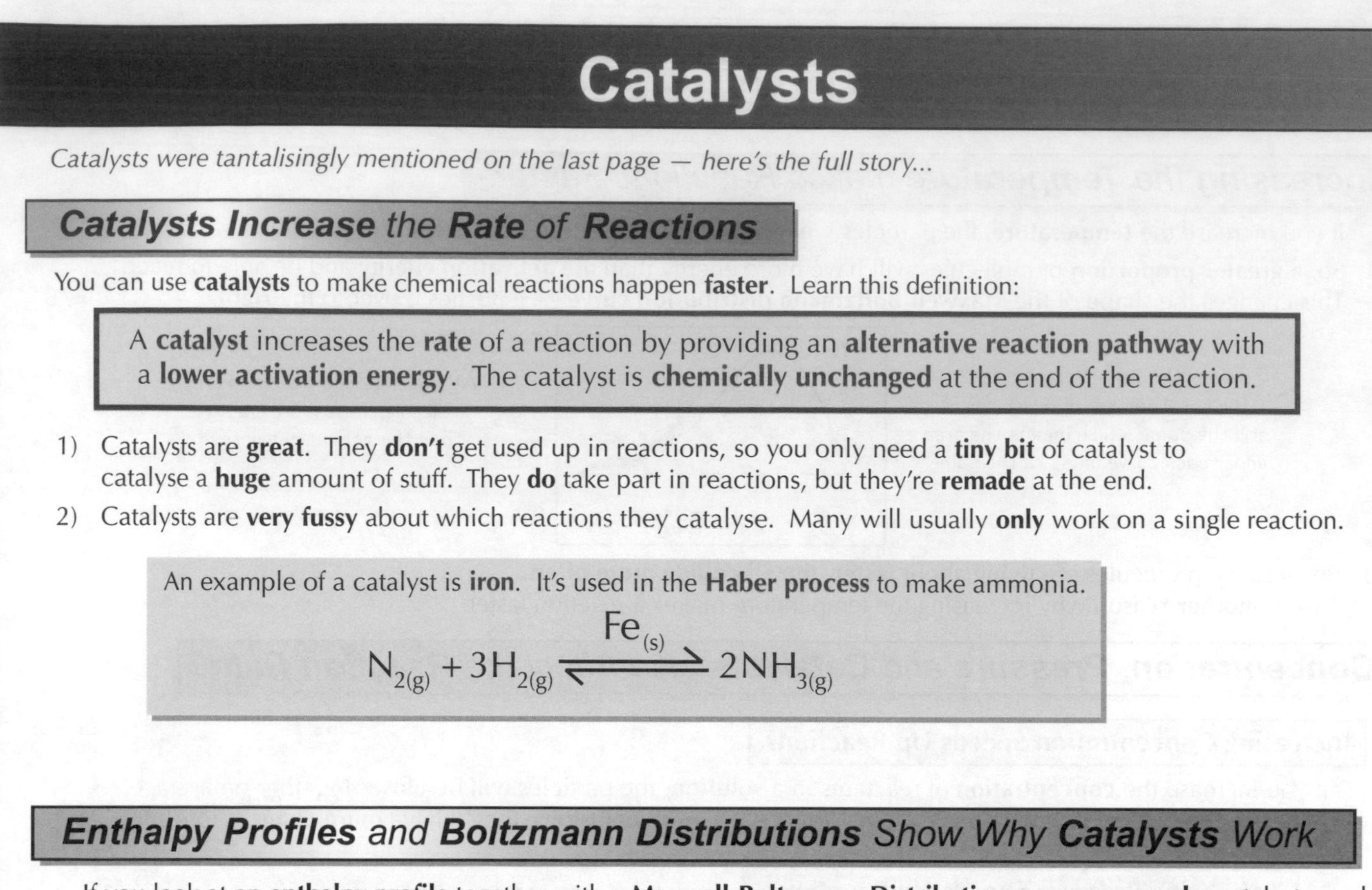

Enthalpy Profiles and Boltzmann Distributions Show Why Catalysts Work

If you look at an **enthalpy profile** together with a **Maxwell-Boltzmann Distribution**, you can see **why** catalysts work.

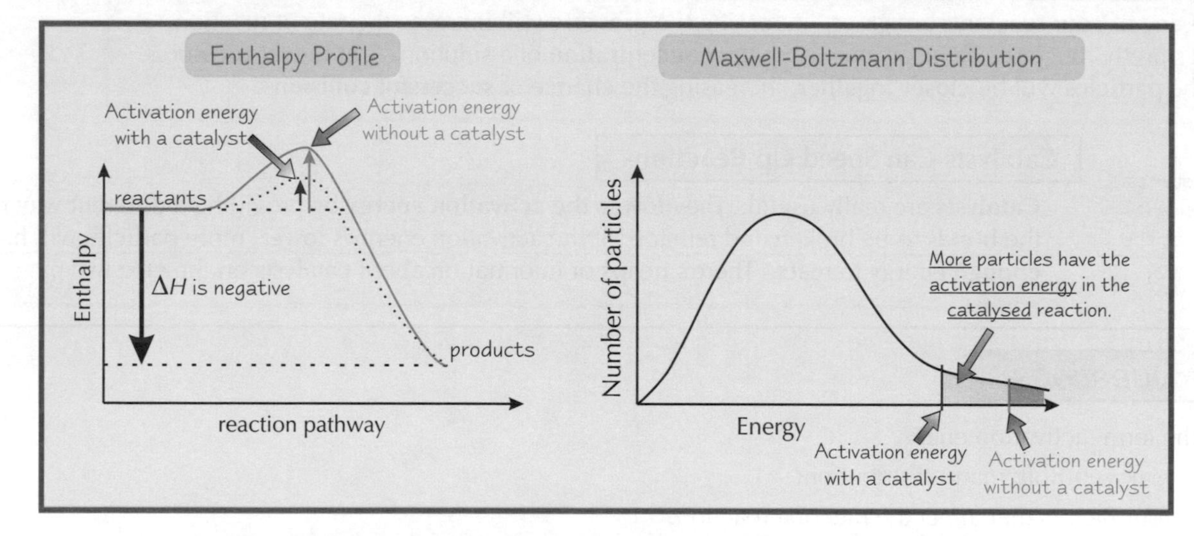

The catalyst **lowers the activation energy**, meaning there's **more particles** with **enough energy** to react when they collide. So, in a certain amount of time, **more particles react**.

Enzymes are Often Used in Industry Too

1) Enzymes are **biological catalysts** — they're **proteins** that catalyse certain **biochemical reactions**.
2) People have used enzymes for thousands of years — since way before we even knew what enzymes were. For example, enzymes produced by **yeast** are used to make bread and alcohol. More recent uses of enzymes include being added to **washing powders** to help break down stains and being used to partly digest **baby food**.
3) Many enzymes operate in conditions close to **room temperature and pressure**, so they're useful in **industry** because they can reduce the need for high temperature, fuel-guzzling processes.
4) Enzymes tend to be very **picky** about what they catalyse, and most can only be used for very **specific reactions**. This can be useful too — it means they can select one molecule from a mixture and cause that to react without affecting the others. This has been exploited in the production of new drugs.

Catalysts

Catalysts — Good for Industries...

Loads of industries rely on **catalysts**. They can dramatically lower production costs, and help make better products. Here's a few examples —

Iron is used as a catalyst in **ammonia** production. If it wasn't for the catalyst, they'd have to raise the **temperature** loads to make the reaction happen **quick enough**. Not only would this be bad for their fuel bills, it'd **reduce the amount of ammonia** produced.

Using a catalyst can change the properties of a product to make it more useful, e.g. **poly(ethene)**.

	Made without a catalyst	Made with a catalyst (a Ziegler-Natta catalyst to be precise)
Properties of poly(ethene)	less dense, less rigid	more dense, more rigid, higher melting point

Catalysts are used **loads** in the **petroleum industry** too.
They're used for **cracking**, **isomerisation**, and **reforming** alkanes (see p56-57).

...And for the Environment

1) Using catalysts means that lower temperatures and pressures can be used. So energy is saved, meaning **less CO_2** is released, and fossil fuel reserves are preserved. They can also **reduce waste** by allowing a different reaction to be used with a better **atom economy**. (See page 52 for more on atom economy.)

   For example, making the painkiller ibuprofen by the traditional method involves 6 steps and has an atom economy of 32%. Using catalysts it can be made in **3 steps** with an **atom economy of 77%**.

2) **Catalytic converters** on cars are made from **alloys of platinum, palladium and rhodium**. They reduce the pollution released into the atmosphere by speeding up the reaction **$2CO + 2NO \rightarrow 2CO_2 + N_2$**.

But catalysts don't last forever. All catalysts eventually need to be disposed of. The trouble is, many contain nasty **toxic** compounds, which may leach into the soil if they're sent directly to **landfill**. So it's important to try to recycle them, or convert them to non-leaching forms.

If catalysts contain **valuable metals**, such as platinum, it's worth recovering and recycling it — and there's special companies eager to do this. The decision whether to recycle the catalyst or to send it to landfill is made by balancing the **economic and environmental factors**.

Practice Questions

Q1 Explain what a catalyst is.

Q2 Draw an enthalpy profile diagram and a Maxwell-Boltzmann distribution diagram to show how a catalyst works.

Q3 Describe three reasons why catalysts are useful for industry.

Exam Questions

Q1 Sulfuric acid is manufactured by the contact process. In one of the stages, sulfur dioxide is converted into sulfur trioxide. A vanadium(V) oxide catalyst is used.

$$2SO_{2(g)} + O_{2(g)} \xrightarrow{V_2O_{5(s)}} 2SO_{3(g)} \qquad \Delta H = -197 \text{ kJ mol}^{-1}$$

a) Draw and label an enthalpy profile diagram for the catalysed reaction. Label the activation energy. [3 marks]

b) On your diagram from part a), sketch a profile for the uncatalysed reaction. [1 mark]

c) Explain how catalysts work. [2 marks]

Q2 The decomposition of hydrogen peroxide, H_2O_2, into water and oxygen is catalysed by manganese(IV) oxide, MnO_2.

a) Write an equation for the reaction. [2 marks]

b) Sketch a Maxwell-Boltzmann distribution for the reaction.
Mark on the activation energy for the catalysed and uncatalysed process. [3 marks]

c) Referring to your diagram from part b), explain how manganese(IV) oxide acts as a catalyst. [3 marks]

I'm a catalyst — I like to speed up arguments without getting too involved...

Whatever you do, do not confuse the Maxwell-Boltzmann diagram for catalysts with the one for a temperature change. Catalysts lower the activation energy without changing the shape of the curve. BUT, the shape of the curve does change with temperature. Get these mixed up and you'll be the laughing stock of the Examiners' tea room.

Dynamic Equilibrium

There's a lot of to-ing and fro-ing on this page. Mind your head doesn't start spinning.

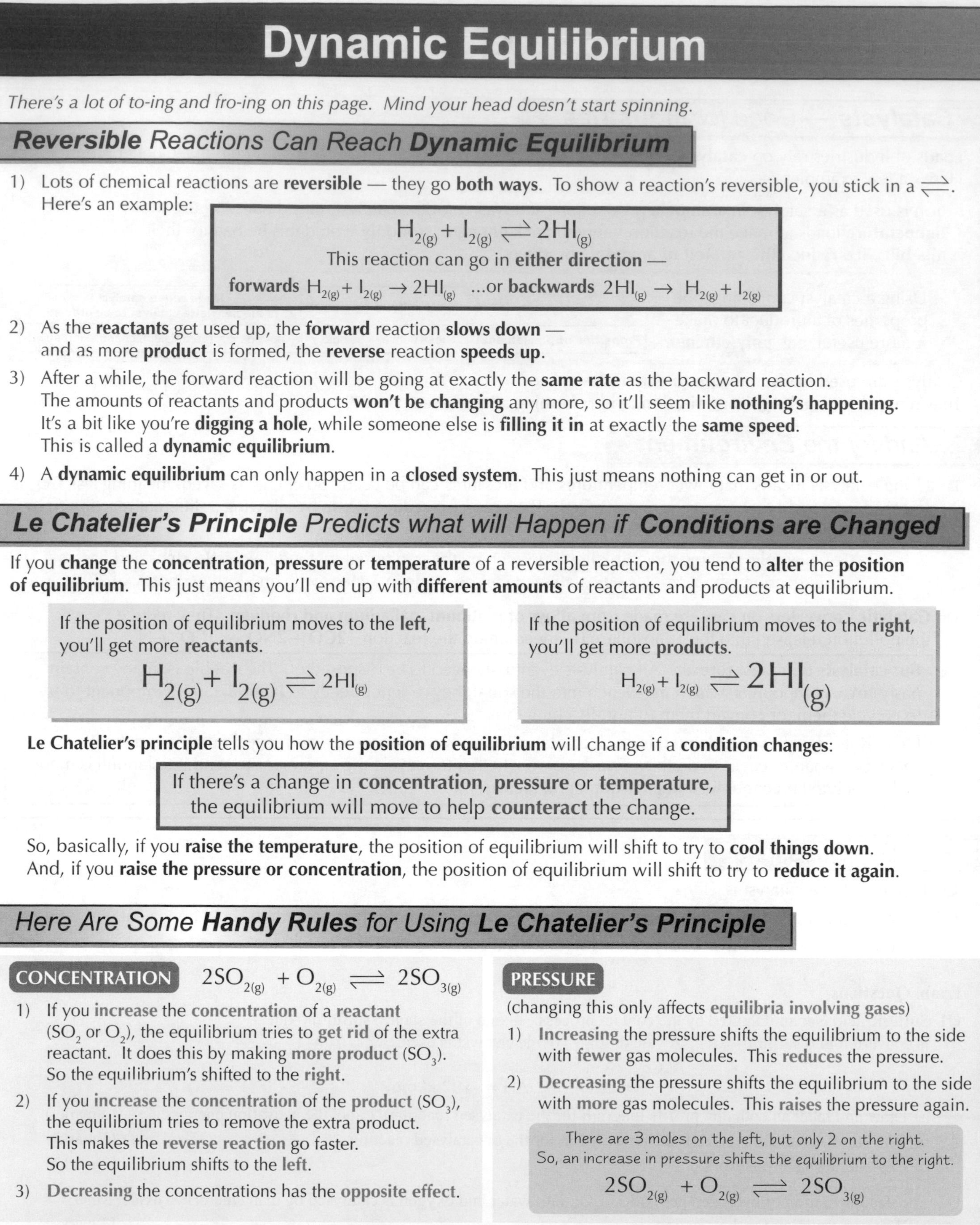

Reversible Reactions Can Reach *Dynamic Equilibrium*

1) Lots of chemical reactions are **reversible** — they go **both ways**. To show a reaction's reversible, you stick in a $\rightleftharpoons$. Here's an example:

$$H_{2(g)} + I_{2(g)} \rightleftharpoons 2HI_{(g)}$$

This reaction can go in **either direction** —

forwards $H_{2(g)} + I_{2(g)} \rightarrow 2HI_{(g)}$ ...or **backwards** $2HI_{(g)} \rightarrow H_{2(g)} + I_{2(g)}$

2) As the **reactants** get used up, the **forward** reaction **slows down** — and as more **product** is formed, the **reverse** reaction **speeds up**.
3) After a while, the forward reaction will be going at exactly the **same rate** as the backward reaction. The amounts of reactants and products **won't be changing** any more, so it'll seem like **nothing's happening**. It's a bit like you're **digging a hole**, while someone else is **filling it in** at exactly the **same speed**. This is called a **dynamic equilibrium**.
4) A **dynamic equilibrium** can only happen in a **closed system**. This just means nothing can get in or out.

Le Chatelier's Principle Predicts what will Happen if *Conditions are Changed*

If you **change** the **concentration**, **pressure** or **temperature** of a reversible reaction, you tend to **alter** the **position of equilibrium**. This just means you'll end up with **different amounts** of reactants and products at equilibrium.

If the position of equilibrium moves to the **left**, you'll get more **reactants**.

$$H_{2(g)} + I_{2(g)} \rightleftharpoons 2HI_{(g)}$$

If the position of equilibrium moves to the **right**, you'll get more **products**.

$$H_{2(g)} + I_{2(g)} \rightleftharpoons 2HI_{(g)}$$

Le Chatelier's principle tells you how the **position of equilibrium** will change if a **condition changes**:

If there's a change in **concentration**, **pressure** or **temperature**, the equilibrium will move to help **counteract** the change.

So, basically, if you **raise the temperature**, the position of equilibrium will shift to try to **cool things down**. And, if you **raise the pressure or concentration**, the position of equilibrium will shift to try to **reduce it again**.

Here Are Some Handy Rules *for Using* Le Chatelier's Principle

CONCENTRATION $2SO_{2(g)} + O_{2(g)} \rightleftharpoons 2SO_{3(g)}$

1) If you **increase** the **concentration** of a **reactant** (SO_2 or O_2), the equilibrium tries to **get rid** of the extra reactant. It does this by making **more product** (SO_3). So the equilibrium's shifted to the **right**.
2) If you **increase** the **concentration** of the **product** (SO_3), the equilibrium tries to remove the extra product. This makes the **reverse reaction** go faster. So the equilibrium shifts to the **left**.
3) **Decreasing** the concentrations has the **opposite effect**.

PRESSURE

(changing this only affects **equilibria involving gases**)

1) **Increasing** the pressure shifts the equilibrium to the side with **fewer** gas molecules. This **reduces** the pressure.
2) **Decreasing** the pressure shifts the equilibrium to the side with **more** gas molecules. This **raises** the pressure again.

There are 3 moles on the left, but only 2 on the right. So, an increase in pressure shifts the equilibrium to the right.

$$2SO_{2(g)} + O_{2(g)} \rightleftharpoons 2SO_{3(g)}$$

TEMPERATURE

1) If you **increase** the temperature, you **add heat**. The equilibrium shifts in the **endothermic (positive ΔH) direction** to absorb this heat.
2) **Decreasing** the temperature **removes heat**. The equilibrium shifts in the **exothermic (negative ΔH) direction** to try to replace the heat.
3) If the forward reaction's **endothermic**, the reverse reaction will be **exothermic**, and vice versa.

This reaction's exothermic in the forward direction. If you increase the temperature, the equilibrium shifts to the left to absorb the extra heat.

Exothermic $\Longrightarrow$

$$2SO_{2(g)} + O_{2(g)} \rightleftharpoons 2SO_{3(g)} \quad \Delta H = -197 \text{ kJ mol}^{-1}$$

$\Longleftarrow$ Endothermic

Dynamic Equilibrium

Catalysts **Don't Affect** The Position of Equilibrium

Catalysts have **NO EFFECT** on the **position of equilibrium**.
They **can't** increase **yield** — but they **do** mean equilibrium is reached **faster**.

Ethanol can be formed from **Ethene and Steam**

1) The industrial production of **ethanol** is a good example of why Le Chatelier's principle is important in **real life**.
2) Ethanol is produced via a **reversible exothermic reaction** between **ethene** and **steam**:

$$C_2H_{4(g)} + H_2O_{(g)} \rightleftharpoons C_2H_5OH_{(g)} \qquad \Delta H = -46 \text{ kJ mol}^{-1}$$

3) The reaction is carried out at a pressure of **60-70 atmospheres** and a temperature of **300 °C**, with a **phosphoric(V) acid** catalyst.

The **Conditions** Chosen are a **Compromise**

1) Because it's an **exothermic reaction**, **lower** temperatures favour the forward reaction. This means **more** ethane and steam is converted to ethanol at lower temperatures — you get a better **yield**.
2) But **lower temperatures** mean a **slower rate of reaction**. You'd be **daft** to try to get a **really high yield** of ethanol if it's going to take you 10 years. So the 300 °C is a **compromise** between **maximum yield** and **a faster reaction**.
3) **Higher pressures** favour the **forward reaction**, so a pressure of **60-70 atmospheres** is used — **high pressure** moves the reaction to the side with **fewer molecules of gas**. **Increasing the pressure** also increases the **rate** of reaction.
4) Cranking up the pressure as high as you can sounds like a great idea so far. But **high pressures** are **expensive** to produce. You need **stronger pipes** and **containers** to withstand high pressure. In this process, increasing the pressure can also cause **side reactions** to occur.
5) So the **60-70 atmospheres** is a **compromise** between **maximum yield** and **expense**. In the end, it all comes down to **minimising costs**.

Mr and Mrs Le Chatelier celebrate another successful year in the principle business

Practice Questions

Q1 Using an example, explain the terms 'reversible' and 'dynamic equilibrium'.

Q2 If the equilibrium moves to the right, do you get more products or reactants?

Q3 A reaction at equilibrium is endothermic in the forward direction.
What happens to the position of equilibrium as the temperature is increased?

Exam Question

Q1 Nitrogen and oxygen gases were reacted together in a closed flask and allowed to reach equilibrium with the nitrogen monoxide formed. The forward reaction is endothermic.

$$N_{2(g)} + O_{2(g)} \rightleftharpoons 2NO_{(g)}$$

a) State Le Chatelier's principle. [1 mark]

b) Explain how the following changes would affect the position of equilibrium of the above reaction:
(i) Pressure is **increased**. [2 marks]
(ii) Temperature is **reduced**. [2 marks]
(iii) Nitrogen monoxide is removed. [1 mark]

c) What would be the effect of a catalyst on the composition of the equilibrium mixture? [1 mark]

Only going forward cos we can't find reverse...

*Equilibria never do what you want them to do. They always **oppose** you. Be sure you know what happens to an equilibrium if you change the conditions. A word about pressure — if there's the same number of gas moles on each side of the equation, then you can raise the pressure as high as you like and it won't make a blind bit of difference to the position of equilibrium.*

Green Chemistry

'Green' things are big news these days — they're everywhere. So it'll be no surprise to find them in AS Chemistry too.

Chemical Industries Could Be More Sustainable

1) Doing something **sustainably** means doing it **without stuffing things up** for the future. Sustainable chemistry (or 'green chemistry') means you don't **use up** all the Earth's **resources**, or put loads of **damaging** chemicals into the environment.
2) Many of the chemical processes used in industry at the moment **aren't** very sustainable. Take the **plastics** industry, for example — the raw materials used often come from non-renewable **crude oil**, and the products themselves are usually **non-biodegradable** or **hard to recycle** when we're finished with them. (See pages 64-65 for more details.)
3) But there are things chemists can do to try and improve things. For example, they can...

Use Renewable Raw Materials

Loads of chemicals are traditionally made from **non-renewable** raw materials (e.g. crude oil fractions, or metal ores). But chemists can often develop **alternative compounds** (or **alternative ways** to make existing ones) involving **renewable** raw materials — e.g. some plastics are now made from **plant products** rather than oil fractions (p65).

Use Renewable Energy Sources

Many chemical processes use a lot of **energy**. Right now, most of that energy comes from **fossil fuels**, which will soon run out. But there are potential **alternatives**...

- **Plant-based fuels** can be used (e.g. bioethanol — see page 59 for more).
- **Solar power** — ways to produce electricity from sunlight are developing rapidly.

There are other renewable energy technologies — like geothermal, wind, wave...

Ensure All the Chemicals Involved are as Non-Toxic as Possible

1) Many common chemicals are **harmful** — either to **humans**, other **living things**, the **environment**, or all three. Where possible, it's generally a good thing to use a **safer** alternative. For example...
   - **Lead** (which can have some nasty effects on your health) used to be used in paint, petrol and for soldering. Alternatives are now used — paint and petrol use lead-free compounds, solder can be made with other metals.
   - Some **foams** used in fire extinguishers are very good at putting out fires, but leave hazardous products behind, including some that deplete the ozone layer (see page 94). Again, alternatives are now available.
   - **Dry cleaners** used to use a solvent based on chlorinated **hydrocarbons**, but these are known to be **carcinogenic** (i.e. they cause cancer). Safer alternatives are now available (liquid 'supercritical' carbon dioxide, as you asked).
2) Sometimes **redesigning** a **process** means you can do without unsafe chemicals completely — e.g. instead of using harmful organic solvents, some reactions can be carried out with one of the **reactants** acting as a solvent.

Make Sure that Products and Waste are Biodegradable or Recyclable

1) Chemists can try to create **recyclable** products (see p64) — a good way to conserve raw materials.
2) **Waste** should be kept to a **minimum**, and preferably be **recyclable** or **biodegradable** (see p65).
3) You can also improve **sustainability** by developing more **efficient processes** — for example, by using **catalysts** (see p86-87), or by picking reactions with higher **atom economy** (see p52).
4) **Laws** can be used to encourage change. For example, when you buy a new TV, the shop now has to agree to recycle your old TV set, with the TV manufacturers paying some of the cost. This creates an incentive to design products that are easier and cheaper to recycle.

Plastics are hard to recycle. Dogs too.

Greener Chemistry Can Have Unexpected Consequences

Pretty much everyone agrees that making the chemical industry more sustainable is a good thing.
But sometimes making things 'greener' can cause unwanted **knock-on effects**. Take biofuels, for example...

- Growing grain for biodiesel (or sugar cane for ethanol) means **less land** is available to grow **food**. So food gets **more expensive** — which will be worst for the **urban poor**, who already struggle to afford food (and can't grow their own).
- Large biofuel companies might buy up the **most fertile** land, forcing small farmers onto land with poorer crop yields.
- The land to grow biofuels often comes from clearing **forests**. Removing loads of trees means less CO_2 is absorbed in photosynthesis, so more stays in the atmosphere — the very problem that the use of biofuels is supposed to tackle...
- And that's not all — destroying existing, varied habitats and replacing them all with vast swathes of the same crop will **reduce biodiversity** and could cause **soil degradation** (loss of nutrients, etc.).

Green Chemistry

International Cooperation Is Needed to Reduce Pollution

1) Pollution doesn't stop at national borders — **rivers** flow from one country to the next, and the **atmosphere** and **oceans** are constantly moving and mixing. This means that eventually **everyone** suffers from **everyone else's** dirty ways.
2) **International cooperation** is important — there are already concerns about countries buying products made using **polluting technologies** from **abroad**, so that they can claim not to be producing the pollution themselves.
3) Various **international treaties** have been agreed. But usually, not all countries sign up because they're worried it will be bad for their **economy** (make things more expensive, cause job losses, and so on).
4) The **Montreal Protocol on Substances that Deplete the Ozone Layer** is probably the most successful 'green chemistry' global treaty to date — virtually everyone's signed up. Countries who signed up to this '**Montreal Protocol**' agreed to phase out production of substances that damaged the ozone layer (see p94).
5) Similarly, most countries have signed the **Stockholm Treaty** on persistent organic pollutants (POPs). POPs are organic chemicals (e.g. some pesticides and fungicides) that **accumulate** in the fatty tissues of living organisms. They're passed up the food chain and are **toxic** to humans and other animals.
6) In 1992, the United Nations held a big conference about the environment and development (the '**Earth Summit**') in Rio de Janeiro. Governments agreed to a set of **27 principles** about sustainable development — the 'Rio Declaration'.

   These principles were all very sensible (e.g. don't cause environmental harm, develop in a sustainable way, and so on) but they **aren't legally binding** — so **no punishment** can be dished out when countries don't keep to the principles.

The summit of the Earth... much harder to get to than the Earth Summit in Rio had been.

You don't need to memorise every detail about the examples on these last two pages, but you should understand the basic principles behind them.

Practice Questions

Q1 List four ways in which the chemical industry can be made more sustainable.

Q2 Give two examples of renewable energy sources.

Q3 List four potential drawbacks of biofuel production.

Q4 Why are international treaties important in controlling pollution?

Exam Questions

Q1 In Brazil, ethanol is produced by fermenting sugar cane.
This ethanol is then used as fuel.

a) Explain the advantages of using ethanol made from sugar cane as a fuel, instead of petrol. [2 marks]

b) Suggest why not all countries produce ethanol for use as a fuel in this way. [1 mark]

c) Describe two possible negative effects of growing sugar cane to make ethanol. [2 marks]

Q2 Much research is currently done on new catalysts.

a) Explain why catalysts are important in making chemical processes 'greener'. [2 marks]

b) The discovery of a new catalyst has made it possible to make ethanoic acid very efficiently by reacting methanol with carbon monoxide:

$$CH_3OH + CO \rightarrow CH_3COOH$$

Describe one way in which this process could be considered 'green'. [2 marks]

Like the contents of my fridge, Chemistry's going greener by the day...

It's important stuff, all this. It'll be important for your exam, obviously, but it's my bet that you'll come across this stuff long after you've taken your exam as well, which makes it doubly useful. On a different note... isn't it weird how you can sign up for an AS level in Chemistry, and only then be told that you'll be studying international politics too...

The Greenhouse Effect

Now I'm sure you know this already but it's good to be sure — the greenhouse effect, global warming and climate change are all different things. They're linked (and you need to know how) — but they are not the same. Ahem.

The Greenhouse Effect Keeps Us Alive

1) Some of the **electromagnetic radiation** from the Sun reaches the Earth and is **absorbed**. The Earth then **re-emits** it as **infrared radiation** (heat).
2) Various gases in the troposphere (the lowest layer of the atmosphere) **absorb** some of this infrared radiation... and **re-emit** it in **all directions** — including back towards Earth, keeping us warm. This is called the '**greenhouse effect**' (even though a real greenhouse doesn't actually work like this, annoyingly).

3) The main greenhouse gases are **water vapour**, **carbon dioxide** and **methane**. Their molecules **absorb IR radiation** to make the bonds in the molecule **vibrate more**. This extra energy is passed on to other molecules in the air by **collisions**, giving the other molecules more kinetic energy and raising the overall temperature.
4) The contribution of any particular gas to the greenhouse effect depends on:
   - how much radiation one molecule of the gas absorbs
   - how much of that gas there is in the atmosphere (concentration in ppm, say)

   For example, one methane molecule traps far more heat than one carbon dioxide molecule, but there's much **less methane** in the atmosphere, so its overall contribution to the greenhouse effect is smaller.

An Enhanced Greenhouse Effect Causes Global Warming

1) Over the last 150 years or so, the world's **human population** has shot up and we've become more **industrialised**. We've been **burning fossil fuels**, releasing **tons** of CO_2, and we've been **chopping down forests** which used to absorb CO_2 by photosynthesis.

2) **Methane** levels have risen as we've grown more food. **Cows** produce large amounts of methane (from both ends). Paddy fields, in which rice is grown, kick out a fair bit of it too.
3) These **human activities** have caused a rise in greenhouse gas concentrations, which **enhances** the greenhouse effect. **More heat** is being trapped and the Earth is **getting warmer** — this is **global warming**.

Global warming won't just make everywhere a bit warmer and affect the skiing — the warmer oceans will expand and massive ice-sheets in the polar regions will melt, causing **sea levels** to rise and leading to more flooding.

The **climate** in any region of the world depends on a **really complicated** system of ocean currents, winds, etc. Global warming means there's **more heat energy** in the system. This could lead to **stormier**, less predictable weather.

In some places there could be much **less rainfall**, with droughts and crop failures causing famines and forcing entire populations to become refugees.
In other regions, increased rainfall and flooding would bring diseases like cholera.

There's Scientific Evidence for the Increase in Global Warming

1) Scientists have collected data to confirm whether or not climate change is happening, e.g. from analysing air samples and sea water samples.
2) The evidence shows that the Earth's average temperature has increased **dramatically** in the last 50 years, and that CO_2 levels have increased at the same time.

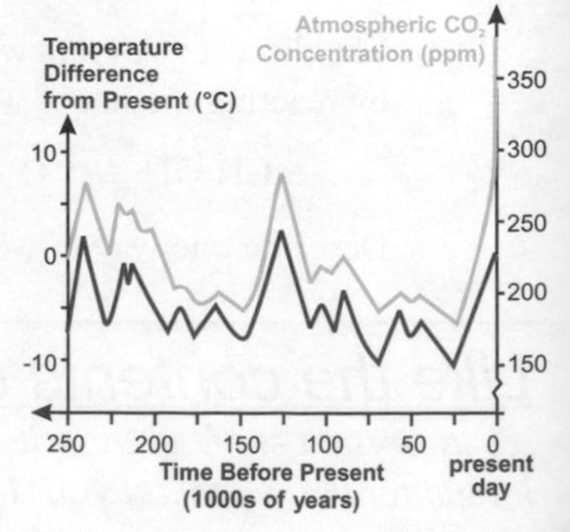

3) The **correlation** between CO_2 and temperature is pretty clear, but there's been debate about whether rising carbon dioxide levels have **caused** the recent temperature rise. Just showing a correlation doesn't prove that one thing causes another — there has to be a plausible mechanism for how one change causes the other (in this case, the explanation is the enhanced greenhouse effect).
4) There's now a consensus among climate scientists that the link **is** causal, and that recent warming is **anthropogenic** — **human activities** are to blame.

The Greenhouse Effect

Scientists are **Monitoring Global Warming...**

1) Scientific evidence gathered by the Intergovernmental Panel on Climate Change (IPCC) persuaded most of the world's governments that global warming is happening. There's now **global agreement** that climate change could be very damaging for millions of people, the environment and economies, and that we should try to **limit** it.
2) In 1997 the **Kyoto protocol** was signed — industrialised countries (including the UK) promised to reduce their greenhouse gas emissions to agreed levels. Many chemists are now involved in **monitoring** greenhouse gas emissions to see if countries will meet the targets (it looks like many won't).
3) Chemists also continue to monitor the environment to see how it's changing now. The data they collect and analyse is used in **climate models** (a big load of equations run on a computer to simulate how the climate system works).
4) Climate scientists use these models to predict future changes. It's a big job — when **new factors** affecting the climate are discovered by other scientists, the modellers have to 'tweak' their models to take this into account.

...and **Investigating** Ways to **Limit It**

Scientists are investigating various ways to help **reduce** carbon dioxide emissions. These include:

1) **Carbon capture and storage** (CCS). This means removing waste CO_2 from, say, power stations, and either
   - injecting it as a **liquid** into the **deep ocean**, or
   - storing it deep **underground** — one possibility is to use old oil- or gas-fields under the sea-bed, or
   - reacting it with metal oxides to form stable, easily stored **carbonate minerals**, e.g. calcium carbonate.
2) Developing alternative fuels. See pages 58-59 for more on this.

Practice Questions

Q1 What type of electromagnetic radiation does the Earth emit?

Q2 What's the difference between the greenhouse effect and global warming?

Q3 Give three reasons why climate change is seen as a problem.

Exam Questions

Q1 a) Name the three main greenhouse gases. [3 marks]

b) Explain how greenhouse gases keep the temperature in the lower layer of the Earth's atmosphere higher than it would otherwise be. [3 marks]

c) What factors affect the contribution a gas makes to the greenhouse effect? [2 marks]

Q2 The concentration of carbon dioxide in the Earth's atmosphere has increased over the last 50 years.

a) Give two reasons for this increase. [2 marks]

b) How do governments know that global warming is happening? [1 mark]

c) Describe two methods that chemists are developing as a way of reducing carbon dioxide emissions. [2 marks]

Global Warming probably just isn't funny...

You may be sick of global warming, because it's all over the news these days. But how scientists gathered all the evidence to back up the theory that global warming is caused by human activity is a great example of How Science Works. What's more, the evidence was used to instigate an international treaty — a beauty of an example of how science informs decision-making.

The Ozone Layer and Air Pollution

Three pages on air pollution coming up, so take a deep breath...
unless you're hanging around somewhere with a lot of air pollution, that is...

The Earth has a Layer of **Ozone** at the Edge of the **Stratosphere**

The **ozone layer** is in a layer of the atmosphere called the **stratosphere**. It contains most of the atmosphere's **ozone molecules**, O_3. Ozone is formed when **UV radiation** from the Sun hits oxygen molecules.

If the right amount of **UV radiation** is absorbed by an oxygen molecule, the oxygen molecule splits into separate atoms or **free radicals**. The free radicals then **combine** with other oxygen molecules to form **ozone molecules**, O_3.

$$O_2 + h\nu \rightarrow O\bullet + O\bullet \longrightarrow O_2 + O\bullet \rightarrow O_3$$

a quantum of UV radiation

The Ozone Layer is Constantly Being **Replaced**

1) UV radiation can also **reverse** the formation of ozone.

$$O_3 + h\nu \rightarrow O_2 + O\bullet$$

The radical produced then forms more ozone with an O_2 molecule, as shown above.

2) So, the ozone layer is continuously being **destroyed** and **replaced** as UV radiation hits the molecules. An **equilibrium** is set up, so the concentrations stay fairly constant:

$$O_2 + O\bullet \rightleftharpoons O_3$$

The Ozone Layer **Protects** the Earth

1) The **UV radiation** from the Sun is made up of **different frequencies**. These are grouped into **three bands**:

UVA UVB UVC
INCREASING FREQUENCY AND ENERGY

2) The ozone layer removes all the high energy **UVC radiation** and about 90% of the **UVB**. These types of UV radiation are harmful to humans and most other life on Earth.
3) **UVB** can damage the DNA in cells and cause **skin cancer**. It's the main cause of **sunburn** too. **UVA** can also lead to **skin cancer**. Both types of UV break down collagen fibres in the skin causing it to **age faster**.
4) When the skin's exposed to UV, it **tans**. This helps protect **deeper tissues** from the effects of the radiation.
5) **BUT...** UV radiation isn't all bad — in fact it's **essential** for us humans. We need it to produce **vitamin D**.

CFCs and **Nitrogen Oxides** Break Ozone Down

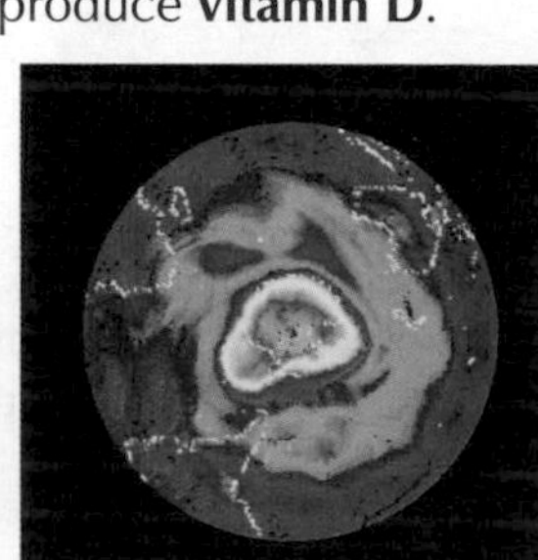

Here's a satellite map showing the 'hole' in the ozone layer over Antarctica. The 'hole' is shown by the white and pink area.

LABORATORY FOR ATMOSPHERES, NASA GODDARD SPACE FLIGHT CENTER/SCIENCE PHOTO LIBRARY

1) In the 1970s and 1980s, scientists discovered that the **ozone layer** above **Antarctica** was getting **thinner** — in fact, it was decreasing very rapidly. The ozone layer over the **Arctic** has been found to be thinning too. These 'holes' in the ozone layer are bad because they allow more harmful **UVB radiation** to reach the Earth.
2) **CFCs** (see p72) absorb UV radiation and split to form **chlorine free radicals**. These free radicals **destroy ozone molecules** and are then **regenerated** to destroy more ozone. One chlorine atom can destroy 10 000 ozone molecules before it forms a stable compound.
3) **NO• free radicals** from **nitrogen oxides** destroy ozone too. Nitrogen oxides are produced by **car and aircraft engines** and **thunderstorms**. NO• free radicals affect ozone in the **same way** as chlorine radicals.
4) The reactions can be represented by these equations, where **R** represents either Cl• or NO•. The free radicals acts as **catalysts** for the destruction of the ozone.

$$\mathbf{R} + O_3 \rightarrow \mathbf{RO} + O_2$$
$$\mathbf{RO} + O\bullet \rightarrow \mathbf{R} + O_2$$

Formed when UV breaks down O_2. The harmful radical is regenerated.

NO• and Cl• aren't the only culprits — free radicals are produced from other halogenoalkanes too.

The overall reaction is: $O_3 + O\bullet \rightarrow 2O_2$

The Ozone Layer and Air Pollution

CFCs and nitrogen oxides breaking the ozone layer down isn't the only air pollution problem you need to know about...

Burning **Hydrocarbons** can Produce **Carbon Monoxide**

Fuels from crude oil are used all the time, for things such as transport and in power stations.

1) When pure alkanes burn **completely**, all you get is **carbon dioxide** and **water**.
2) But if there's **not enough oxygen**, hydrocarbons combust **incompletely**, and you get **carbon monoxide** gas produced instead of carbon dioxide. This can happen in internal combustion engines (as used in most cars on the planet).

Here's how carbon monoxide forms when methane burns without enough oxygen:

$$CH_{4\,(g)} + 1\tfrac{1}{2}O_{2\,(g)} \rightarrow CO_{(g)} + 2H_2O_{(g)}$$

And here's the equation for incomplete combustion of octane:

$$C_8H_{18\,(g)} + 8\tfrac{1}{2}O_{2\,(g)} \rightarrow 8CO_{(g)} + 9H_2O_{(g)}$$

3) This is bad news — carbon monoxide gas is poisonous. Carbon monoxide molecules bind to the same sites on **haemoglobin molecules** in red blood cells as oxygen molecules. So **oxygen** can't be carried around the body.

And if that's Not Bad Enough... **Burning Fuels** Produces Other **Pollutants** Too

Carbon monoxide's not the only pollutant gas that comes out of a car exhaust.

1) Engines **don't burn** all the fuel molecules. Some of these come out as **unburnt hydrocarbons**.
2) **Oxides of nitrogen** (NO_x) are produced when the high pressure and temperature in a car engine cause the nitrogen and oxygen atoms in the air to react together. Oxides of nitrogen don't just contribute to the breaking down of the ozone layer...
3) The hydrocarbons and nitrogen oxides react with sunlight to form **ground-level ozone** (O_3), which is a major component of **smog**. Specifically, it's part of **photochemical smog** — the dangerous chemicals that form when certain pollutant gases react with sunlight.
4) **Ground-level ozone** irritates people's eyes, aggravates respiratory problems and even damages our lungs (ozone isn't very nice stuff, unless it is high up in the atmosphere as part of the ozone layer).

Luckily, carbon monoxide, unburnt hydrocarbons and oxides of nitrogen can be removed by **catalytic converters** on cars. Unluckily, you need to know some of the chemistry behind them...

Catalytic Converters Reduce Harmful **Exhaust Emissions**

a cat a list

1) Catalytic converters sit quietly in a car **exhaust** and get rid of **pollutant gases** like carbon monoxide, oxides of nitrogen and unburnt hydrocarbons by changing them to **harmless gases**, like water vapour and nitrogen, or to **less harmful** ones like carbon dioxide.
2) **Solid** heterogeneous catalysts can provide a **surface** for a reaction to take place on. Here's how it works —
   - **Reactant molecules** arrive at the **surface** and **bond** with the solid catalyst. This is called **a<u>d</u>sorption**.
   - The bonds between the **reactant's** atoms are **weakened** and **break up**. This forms **radicals**. These radicals then **get together** and make **new molecules**.
   - The new molecules are then detached from the catalyst. This is called **desorption**.

This example shows you how a catalytic converter changes the harmful gases **nitrogen monoxide, NO**, and **carbon monoxide, CO**, to **nitrogen** and **carbon dioxide**.

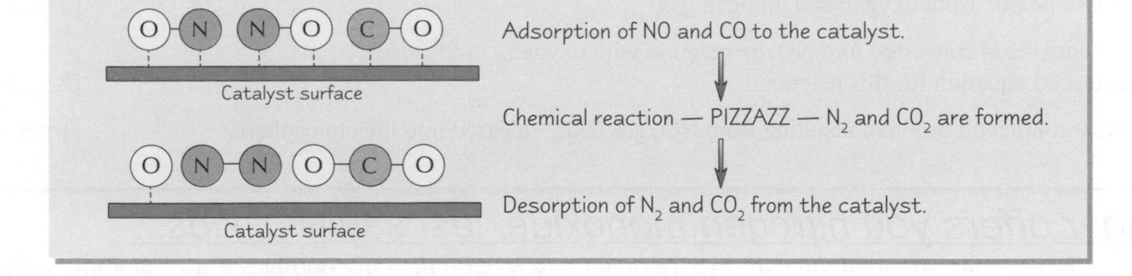

Remember — the adsorption **mustn't** be **too strong** or it won't **let go** of the atoms. **BUT** — it needs to be **strong enough** to **weaken** the bonds between the reactant molecules so that the new molecules can form.

The Ozone Layer and Air Pollution

Infrared Spectroscopy is Used to *Monitor Air Pollution*

You can use **infrared spectroscopy** to measure how much of a **polluting gas** is present in the **air**.
(If you've forgotten what infrared spectroscopy is, go back and look at page 75.)
Here's an outline of how infrared spectroscopy is used to check how much **carbon monoxide** there is:

1) A **sample of air** is drawn into the spectrometer. A beam of **infrared radiation** of a certain frequency is passed through the sample. Any carbon monoxide that is present will **absorb** some of this radiation.
2) At the same time, a beam of infrared radiation of the **same frequency** is passed through a sample of a gas that **doesn't absorb any infrared**, like N_2. This acts as a kind of **control** reading.
3) The **difference** in the amount of infrared energy absorbed by the gases in the two chambers is a measure of the **amount** of carbon monoxide present in the air sample.

You can use the same technique to monitor the levels of any polluting gas that can absorb infrared, like NO, SO_2 or CH_4. Only molecules containing at least two different atoms will absorb infrared radiation.

Practice Questions

Q1 What is ozone, and where is the ozone layer?

Q2 Which has higher energy — UVA, UVB, or UVC?

Q3 Write out equations to show how ozone is destroyed, using R to represent the radical.

Q4 Write a chemical equation for the incomplete combustion of methane gas in air.

Q5 What exhaust gases contribute to photochemical smog?

Q6 Describe how the catalytic converter in a car exhaust works.

Exam Questions

Q1 The 'ozone layer' lies mostly between 15 and 30 km above the Earth's surface.

a) Explain how ozone forms in this part of the atmosphere. [3 marks]

b) What are the benefits to humans of the ozone layer? [2 marks]

c) How does the ozone layer absorb harmful radiation without being permanently destroyed? [3 marks]

Q2 Nitrogen monoxide gas is a pollutant formed when internal combustion engines burn fuels.

a) Write a balanced chemical equation for the formation of nitrogen monoxide from oxygen and nitrogen gas. [2 marks]

b) Nitrogen monoxide is converted into NO_2 by reaction with oxygen. Write a balanced equation for this reaction. [2 marks]

c) Name an environmental problem resulting from NO_2 gas being released into the atmosphere. [1 mark]

If a stranger offers you nitrogen monoxide, just say NO, kids,,,

That's right, it's yet more pages about all the things we're doing to screw up the environment. Lucky all those chemists are there to invent catalytic converters and stuff to save us from ourselves. I think the best plan is to get rid of all cars and walk everywhere. Except when I've got a lot of stuff to carry. Or when I'm in a hurry. Or when I'm feeling a bit lazy. Or...

Practical and Investigative Skills

You're going to have to do some practical work too — and once you've done it, you have to make sense of your results...

Make it a **Fair Test** — Control your **Variables**

You probably know this all off by heart but it's easy to get mixed up sometimes. So here's a quick recap:

Variable — A variable is a **quantity** that has the **potential to change**, e.g. mass.
There are two types of variable commonly referred to in experiments:

- **Independent variable** — the thing that you **change** in an experiment.
- **Dependent variable** — the thing that you **measure** in an experiment.

When drawing graphs, the dependent variable should go on the y-axis, the independent on the x-axis.

So, if you're investigating the effect of **temperature** on rate of reaction using the apparatus on the right, the variables will be:

Independent variable	Temperature
Dependent variable	Amount of oxygen produced — you can measure this by collecting it in a gas syringe
Other variables — you MUST keep these the same	Concentration and volume of solutions, mass of solids, pressure, the presence of a catalyst and the surface area of any solid reactants

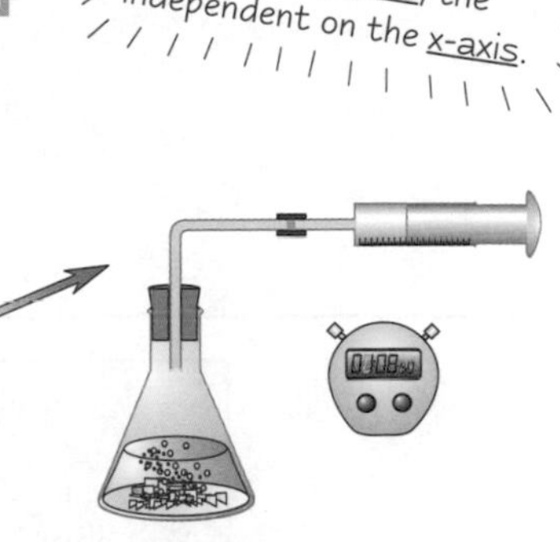

Know Your Different Sorts of **Data**

Experiments always involve some sort of measurement to provide **data**.
There are different types of data — and you need to know what they are.

Discrete — you get discrete data by **counting**. E.g. the number of bubbles produced in a reaction would be discrete. You can't have 1.25 bubbles. That'd be daft. Shoe size is another good example of a discrete variable.

Continuous — a continuous variable can have **any value** on a scale. For example, the volume of gas produced or the mass of products from a reaction. You can never measure the exact value of a continuous variable.

Categoric — a categoric variable has values that can be sorted into **categories**. For example, the colours of solutions might be blue, red and green. Or types of material might be wood, steel, glass.

Ordered (ordinal) — Ordered data is similar to categoric, but the categories can be **put in order**. For example, if you classify reactions as 'slow', 'fairly fast' and 'very fast' you'd have ordered data.

Organise Your Results in a **Table** — And Watch Out For **Anomalous** Ones

Before you start your experiment, make a **table** to write your results in.
You'll need to repeat each test at least three times to check your results are reliable.

This is the sort of table you might end up with when you investigate the effect of **temperature** on **reaction rate**. (You'd then have to do the same for **different temperatures**.)

Temperature	Time (s)	Volume of gas evolved (cm^3) Run 1	Volume of gas evolved (cm^3) Run 2	Volume of gas evolved (cm^3) Run 3	Average volume of gas evolved (cm^3)
20 °C	**10**	8	7	8	**7.7**
	20	17	19	20	**18.7**
	30	28	20	30	**29**

Find the average of each set of repeated values.
You need to add them all up and divide by how many there are.
E.g.: $(8 + 7 + 8) \div 3 = 7.7\ cm^3$

Watch out for **anomalous results**. These are ones that don't fit in with the other values and are likely to be wrong. They're likely to be due to random errors — here the syringe plunger may have got stuck.

Ignore anomalous results when you calculate the average.

Practical and Investigative Skills

Graphs: **Line, Bar or Scatter** — Use the **Best Type**

You'll usually be expected to make a **graph** of your results. Not only are graphs **pretty**, they make your data **easier to understand** — so long as you choose the right type.

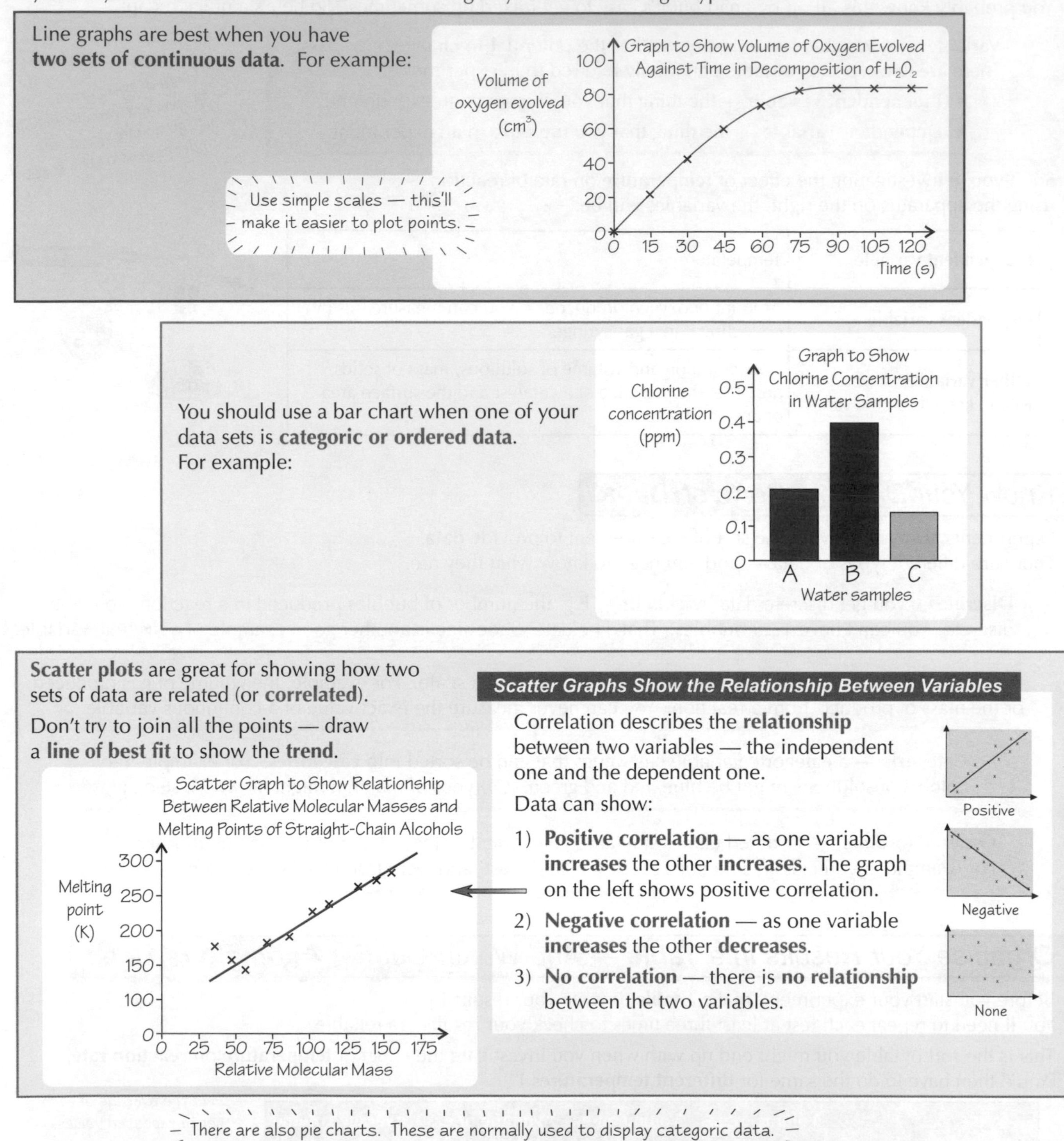

Remember These **Important Points** When **Drawing Graphs**

Whatever type of graph you make, you'll ONLY get full marks if you:

- Choose a sensible scale — don't do a tiny graph in the corner of the paper.
- Label both axes — including units.
- Plot your points accurately — using a sharp pencil.

Practical and Investigative Skills

Correlation **Doesn't Necessarily** Mean **Cause** — Don't Jump to Conclusions

1) Ideally, only **two** quantities would **ever** change in any experiment — everything else would remain **constant**.
2) But in experiments or studies outside the lab, you **can't** usually control all the variables. So even if two variables are correlated, the change in one may **not** be causing the change in the other. Both changes might be caused by a **third variable**.

Watch out for bias too — for instance, a bottled water company might point these studies out to people without mentioning any of the doubts.

Example

Some studies have found a correlation between **drinking chlorinated tap water** and the risk of developing certain cancers. So some people argue that this means water shouldn't have chlorine added.

BUT it's hard to control all the variables between people who drink tap water and people who don't. It could be many lifestyle factors.

Or, the cancer risk could be affected by something else in tap water — or by whatever the non-tap water drinkers drink instead...

Don't Get **Carried Away** When Drawing Conclusions

The **data** should always **support** the conclusion. This may sound obvious but it's easy to **jump** to conclusions. Conclusions have to be **specific** — not make sweeping generalisations.

Example

The rate of an enzyme-controlled reaction was measured at **10 °C, 20 °C, 30 °C, 40 °C, 50 °C and 60 °C**. All other variables were kept constant, and the results are shown in this graph.

A science magazine **concluded** from this data that enzyme X works best at **40 °C**. The data **doesn't** support this.

The enzyme **could** work best at 42 °C or 47 °C but you can't tell from the data because **increases** of **10 °C** at a time were used.

The rate of reaction at in-between temperatures **wasn't** measured.

All you know is that it's faster at **40 °C** than at any of the other temperatures tested.

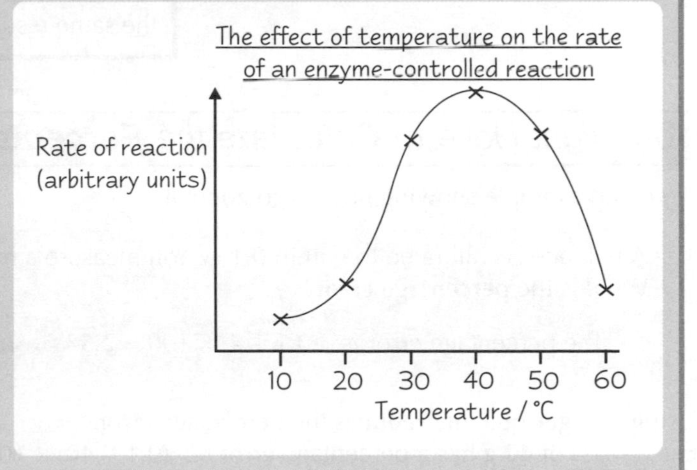

Example

The experiment above **ONLY** gives information about this particular enzyme-controlled reaction. You can't conclude that **all** enzyme-controlled reactions happen faster at a particular temperature — only this one. And you can't say for sure that doing the experiment at, say, a different constant pressure, wouldn't give a different optimum temperature.

Practical and Investigative Skills

*You Need to Look **Critically** at Your Results*

There are a few bits of lingo that you need to understand.
They'll be useful when you're evaluating how convincing your results are.

VALID RESULTS

Valid results answer the original question. For example, if you haven't **controlled all the variables** your results won't be valid, because you won't be testing just the thing you wanted to.

ACCURATE RESULTS

Accurate results are those that are **really close** to the **true** answer.

PRECISE RESULTS

These are results taken using **sensitive instruments** that measure in **small increments**, e.g. pH measured with a meter (pH 7.692) will be **more precise** than pH measured with paper (pH 7).

It's possible for results to be precise **but not** accurate, e.g. a balance that weighs to 1/1000 th of a gram will give precise results but if it's not **calibrated** properly the results won't be accurate.

RELIABLE RESULTS

Reliable means the results can be **consistently reproduced** in independent experiments. And if the results are reproducible they're more likely to be **true**. If the data isn't reliable for whatever reason you **can't draw** a valid **conclusion**.

For experiments, the **more repeats** you do, the **more reliable** the data. If you get the **same result** twice, it could be the correct answer. But if you get the same result **20 times**, it'd be much more reliable. And it'd be even more reliable if everyone in the class got about the same results using different apparatus.

*You Might Have to **Calculate** the **Percentage Error** of a Measurement*

Here's an example showing how to go about it:

A balance is calibrated to within 0.1 g. You measure a mass as **4 g**. What is the **percentage error**?

The percentage error is: $(0.1 \div 4) \times 100 =$ **2.5%**

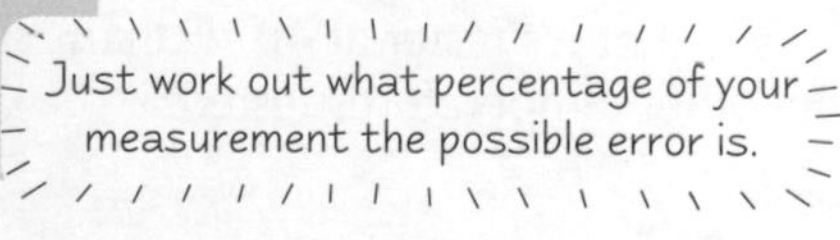

Using a **larger** quantity **reduces** the percentage error.
E.g. a mass of **40 g** has a percentage error of: $(0.1 \div 40) \times 100 =$ **0.25%**

*Work **Safely** and **Ethically** — Don't Blow Up the Lab or Harm Small Animals*

In any experiment you'll be expected to show that you've thought about the **risks and hazards**. It's generally a good thing to wear a lab coat and goggles, but you may need to take additional safety measures, depending on the experiment. For example, anything involving nasty gases will need to be done in a fume cupboard.

You need to make sure you're working **ethically** too. This is most important if there are other people or animals involved. You have to put their welfare first.

A2-Level Chemistry

Exam Board: OCR A

Benzene

We begin A2 chemistry with a fantastical tale of the discovery of the magical rings of Benzene. Our story opens in a shire where four hobbits are getting up to mischief... Actually no, that's something else.

Benzene has a Ring Of Carbon Atoms

Benzene has the formula C_6H_6. It has a cyclic structure, with its six carbon atoms joined together in a ring. There are two ways of representing it — the **Kekulé model** and the **delocalised model**.

The Kekulé Model Came First

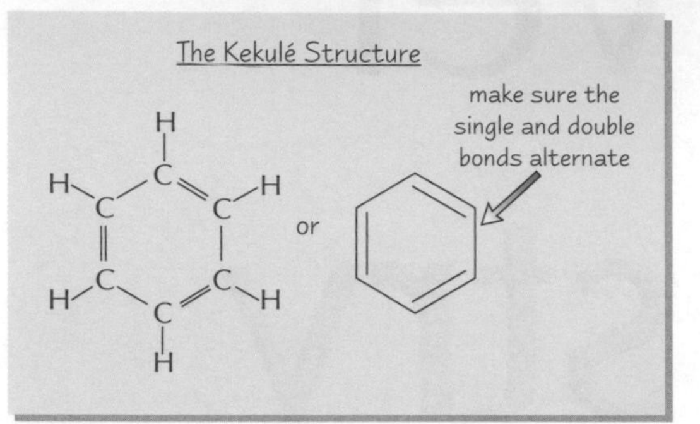

1) This was proposed by German chemist Friedrich August Kekulé in 1865. He came up with the idea of a **ring** of C atoms with **alternating single** and **double** bonds between them.
2) He later adapted the model to say that the benzene molecule was constantly **flipping** between two forms (**isomers**) by switching over the double and single bonds.

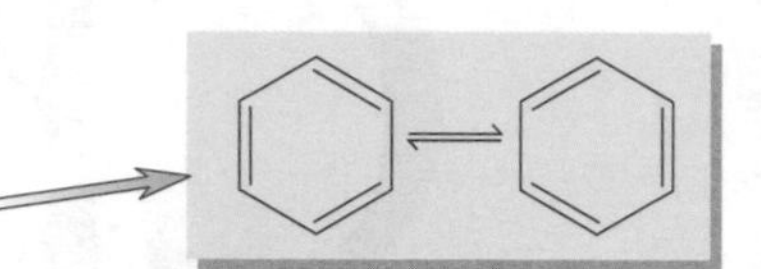

3) If the Kekulé model was correct, you'd expect there to always be three bonds with the length of a **C–C bond** (147 pm) and three bonds with the length of a **C=C bond** (135 pm).
4) However **X-ray diffraction studies** have shown that all the carbon-carbon bonds in benzene have the **same length** of 140 pm — i.e. they are **between** the length of a single bond and a double bond.
5) So the Kekulé structure **can't** be completely right, but it's still used today as it's useful for drawing reaction mechanisms.

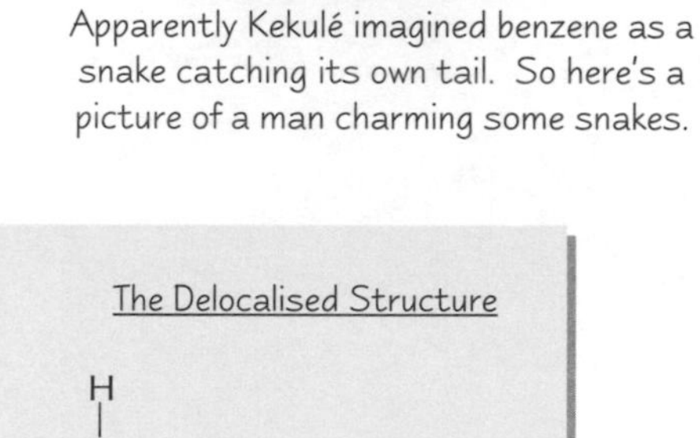

Apparently Kekulé imagined benzene as a snake catching its own tail. So here's a picture of a man charming some snakes.

The Delocalised Model Replaced Kekulé's Model

The bond-length observations are explained with the delocalised model.

1) The delocalised model says that the **p-orbitals** of all six carbon atoms **overlap** to create **π-bonds**.
2) This creates two **ring-shaped** clouds of electrons — one above and one below the plane of the six carbon atoms.
3) All the bonds in the ring are the **same length** because all the bonds are the same.
4) The electrons in the rings are said to be **delocalised** because they don't belong to a specific carbon atom. They are represented as a circle inside the ring of carbons rather than as double or single bonds.

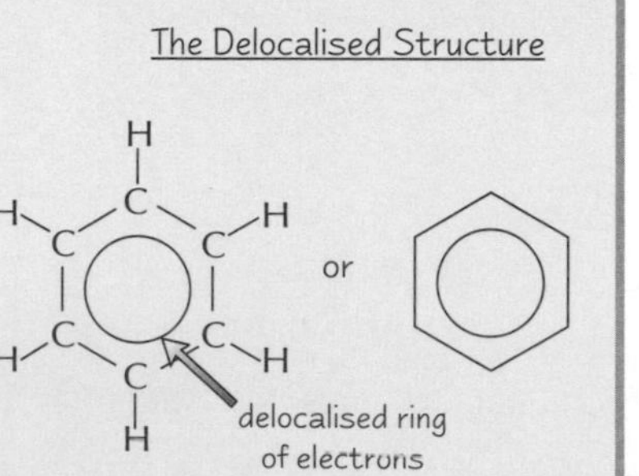

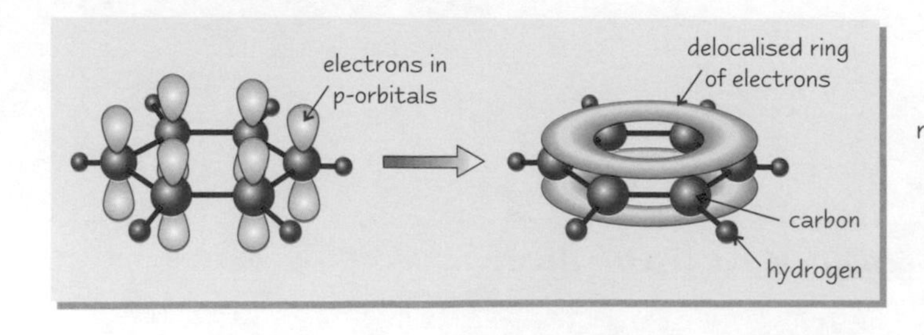

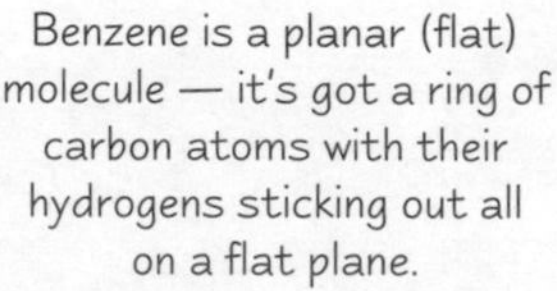

Benzene is a planar (flat) molecule — it's got a ring of carbon atoms with their hydrogens sticking out all on a flat plane.

Benzene

Enthalpy Changes Give More Evidence for Delocalisation

1) Cyclohexene has **one** double bond. When it's hydrogenated, the enthalpy change is **–120 kJmol⁻¹**. If benzene had three double bonds (as in the Kekulé structure), you'd expect it to have an enthalpy of hydrogenation of –360 kJmol⁻¹.

2) But the **experimental** enthalpy of hydrogenation of benzene is **–208 kJmol⁻¹** — far **less exothermic** than expected.

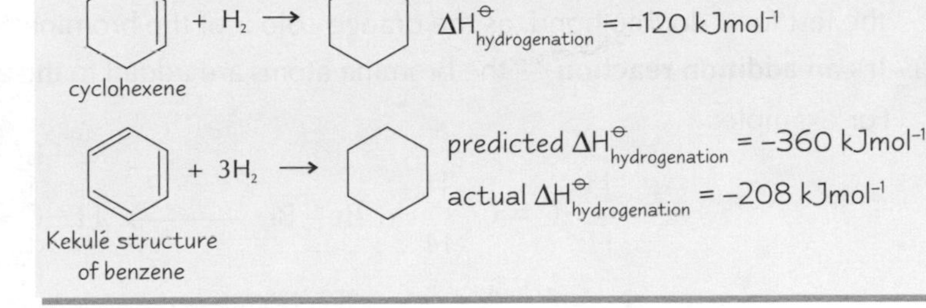

3) Energy is put in to break bonds and released when bonds are made. So **more energy** must have been put in to break the bonds in benzene than would be needed to break the bonds in the Kekulé structure.

4) This difference indicates that benzene is **more stable** than the Kekulé structure would be. This is thought to be due to the **delocalised ring of electrons**.

See page 168 for more about enthalpy changes.

Aromatic Compounds are Derived from Benzene

Compounds containing a **benzene ring** are called **arenes** or **'aromatic compounds'**. Don't be confused by the term 'aromatic' — although some of them are smelly, it's really just a term for compounds that have this structure. Arenes are **named** in two ways. There's no easy rule — you just have to learn these examples:

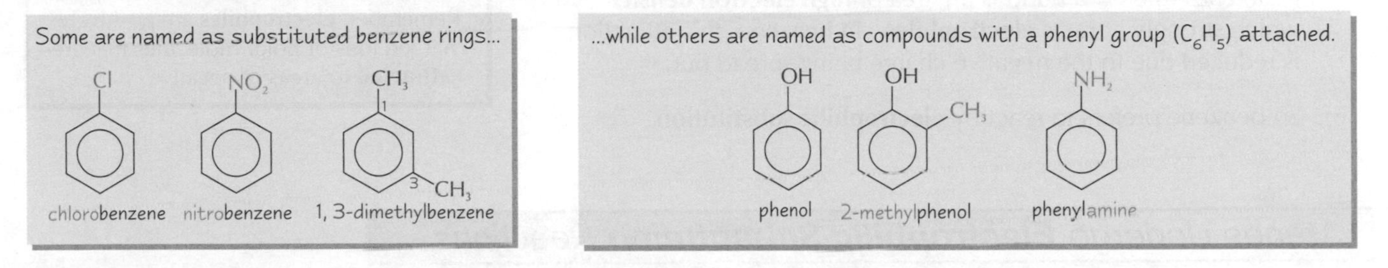

Practice Questions

Q1 Draw the Kekulé and delocalised models of benzene.

Q2 Give two pieces of evidence for the delocalised electron ring in benzene.

Q3 What type of bonds exist between C atoms in the delocalised model?

Exam Questions

1 a) The diagram represents the compound methylbenzene. What is its chemical formula? [1 mark]

CH_3

b) What name is given to compounds that contain a ring like this? [1 mark]

c) Name compounds A, B and C, shown on the right. [3 marks]

A: Cl, Cl   B: NO_2 C: OH, CH_3, CH_3

2 a) In 1865, Friedrich Kekulé proposed the structure shown for a benzene molecule. What does this model imply about the C-C bond lengths in the molecule? [1 mark]

b) What technique has been used to show that the bond lengths suggested by the Kekulé structure are incorrect? [1 mark]

c) How does this technique show that Kekulé's structure is incorrect? [1 mark]

Everyone needs a bit of stability in their life...

The structure of benzene is bizarre — even top scientists struggled to find out what its molecular structure looked like. Make sure you can draw all the different representations of benzene given on this page, including the ones showing the Cs and Hs. Yes, and don't forget that there's a hydrogen at every point on the ring — it's easy to forget they're there.

Reactions of Benzene

Benzene is an alkene but it often doesn't behave like one — whenever this is the case, you can pretty much guarantee that our kooky friend Mr Delocalised Electron Ring is up to his old tricks again...

Alkenes usually like Addition Reactions, but Not Benzene

1) **Alkenes** react easily with **bromine** water at room temperature. The reaction is the basis of the test for a double bond, as the orange colour of the bromine water is lost.
2) It's an **addition reaction** — the bromine atoms are added to the alkene.

For example:

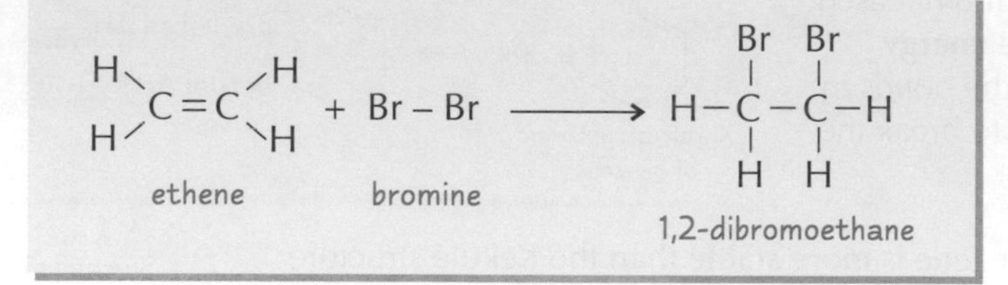

3) If the Kekulé structure (see page 102) were correct, you'd expect a **similar reaction** between benzene and bromine. In fact, to make it happen you need **hot benzene** and **ultraviolet light** — and it's still a real **struggle**.
4) This difference between benzene and other alkenes is explained by the **delocalised electron rings** above and below the plane of carbon atoms. They make the benzene ring very **stable**, and **spread out** the negative charge. So benzene is very **unwilling** to undergo **addition reactions** which would destroy the stable ring.
5) In alkenes, the C=C bond is an area of **high electron density** which strongly attracts **electrophiles**. In benzene, this attraction is reduced due to the negative charge being spread out.
6) So benzene prefers to react by **electrophilic substitution**.

Remember, **electrophiles** are positively charged ions or polar molecules that are **attracted** to areas of negative charge.

*Arenes Undergo **Electrophilic Substitution** Reactions...*

*1) With **Halogens** using a **Halogen Carrier***

1) Benzene will react with bromine, Br–Br, in the presence of aluminium chloride, $AlCl_3$.
2) Br–Br is the **electrophile**.
3) $AlCl_3$ acts as a **halogen carrier** (see below) which makes the **electrophile stronger**.
   Without the halogen carrier, the electrophile doesn't have a strong enough positive charge to attack the stable benzene ring.
4) A Br atom is **substituted** in place of a H atom.
5) Chlorine Cl–Cl will react in just the same way.

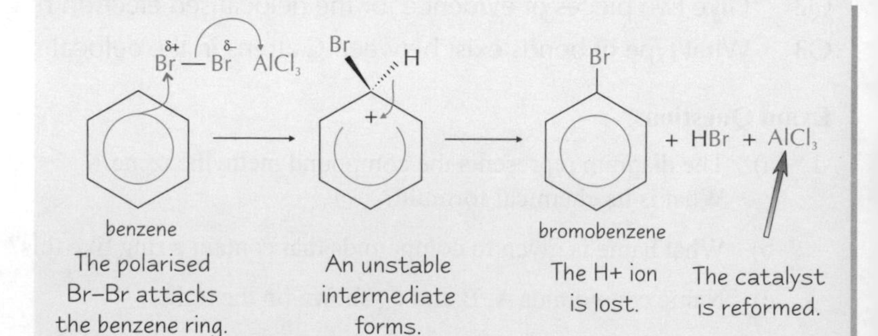

Halogen carriers make the Electrophile Stronger

1) Halogen carriers accept a **lone pair of electrons** from the electrophile.
2) Halogen carriers include **aluminium halides, iron halides** and **iron**.
3) **E.g.** Aluminium chloride combines with the bromine molecule like this in the example above: $AlCl_3 + Br-Br \rightarrow AlCl_3Br^{\delta-}-Br^{\delta+}$

Reactions of Benzene

2) With **Nitric Acid** and a **Catalyst**

When you warm **benzene** with **concentrated nitric** and **sulfuric acid**, you get **nitrobenzene**.

Sulfuric acid works as a **catalyst** — it helps make the nitronium ion, $\mathbf{NO_2^+}$, which is the **electrophile**, but is regenerated at the end of the reaction mechanism.

$$HNO_3 + H_2SO_4 \rightarrow H_2NO_3^+ + HSO_4^-$$

$$H_2NO_3^+ \rightarrow NO_2^+ + H_2O$$

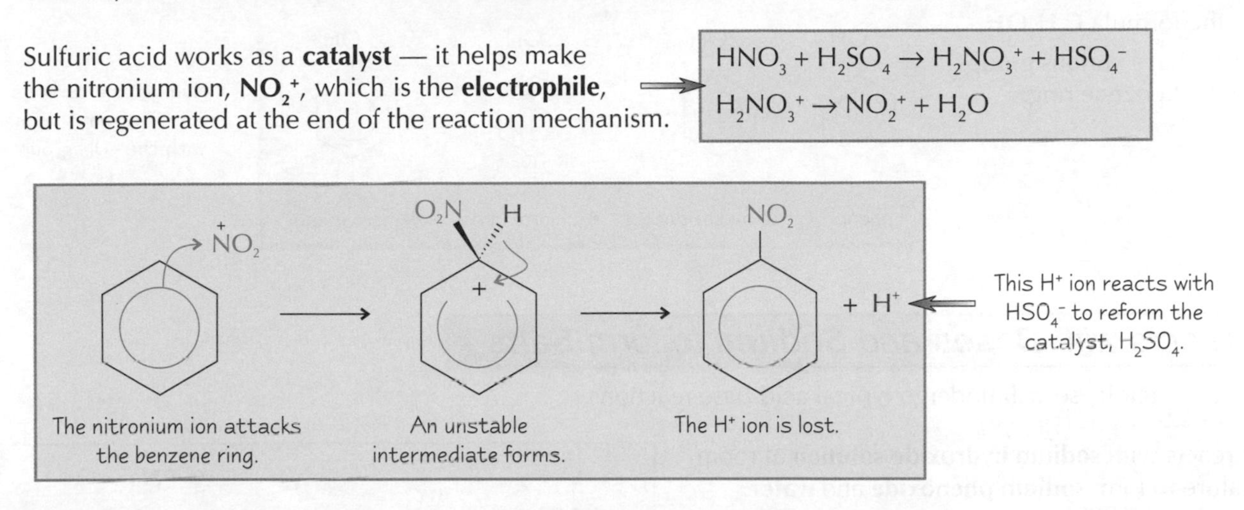

If you only want one NO_2 group added (**mononitration**), you need to keep the temperature **below 55 °C**. Above this temperature you'll get lots of substitutions.

Practice Questions

Q1 What type of reaction does benzene tend to undergo?

Q2 What makes benzene resistant to reaction with bromine?

Q3 Which substances are used as halogen carriers in substitution reactions of benzene?

Q4 Which two acids are used in the production of nitrobenzene?

Exam Questions

1 Nitrobenzene is a yellow oily substance used in the first step of the production of polyurethane. It is made from benzene by reaction with concentrated nitric and sulfuric acids.

a) Draw the structure of nitrobenzene. [1 mark]

b) 1,3-dinitrobenzene is made by the same process if the temperature is higher. Draw its structure. [1 mark]

c) What kind of reaction is this? [1 mark]

d) Outline a mechanism for this reaction. [3 marks]

2 In the Kekulé model of benzene, there are 3 double bonds. Cyclohexene has 1 double bond.

a) Describe what you would see if benzene and cyclohexene were each mixed with bromine water. [1 mark]

b) Explain the difference that you would observe. [2 marks]

c) To make bromine and benzene react, they are heated together with iron(III) chloride.

i) What is the function of the iron(III) chloride? [1 mark]

ii) Outline a mechanism for this reaction. [3 marks]

What are you looking at, punk?

Arenes really like Mr Delocalised Electron Ring and they won't give him up for nobody, at least not without a fight. They'd much rather get tangled up in an electrophilic substitution — anything not to bother The Ring. Being associated with The Ring provides Arenes with stability but also serious respect on the mean streets of... err... Organic Chemistryville.

Phenols

A phenol is the aromatic version of an alcohol. Don't drink them though — they'd get your insides a bit too clean.

Phenols *Have Benzene Rings with* ***–OH*** *Groups Attached*

Phenol has the formula $\mathbf{C_6H_5OH}$.
Other phenols have various groups attached to the benzene ring:

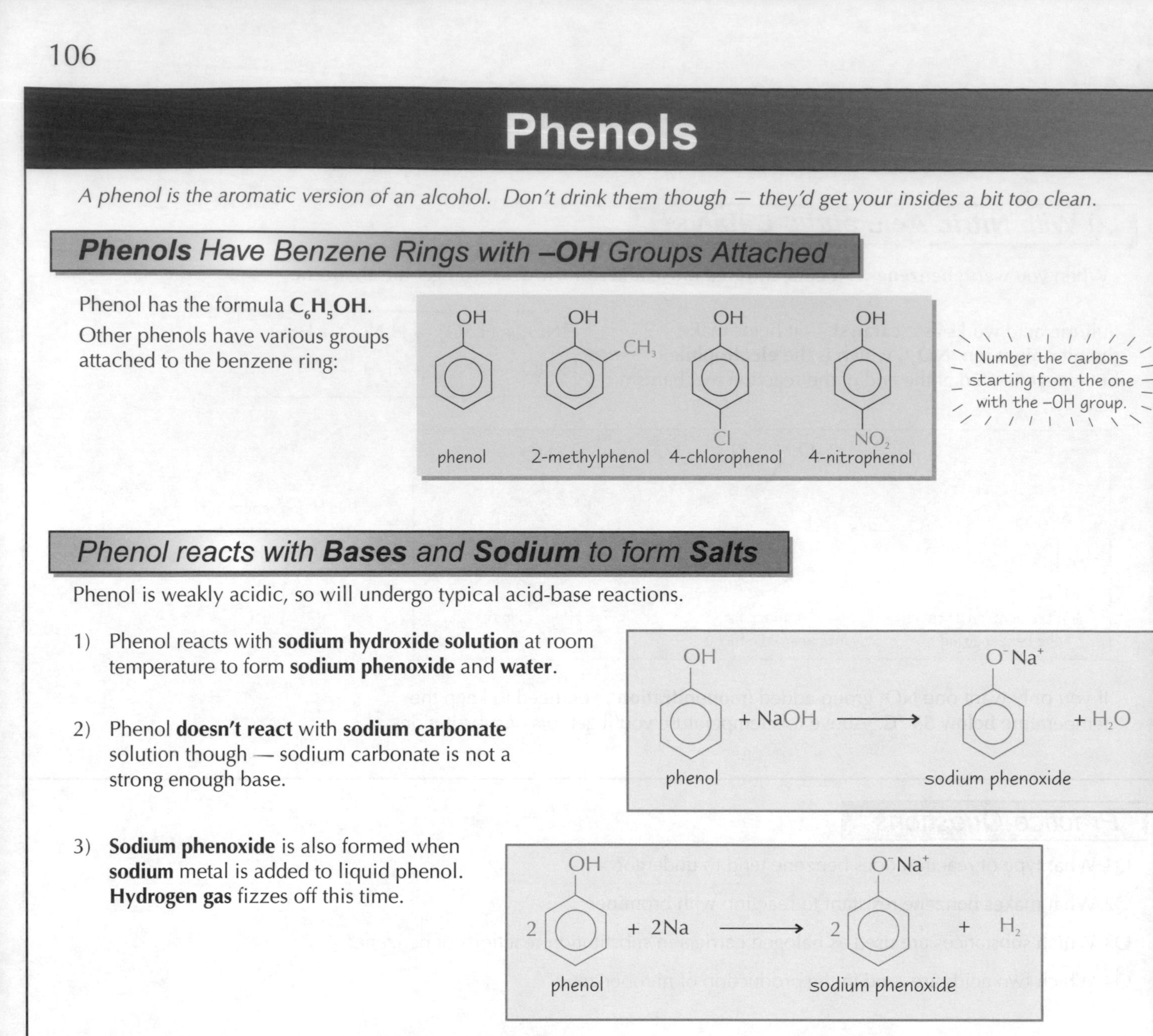

Phenol reacts with ***Bases*** *and* ***Sodium*** *to form* ***Salts***

Phenol is weakly acidic, so will undergo typical acid-base reactions.

1) Phenol reacts with **sodium hydroxide solution** at room temperature to form **sodium phenoxide** and **water**.
2) Phenol **doesn't react** with **sodium carbonate** solution though — sodium carbonate is not a strong enough base.
3) **Sodium phenoxide** is also formed when **sodium** metal is added to liquid phenol. **Hydrogen gas** fizzes off this time.

Phenol Reacts with ***Bromine Water***

1) If you shake phenol with orange bromine water, it will **react**, **decolorising** it.
2) Benzene **doesn't** react with bromine water (see page 104), so phenol's reaction must be to do with the **OH group**.
3) One of the pairs of electrons in a **p-orbital** of the oxygen atom **overlaps** with the delocalised ring of electrons in the benzene ring.
4) This increases the **electron density** of the ring, especially at positions 2, 4 and 6 (for reasons you don't need to know) , making it more likely to be attacked by the bromine molecule in these positions.
5) The hydrogen atoms at 2, 4 and 6 are **substituted** by bromine atoms. The product is called 2,4,6-tribromophenol — it's insoluble in water and **precipitates** out of the mixture. It smells of antiseptic.

This is an electrophilic substitution reaction.

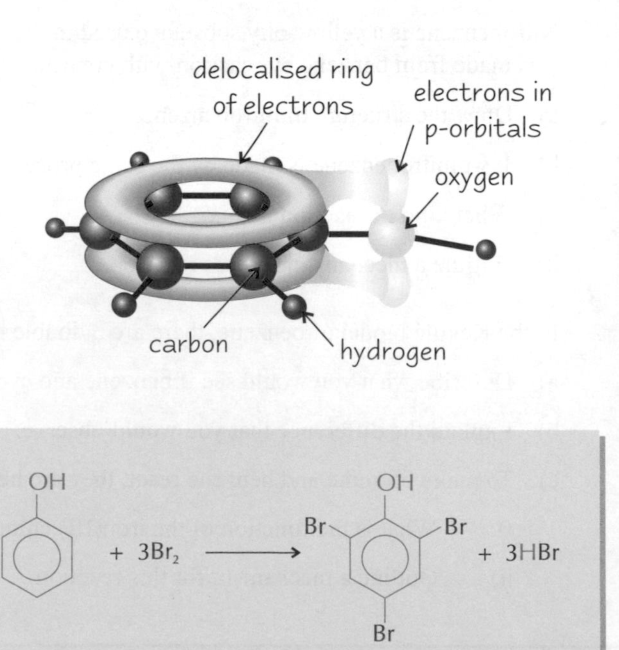

Phenols

Phenol has many Important Uses

Phenol is a major chemical product, with more than 8 million tonnes being produced each year.

1) The first major use was as an **antiseptic** during surgery. Joseph Lister was the first to use it to clean wounds. It wasn't used like this for long though as it was too damaging to tissue. It's still used today in the production of **antiseptics** and **disinfectants** such as TCP™.

2) Another important use is in the production of **polymers**. Kevlar® (see page 122) and polycarbonate are both produced from substances made from phenol. Bisphenol A is used to make polycarbonates, which are used in things like bottles, spectacle lenses and CDs.

3) One of the earliest "plastics" was **Bakelite**™, a polymer of phenol and formaldehyde. It is a resin with good insulating properties and was used to make things like telephones and radio casings. Today, similar compounds are used to make all sorts of objects including saucepan handles, electrical plugs, dominos, billiard balls and chess pieces.

4) Bisphenol A is used in the manufacture of resins called **epoxies**. These have a variety of important uses including **adhesives** and **paints**. They're also really important in electronic circuits where they're used as electrical insulators.

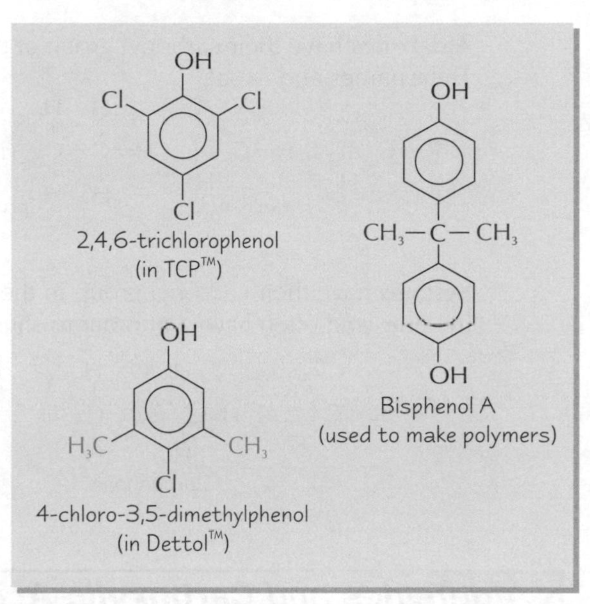

Practice Questions

Q1 Draw the structures of phenol, 4-chlorophenol and 4-nitrophenol.

Q2 Write a balanced equation for the reaction between phenol and sodium hydroxide solution.

Q3 Write a balanced equation for the reaction between phenol and bromine (Br_2).

Q4 Give three uses of phenol.

Exam Questions

1 a) Draw the structure of 2-methylphenol. [1 mark]

b) Write an equation for the reaction of 2-methylphenol with potassium. [1 mark]

c) 1 mole of gas occupies 24 dm^3 at room temperature and pressure.
What mass of 2-methyl phenol would you need to produce 4.8 dm^3 of hydrogen, at room temperature and pressure, by reaction with excess potassium? [3 mark]

2 a) Bromine water can be used to distinguish between benzene and phenol.
Describe what you would observe in each case. [2 marks]

b) Name the product formed when phenol reacts with bromine water. [1 mark]

c) Explain why phenol reacts differently from benzene. [2 marks]

d) What type of reaction occurs between phenol and bromine? [1 mark]

Phenol Destination 4 — more chemicals, more equations, more horror...

You might not like this phenol stuff, but if you were a germ, you'd like it even less. If you're ever looking for TCP™ in the supermarket, try asking for a bottle of 2,4,6-trichlorophenol and see what you get — probably a funny look. Anyway, no time for shopping — you've got to get these pages learned. If you can do all the questions above, you're well on your way.

Aldehydes and Ketones

Aldehydes and ketones are both carbonyl compounds. They've got their carbonyl groups in different positions though.

Aldehydes *and* ***Ketones*** *contain a* ***Carbonyl Group***

Aldehydes and ketones are **carbonyl compounds** — they contain the **carbonyl** functional group, **C=O**.

Aldehydes have their carbonyl group at the **end** of the carbon chain.
Their names end in **–al**.

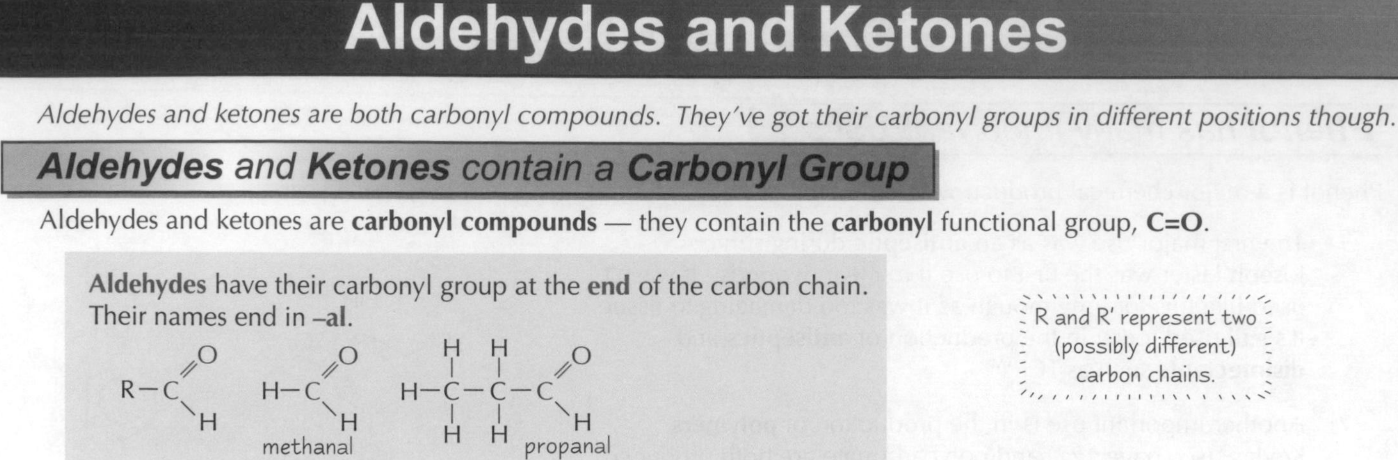

Ketones have their carbonyl group in the **middle** of the carbon chain. Their names end in **–one**, and often have a number to show which **carbon** the carbonyl group is on.

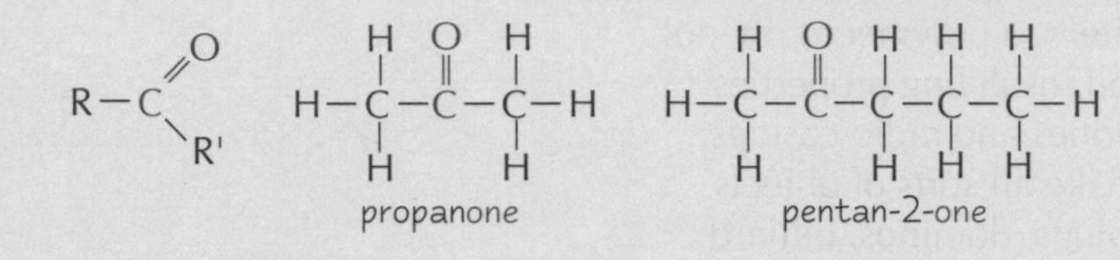

Aldehydes *and* ***Carboxylic Acids*** *are made by Oxidising* ***Primary Alcohols***

You can use **acidified dichromate(VI) ions** ($Cr_2O_7^{2-}$) to **mildly** oxidise alcohols.
Acidified **potassium dichromate(VI)** ($K_2Cr_2O_7$ / H_2SO_4) is often used.

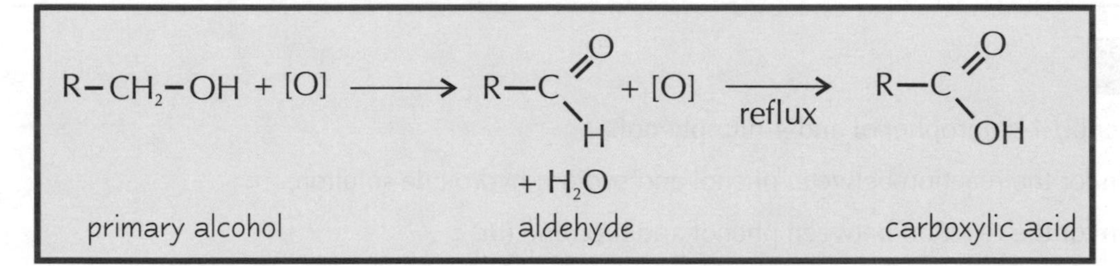

[O] = oxidising agent

The orange dichromate(VI) ion is reduced to the green chromium(III) ion, Cr^{3+}.

You can control how **far** the alcohol is oxidised by controlling the **reaction conditions**:

Oxidising Primary Alcohols

1) Gently heating ethanol with potassium dichromate(VI) solution and sulfuric acid in a test tube should produce "apple" smelling **ethanal** (an aldehyde). However, it's **really tricky** to control the amount of heat, and the aldehyde is usually oxidised to form "vinegar" smelling **ethanoic acid**.

2) To get just the **aldehyde**, you need to get it out of the oxidising solution **as soon** as it's formed. You can do this by gently heating excess alcohol with a **controlled** amount of oxidising agent in **distillation apparatus**, so the aldehyde (which boils at a lower temperature than the alcohol) is distilled off **immediately**.

Reflux Apparatus

water out
Liebig condenser
water in
round bottomed flask
anti-bumping granules (added to make boiling smoother)
heat

3) To produce the **carboxylic acid**, the alcohol has to be **vigorously oxidised**. The alcohol is mixed with excess oxidising agent and heated under **reflux**. Heating under reflux means you can increase the **temperature** of an organic reaction to boiling without losing **volatile** solvents, reactants or products. Any vaporised compounds are cooled, condense and drip back into the reaction mixture. Handy, hey.

Aldehydes and Ketones

Ketones are made by Oxidising **Secondary Alcohols**

1) Refluxing a secondary alcohol, e.g. propan-2-ol, with acidified dichromate(VI) will produce a **ketone**.

$$R_1-CH(OH)-R_2 + [O] \xrightarrow{\text{reflux}} R_1R_2C{=}O + H_2O$$

2) Ketones can't be oxidised easily, so even prolonged refluxing won't produce anything more.

Tertiary Alcohols **Can't** be Oxidised Easily

Tertiary alcohols are **resistant** to oxidation. They **don't react** with potassium dichromate(VI) at all — the solution stays orange.

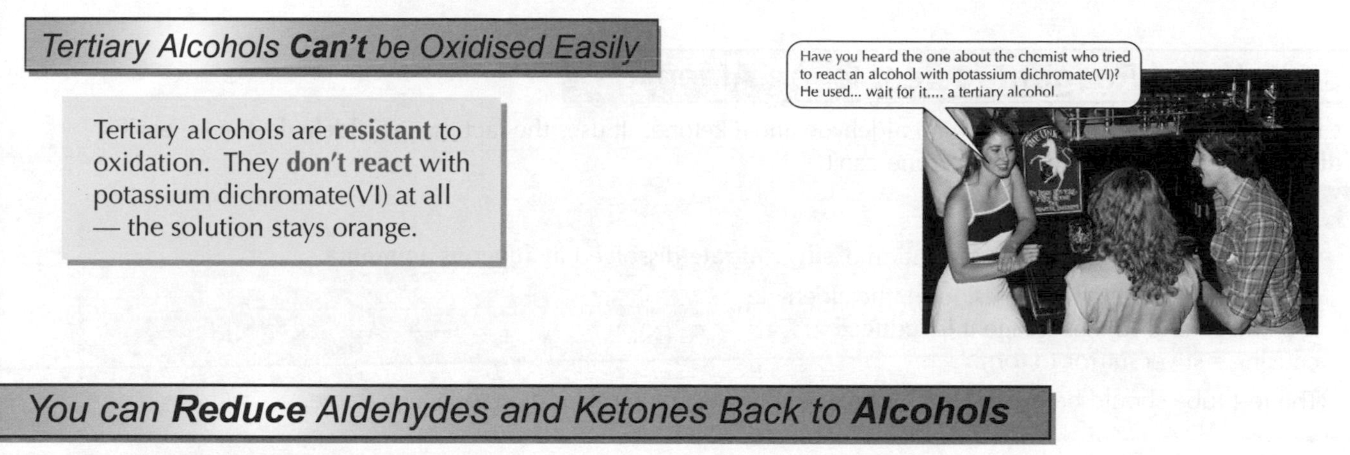

You can **Reduce** Aldehydes and Ketones Back to **Alcohols**

Using a **reducing agent** [H] you can:

1) reduce an **aldehyde** to a **primary alcohol**. 2) reduce a **ketone** to a **secondary alcohol**.

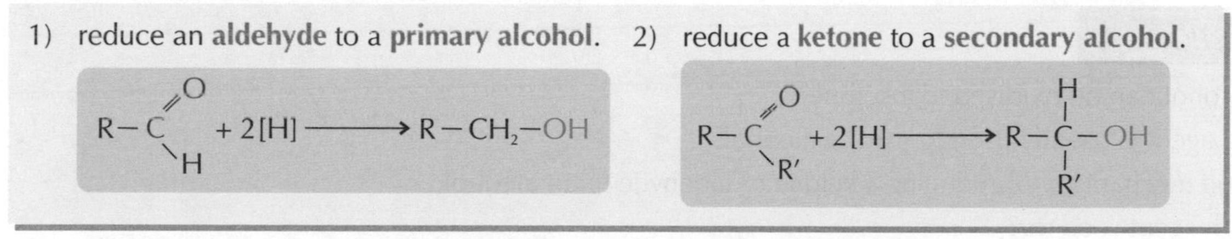

$$R-CHO + 2[H] \longrightarrow R-CH_2-OH$$

$$R-CO-R' + 2[H] \longrightarrow R-CH(OH)-R'$$

You'd usually use **NaBH$_4$** (sodium tetrahydridoborate(III) or sodium borohydride) dissolved in water with methanol as the reducing agent.

The **Reduction** of Aldehydes and Ketones is **Nucleophilic Addition**

The reducing agent, e.g. NaBH$_4$, supplies **hydride ions**, H$^-$.
The extra pair of electrons on the H$^-$ make this a **nucleophile**, which will attack the δ+ carbon on the **carbonyl group** of an aldehyde or ketone:

$$:H^- + RR'C^{\delta+}{=}O^{\delta-} \longrightarrow R-CH(O^-)-R'$$

Addition of water then gives...

$$R-CH(O^-)-R' + H^{\delta+}-O^{\delta-}-H^{\delta+} \longrightarrow R-CH(OH)-R' + OH^-$$

Aldehydes and Ketones

Knowing what aldehyde and ketone molecules look like won't help you decide which is which if you've got a test tube of each. There are chemical tests that will though. You can also just test for a carbonyl group.

Brady's Reagent Tests for a Carbonyl Group

Brady's reagent is **2,4-dinitrophenylhydrazine** (2,4-DNPH) dissolved in methanol and concentrated sulfuric acid.
The **2,4-dinitrophenylhydrazine** forms a **bright orange precipitate** if a carbonyl group is present.
This only happens with **C=O groups**, not with ones like COOH, so it only tests for **aldehydes** and **ketones**.

The orange precipitate is a **derivative** of the carbonyl compound. Each different carbonyl compound produces a crystalline derivative with a **different melting point**.
So if you measure the melting point of the crystals and compare it against the **known** melting points of the derivatives, you can **identify** the carbonyl compound.

Use Tollens' Reagent to Test for an Aldehyde

This test lets you distinguish between an aldehyde and a ketone. It uses the fact that an **aldehyde** can be **easily oxidised** to a carboxylic acid, but a ketone can't.

TOLLENS' REAGENT

Tollens' reagent is a **colourless** solution of **silver nitrate** dissolved in **aqueous ammonia**.

When heated together in a test tube, the aldehyde is **oxidised** and Tollens' reagent is **reduced** causing a silver mirror to form.

colourless / silver

$$Ag(NH_3)_2{}^+{}_{(aq)} + e^- \longrightarrow Ag_{(s)} + 2NH_{3(aq)}$$

The test tube should be heated in a beaker of hot water, rather than directly over a flame.

Practice Questions

Q1 What type of alcohol can be oxidised to a ketone?

Q2 Which oxidising agent is usually used to oxidise alcohols?

Q3 Draw the reaction mechanism for reducing a ketone or aldehyde to an alcohol.

Q4 What is Tollens' reagent used for?

Exam Questions

1 Compound X is an alcohol. When heated under reflux with acidified potassium dichromate(VI), compound Y is made. Compound Y does not give a silver mirror when heated with Tollens' reagent. Compound Y has a neutral pH.

a) What type of compound is substance Y? Explain your answer. [3 marks]

b) What type of alcohol is X? Explain your answer. [1 mark]

c) Suggest a further test that could be used to identify compound Y. [2 marks]

2 Butan-2-one is a compound that occurs in some fruits, but is manufactured on a large scale for use as a solvent in things like paints and white board pens.

a) Draw the structure of butan-2-one. [2 marks]

b) Draw and name the structure of the alcohol that butan-2-one is made from. [2 marks]

3 Propanoic acid, C_2H_5COOH, is added to animal feed and bread as an anti-fungal agent.

a) Briefly describe a method that could be used to produce propanoic acid from propan-1-ol. [1 mark]

b) How would you change the experiment if you wanted to produce propanal? [1 mark]

c) Describe two tests that you could use to distinguish between propan-1-ol, propanal and propanoic acid. [3 marks]

Round bottomed flasks you make the rocking world go round...

You've got to be a dab hand at recognising different functional groups from a mile off. Make sure you know how aldehydes differ from ketones and what you get when you oxidise them both. And how to reduce them. And don't forget all the details of those pesky tests. Phew, it's hard work some of this chemistry. Don't worry, it'll only make you stronger.

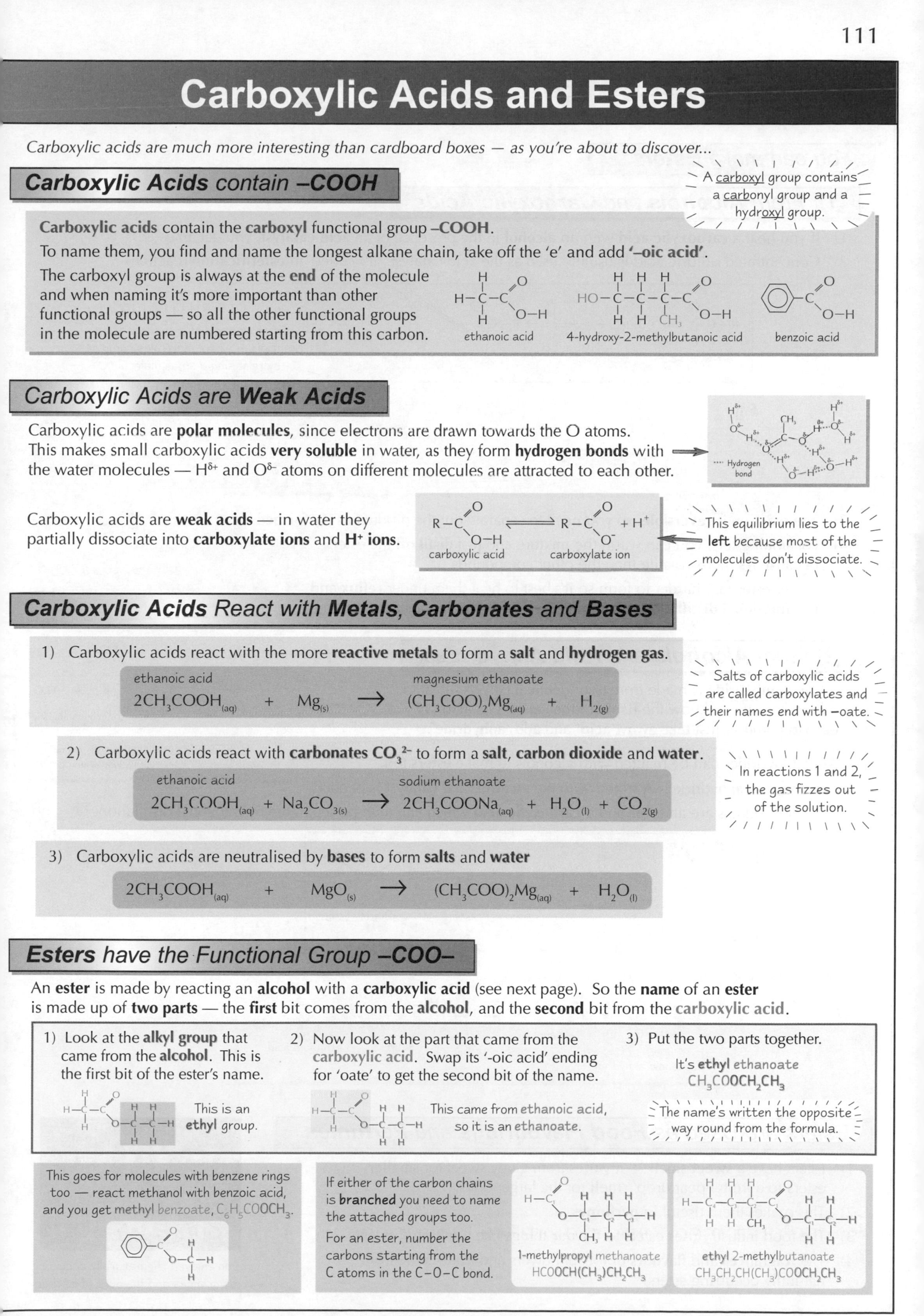

Carboxylic Acids and Esters

Carboxylic acids are much more interesting than cardboard boxes — as you're about to discover...

Carboxylic Acids contain –COOH

A carboxyl group contains a carbonyl group and a hydroxyl group.

Carboxylic acids contain the **carboxyl** functional group **–COOH**.

To name them, you find and name the longest alkane chain, take off the 'e' and add **'–oic acid'**.

The carboxyl group is always at the **end** of the molecule and when naming it's more important than other functional groups — so all the other functional groups in the molecule are numbered starting from this carbon.

Carboxylic Acids are Weak Acids

Carboxylic acids are **polar molecules**, since electrons are drawn towards the O atoms. This makes small carboxylic acids **very soluble** in water, as they form **hydrogen bonds** with the water molecules — $H^{\delta+}$ and $O^{\delta-}$ atoms on different molecules are attracted to each other.

Carboxylic acids are **weak acids** — in water they partially dissociate into **carboxylate ions** and **H⁺ ions**.

$$RCOOH \rightleftharpoons RCOO^- + H^+$$

This equilibrium lies to the **left** because most of the molecules don't dissociate.

Carboxylic Acids React with Metals, Carbonates and Bases

1) Carboxylic acids react with the more **reactive metals** to form a **salt** and **hydrogen gas**.

ethanoic acid → magnesium ethanoate

$$2CH_3COOH_{(aq)} + Mg_{(s)} \rightarrow (CH_3COO)_2Mg_{(aq)} + H_{2(g)}$$

Salts of carboxylic acids are called carboxylates and their names end with –oate.

2) Carboxylic acids react with **carbonates CO_3^{2-}** to form a **salt**, **carbon dioxide** and **water**.

ethanoic acid → sodium ethanoate

$$2CH_3COOH_{(aq)} + Na_2CO_{3(s)} \rightarrow 2CH_3COONa_{(aq)} + H_2O_{(l)} + CO_{2(g)}$$

In reactions 1 and 2, the gas fizzes out of the solution.

3) Carboxylic acids are neutralised by **bases** to form **salts** and **water**

$$2CH_3COOH_{(aq)} + MgO_{(s)} \rightarrow (CH_3COO)_2Mg_{(aq)} + H_2O_{(l)}$$

Esters have the Functional Group –COO–

An **ester** is made by reacting an **alcohol** with a **carboxylic acid** (see next page). So the **name** of an **ester** is made up of **two parts** — the **first** bit comes from the **alcohol**, and the **second** bit from the **carboxylic acid**.

1) Look at the **alkyl group** that came from the **alcohol**. This is the first bit of the ester's name.

This is an **ethyl** group.

2) Now look at the part that came from the **carboxylic acid**. Swap its '-oic acid' ending for 'oate' to get the second bit of the name.

This came from ethanoic acid, so it is an ethanoate.

3) Put the two parts together.

It's **ethyl** ethanoate
$CH_3COOCH_2CH_3$

The name's written the opposite way round from the formula.

This goes for molecules with benzene rings too — react methanol with benzoic acid, and you get methyl benzoate, $C_6H_5COOCH_3$.

If either of the carbon chains is **branched** you need to name the attached groups too.

For an ester, number the carbons starting from the C atoms in the C–O–C bond.

1-methylpropyl methanoate
$HCOOCH(CH_3)CH_2CH_3$

ethyl 2-methylbutanoate
$CH_3CH_2CH(CH_3)COOCH_2CH_3$

Carboxylic Acids and Esters

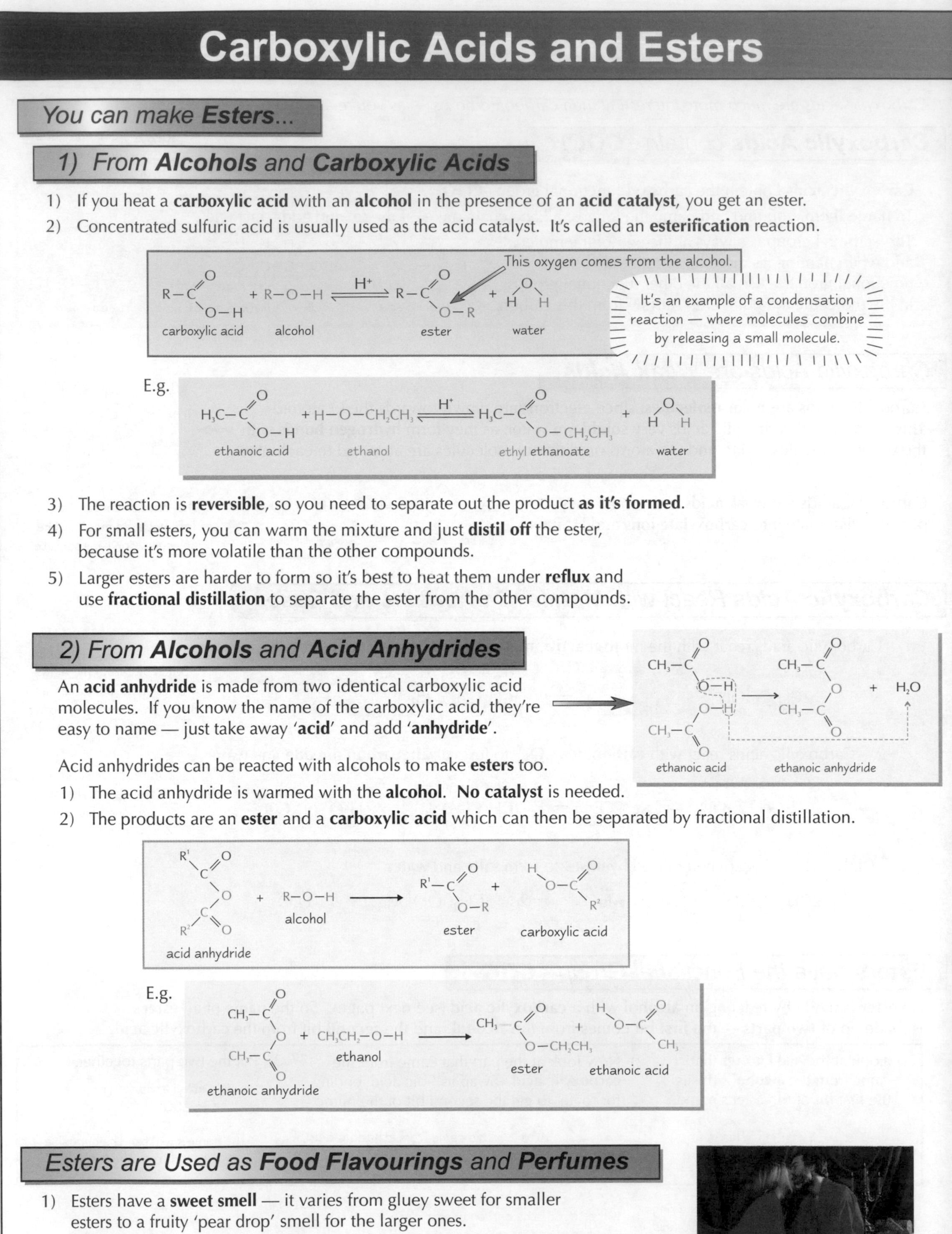

You can make Esters...

1) From Alcohols and Carboxylic Acids

1) If you heat a **carboxylic acid** with an **alcohol** in the presence of an **acid catalyst**, you get an ester.
2) Concentrated sulfuric acid is usually used as the acid catalyst. It's called an **esterification** reaction.

$$\text{R–COOH (carboxylic acid)} + \text{R–O–H (alcohol)} \overset{H^+}{\rightleftharpoons} \text{R–COO–R (ester)} + H_2O \text{ (water)}$$

This oxygen comes from the alcohol.

It's an example of a condensation reaction — where molecules combine by releasing a small molecule.

E.g.

$$H_3C\text{–COOH (ethanoic acid)} + H\text{–O–}CH_2CH_3 \text{ (ethanol)} \overset{H^+}{\rightleftharpoons} H_3C\text{–COO–}CH_2CH_3 \text{ (ethyl ethanoate)} + H_2O \text{ (water)}$$

3) The reaction is **reversible**, so you need to separate out the product **as it's formed**.
4) For small esters, you can warm the mixture and just **distil off** the ester, because it's more volatile than the other compounds.
5) Larger esters are harder to form so it's best to heat them under **reflux** and use **fractional distillation** to separate the ester from the other compounds.

2) From Alcohols and Acid Anhydrides

An **acid anhydride** is made from two identical carboxylic acid molecules. If you know the name of the carboxylic acid, they're easy to name — just take away '**acid**' and add '**anhydride**'.

$$2\,CH_3\text{–COOH (ethanoic acid)} \rightarrow (CH_3CO)_2O \text{ (ethanoic anhydride)} + H_2O$$

Acid anhydrides can be reacted with alcohols to make **esters** too.

1) The acid anhydride is warmed with the **alcohol**. **No catalyst** is needed.
2) The products are an **ester** and a **carboxylic acid** which can then be separated by fractional distillation.

$$R^1\text{CO–O–CO}R^2 \text{ (acid anhydride)} + \text{R–O–H (alcohol)} \longrightarrow R'\text{–COO–R (ester)} + H\text{–O–CO–}R^2 \text{ (carboxylic acid)}$$

E.g.

$$CH_3\text{CO–O–CO}CH_3 \text{ (ethanoic anhydride)} + CH_3CH_2\text{–O–H (ethanol)} \longrightarrow CH_3\text{–COO–}CH_2CH_3 \text{ (ester)} + H\text{–O–CO–}CH_3 \text{ (ethanoic acid)}$$

Esters are Used as Food Flavourings and Perfumes

1) Esters have a **sweet smell** — it varies from gluey sweet for smaller esters to a fruity 'pear drop' smell for the larger ones.
2) This makes them useful in **perfumes**.
3) The food industry uses esters to **flavour** things like drinks and sweets.
4) The **fragrances** and **flavours** of lots of flowers and fruits come from naturally occurring esters.

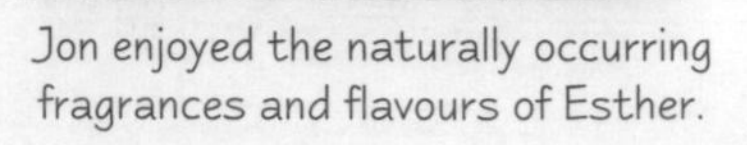

Jon enjoyed the naturally occurring fragrances and flavours of Esther.

Carboxylic Acids and Esters

Esters are **Hydrolysed** to Form **Alcohols**

There are two ways to hydrolyse esters — using **acid hydrolysis** or **base hydrolysis**.
With both types you get an **alcohol**, but the second product in each case is different.

ACID HYDROLYSIS

Acid hydrolysis splits the ester into a **carboxylic acid** and an **alcohol**.
You have to **reflux** the ester with a **dilute acid**, such as hydrochloric or sulfuric.

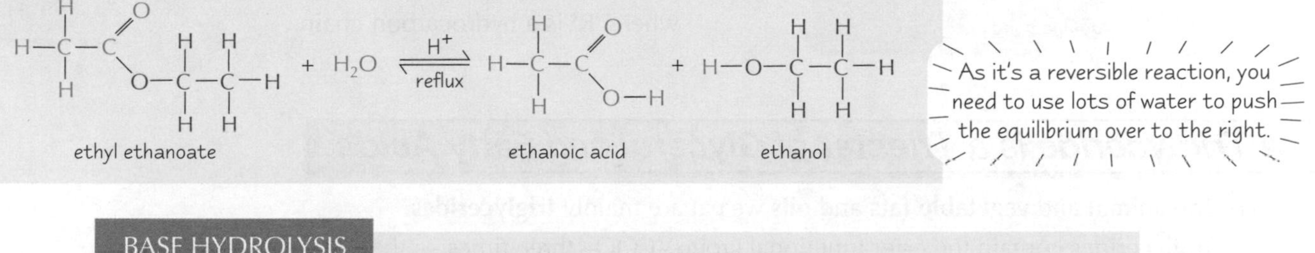

BASE HYDROLYSIS

This time you have to **reflux** the ester with a **dilute alkali**, such as sodium hydroxide.
You get a **carboxylate salt** and an **alcohol**.

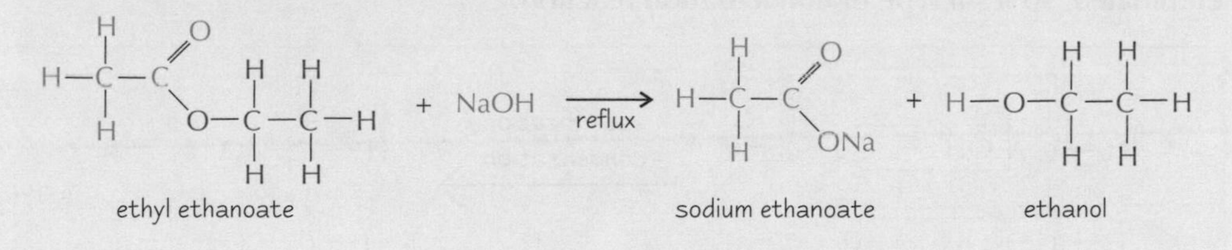

Practice Questions

Q1 What is the functional group of: a) a carboxylic acid, b) an ester?

Q2 Explain why carboxylic acids dissolve in water.

Q3 What products are formed when carboxylic acids react with: a) metals, b) carbonates, c) bases?

Q4 Draw the structures of: a) propanoic acid, b) 3-methylbutanoic acid, c) benzoic acid.

Q5 Draw the structure of the ester 1-methylpropyl methanoate.

Exam Questions

1 Compound C, shown on the right, is found in raspberries.

a) Name compound C. [1 mark]

b) Suggest a use for compound C. [1 mark]

c) Compound C is refluxed with dilute sulfuric acid.
Name the products formed and draw their structures. What kind of reaction is this? [5 marks]

d) If compound C is refluxed with excess sodium hydroxide, a similar reaction occurs.
Give a difference between this reaction and the reaction described in (c). [1 mark]

2 Propyl ethanoate is a pear-scented oil. It is used as a solvent and as a flavouring.

a) Name the alcohol that it can be made from. [1 mark]

b) Name two different substances that could be added to the alcohol to produce the propyl ethanoate. [2 marks]

c) Draw the structures of the two substances you have chosen in b). [2 marks]

Carboxylic acid + alcohol produces ester — well, that's life...

I bet your apple flavoured sweets have never been near a nice rosy apple — it's all the work of esters. And for that matter, I reckon prawn cocktail crisps have never met a prawn, or a cocktail either. None of it's real. And as for potatoes...

Fatty Acids and Fats

OK, brace yourself, as you're about to experience three solid pages of pure, unadulterated lardfest.

Fatty Acids are **Carboxylic Acids**

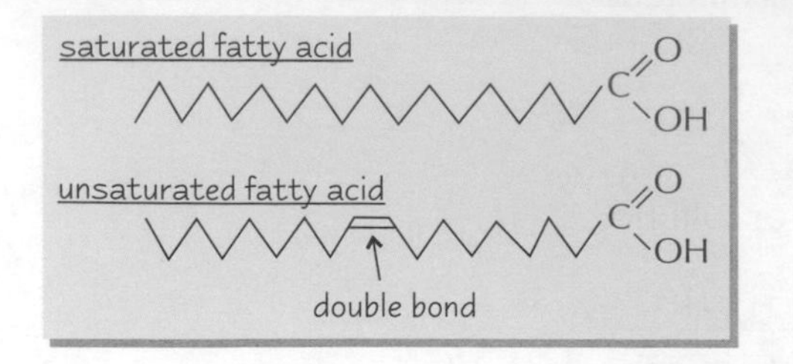

Fatty acids have a long hydrocarbon chain with a **carboxylic acid** group at the end. If the hydrocarbon chain contains **no double bonds** then the fatty acid is **saturated**, but if it contains one or more double bonds then it's **unsaturated**.

Fatty acids can also be written like this where 'R' is a hydrocarbon chain.

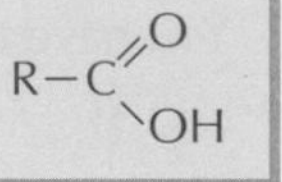

A **Triglyceride** is a **Triester** of **Glycerol** and **Fatty Acids**

1) The **animal** and **vegetable fats** and **oils** we eat are mainly **triglycerides**.
2) Triglycerides contain the ester functional group –COO– three times — they are **triglyceryl esters**. They're made by reacting **glycerol** (**propane-1,2,3-triol**) with **fatty acids**.
3) The **three -OH groups** on the glycerol molecules link up to **fatty acids** to produce triglyceride molecules. Water is eliminated, so it's a type of **condensation** reaction.

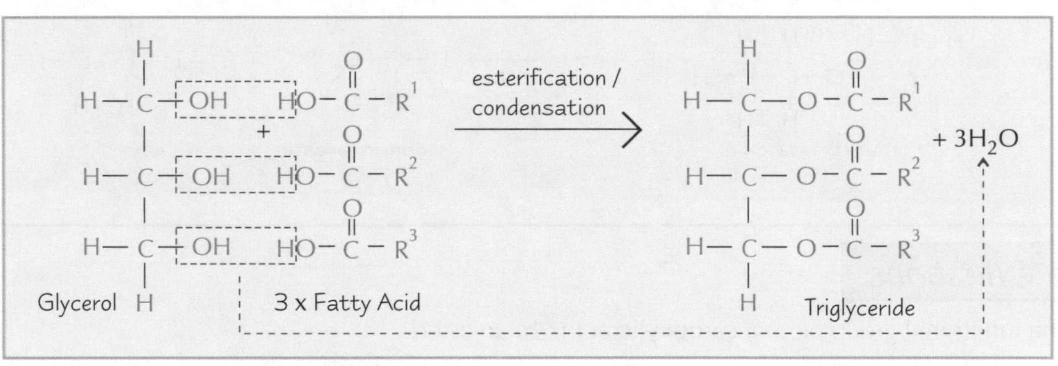

You Can **Name Fatty Acids** Using **Systematic** Or **Shorthand** Names

Fatty acids are pretty **complex molecules**, which means that describing their structure clearly without having to draw them can be a bit tricky.
The best way is to use either their **systematic name** or **shorthand name**.

Most of these acids also have a common name (like 'oleic acid') but that's no help if you want to know about their structure.

Naming Fatty Acids Using **Systematic Names**

1) Count **how many carbon atoms** are in the molecule. This tells you the **stem** of its name.
2) Add some **numbers** to say where any **double bonds** are. Count from the **-COOH** end of the chain and when you get to the first carbon of a double bond write the number down.
3) Stick the **suffix** that tells you it's a **carboxylic acid** on the end of the name. That's **-anoic acid** if it's saturated, or **-enoic acid** if it's unsaturated.
4) If the molecule has **more than one double bond** add a bit to the suffix to say how many it has (two is **-dienoic acid**, three is **-trienoic**, and so on).

For example:
tetra = 4
hexa = 6
octa = 8
deca = 10
tetradeca = 14

Example 1

COOH

A saturated fatty acid with **14** carbon atoms and **no** double bonds is called **tetradecanoic acid**.

Example 2

COOH

This unsaturated fatty acid has **16** carbons and **2** double bonds — one starting from **carbon 3** and one starting from **carbon 7**. So it's called **hexadeca-3,7-dienoic acid**.

Fatty Acids and Fats

Naming Fatty Acids Using *Shorthand Names*

1) First write down the **number of carbon atoms** in the fatty acid.
2) Then count the **number of double bonds** in the molecule.
   Write this **after** the number of carbons, separating them with a comma.
3) Work out **where** in the carbon chain the **double bonds** are (if there are any).
   Add these numbers, in **brackets**, to the **end** of the name.

This is a quicker method than systematic naming because it only uses numbers.

Example 1

COOH

This fatty acid has **14** carbons and **no** double bonds. So its shorthand name is **14, 0**.

Example 2

COOH

This fatty acid has **16** carbons and **2** double bonds — one starting from **carbon 3** and one starting from **carbon 7**. So its shorthand name is **16, 2(3, 7)**.

Fatty acids come in *Cis* and *Trans* forms

Fatty acids can exist as **geometric isomers** — molecules with the same structural formula but different arrangements in space. This is because the C=C bond is **rigid** — the molecule can't rotate about this bond.

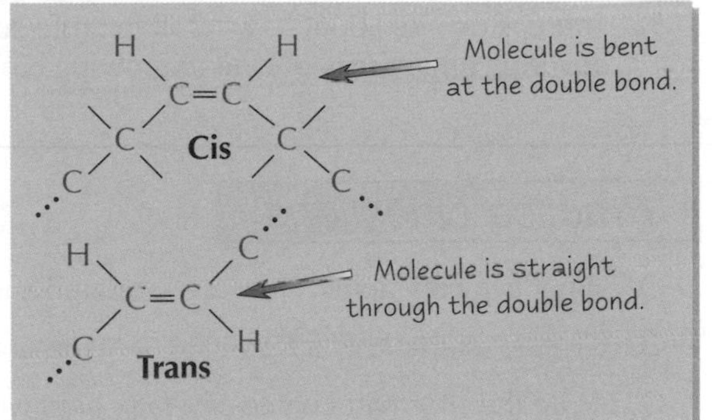

1) Almost all naturally-occurring fatty acids that are unsaturated have the **cis configuration**. This means that the hydrogens each side of the double bond are on the same side. This results in a **bent molecule**, or with several double bonds, a **curved molecule**.
2) In **trans fatty acids**, the **hydrogens** are on **opposite sides**.
   This gives long, straight molecules, similar to saturated fatty acids. They're almost always the product of human processing — hydrogen is added (**hydrogenation**) to unsaturated vegetable oils to saturate them, raising their melting point and creating solid fats.

See page 189 for more on cis-trans isomerism.

Fats *Aren't Always* Bad (but they Often Are)

1) **Cholesterol** is a soft, waxy material found in cell membranes and transported in your blood stream. It is partly produced by your body and partly absorbed from animal products that you eat, e.g., eggs, meat, dairy products.
2) There are **two types** of cholesterol - 'good' cholesterol and 'bad' cholesterol.
3) **Bad cholesterol** can clog blood vessels, which increases the risk of heart attacks and strokes.
4) **Good cholesterol** removes bad cholesterol, taking it to the liver to be destroyed.
   So high levels of good cholesterol can give protection from heart disease.
5) Recent research has shown that **trans fats increase** the amount of bad cholesterol and decrease the amount of good cholesterol.
6) Trans fats are **triglycerides** made from trans fatty acids. They are almost all man-made and are used in many foods such as biscuits, cakes, chips and crisps. Because of recent health concerns, there have been moves to **reduce their use** and more **clearly label** foods that contain them.
7) Bad cholesterol is also increased by eating **saturated fats** (made from fatty acids with no double bonds). They occur in animal products but much less so in plants.
8) Plant oils such as olive and sunflower oils contain **unsaturated fats**. These can be **polyunsaturated** (several double bonds) or **monounsaturated** (one double bond per chain). Polyunsaturated oils have been shown to reduce "bad" cholesterol and are actually a good thing to eat in moderation to prevent heart disease. They can help counteract obesity if they're used instead of saturated fats.

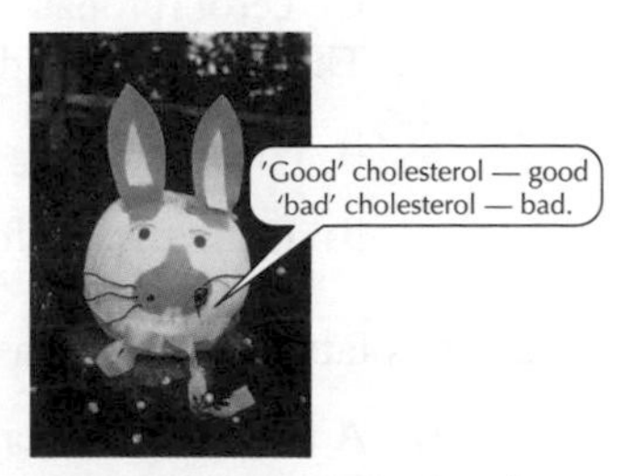

Revision Bunny can help if you're struggling with the key point here.

Fatty Acids and Fats

Fats can be used to make Biodiesel

Biodiesel is a renewable fuel made from **vegetable oil** or **animal fats** that can be used in diesel engines. It is gaining popularity as a viable alternative to crude oil-based diesel.

1) Biodiesel is mainly a mixture of methyl and ethyl **esters of fatty acids**.
2) It's made by reacting **triglycerides** (oils or fats) with **methanol** or **ethanol**.

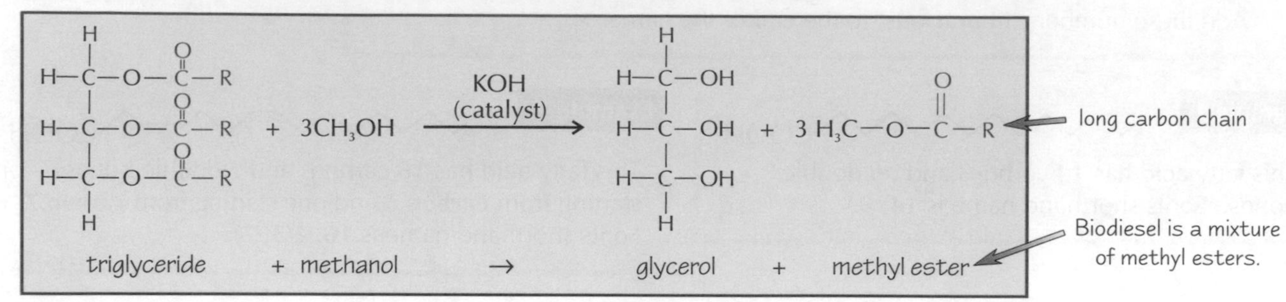

3) The **vegetable oils** used in the process can be **new**, **low grade oil** or **waste oil** from chip shops and restaurants. **Animal fats** that can be used include chicken fat, waste fish oil and lard from meat processing.
4) At present, biodiesel is mainly used **mixed with conventional diesel**, rather than in pure form. **B20 fuel** contains 20% biodiesel and 80% conventional diesel. Diesel engines generally need converting before they're able to run **B100 fuel** (100% biodiesel).
5) There is debate about how feasible **large-scale use** of biodiesel is — to produce significant quantities would mean devoting **huge areas of land** to growing biodiesel crops (e.g. rapeseed and soy beans) rather than **food crops**.

Practice Questions

Q1 What is a fatty acid? How are saturated and unsaturated fatty acids different?

Q2 Give an example of a fatty acid — write out its systematic name and its chemical formula.

Q3 Write the shorthand name of a fatty acid with 14 carbons and one double bond on carbon 6.

Q4 Explain how cis and trans fatty acids are different.

Q5 Give one advantage and one disadvantage of biodiesel over conventional diesel.

Exam Questions

1 Stearic acid is the common name for a fatty acid found in many animal fats.
Stearic acid is used in candle making. Its systematic name is octadecanoic acid.

a) What is the chemical formula of stearic acid? [1 mark]

b) Glycerol (propane-1,2,3-triol) forms a triester with stearic acid.
This triester is widely used in shampoos and cosmetics to give a pearly effect.

i) What name is given to triesters formed from glycerol and fatty acids? [1 mark]

ii) Is the triester formed saturated or unsaturated? Explain how you know. [1 mark]

2 Cis fatty acids are considered to be more healthy than trans fatty acids.

a) A major constituent of olive oil is a cis fatty acid called oleic acid.
What does the term 'cis' tell you about its structure? [1 mark]

b) Explain why trans fatty acids are harmful to human health. [2 marks]

Altogether now... 'Lard, lard, lard'... 'Lard, Lard, Lard'... 'Lard Lard Lard Lard'...

If you're struggling to remember everything, have a go at writing down the key facts like this: Fatty acid = carboxylic acid, triglyceride = triester = "fat". If you're still struggling, as a last resort you could summon Revision Bunny by standing on one leg and saying his name three times. Word of warning though, he's a bit loco ding dong. So don't blame me if things get ugly.

Amines

This is a-mean, fishy smelling topic...

Amines are Organic Derivatives of **Ammonia**

If one or more of the **hydrogens** in **ammonia** (NH_3) is replaced with an organic group, you get an **amine**.

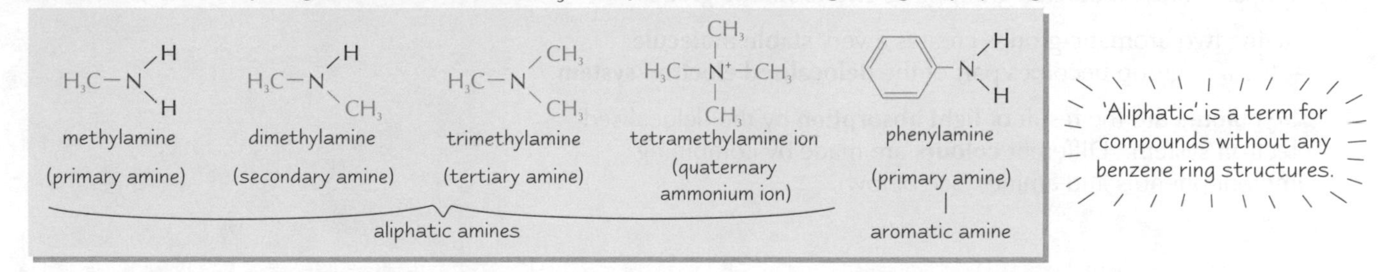

Small amines smell similar to **ammonia**, with a **slightly 'fishy'** twist. **Larger amines** smell very **'fishy'**. (Nice.)

Amines have a **Lone Pair of Electrons** that can Form **Dative Covalent Bonds**

1) Amines will **accept protons (H^+ ions)**.
2) There's a **lone pair of electrons** on the **nitrogen** atom that forms a **dative (coordinate) bond** with an H^+ ion. Dative bonds are covalent bonds where both electrons come from the **same atom**.
3) This means that amines are **bases** — bases can be defined as proton acceptors or electron donors.

$$RNH_2: + H^+ \rightarrow [RNH_3]^+$$

Amines React with **Acids** to Form **Salts**

Amines are **neutralised** by **acids** to make an **ammonium salt**.
For example, **ethylamine** reacts with **hydrochloric** acid to form ethylammonium chloride:

$$CH_3CH_2NH_2 + HCl \rightarrow CH_3CH_2NH_3^+Cl^-$$

Aliphatic Amines can be Made From **Haloalkanes**

Amines can be made by heating a **haloalkane** with an excess of ethanolic **ammonia**.
You'll get a **mixture** of primary, secondary and tertiary amines, and quaternary ammonium salts, as more than one hydrogen is likely to be substituted. You can separate the products using **fractional distillation**.

E.g. $$2\,NH_3 + CH_3CH_2Br \longrightarrow CH_3CH_2NH_2 + NH_4Br$$
ammonia — ethylamine

Aromatic Amines are Made by **Reducing** a **Nitro Compound**

Nitro compounds, such as **nitrobenzene**, are reduced in two steps:

1) Heat a mixture of a **nitro compound**, **tin metal** and **conc. hydrochloric acid** under **reflux** — this makes a salt.
2) Then to get the **aromatic amine**, you have to add **sodium hydroxide**.

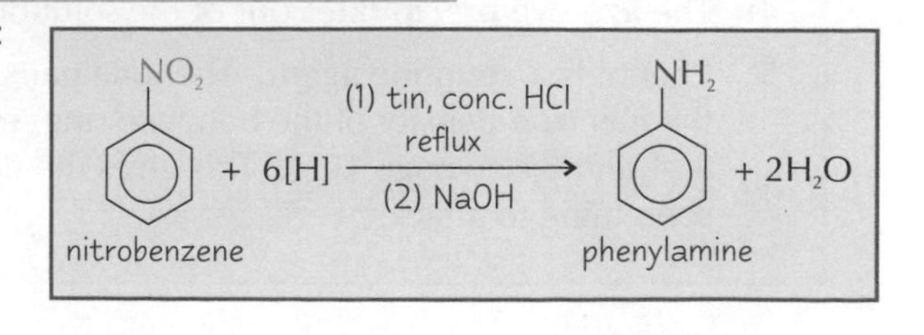

Amines

Aromatic Amines are Used to Make Azo Dyes

1) Azo dyes are man-made dyes that contain the **azo group**, **–N=N–**.
2) In most azo dyes, the azo group links **two aromatic groups**
3) Having two aromatic groups creates a very **stable molecule** — the azo group becomes part of the **delocalised electron system**.
4) The **colours** are the result of **light absorption** by the delocalised electron system. Different **colours** are made by combining different phenols and amines (see below).

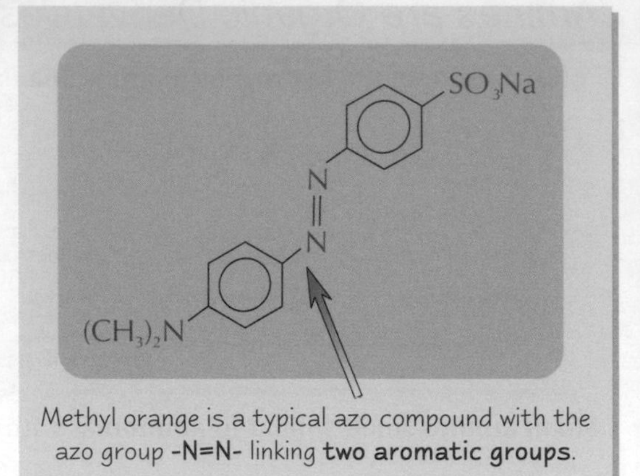

Methyl orange is a typical azo compound with the azo group **-N=N-** linking **two aromatic groups**.

Azo Dyes can be made in a Coupling Reaction

The first step in creating an azo dye is to make a **diazonium salt** — diazonium compounds contain the group $-\overset{+}{N}\equiv N-$. The **azo dye** is then made by **coupling** the diazonium salt with an **aromatic** compound that is susceptible to **electrophilic attack** — like a **phenol**.

Here's the method for creating a yellow-orange azo dye:

React Phenylamine with Nitrous Acid to make a Diazonium Salt

1) **Nitrous acid (HNO_2)** is **unstable**, so it has to be made ***in situ*** from sodium nitrite and hydrochloric acid.

'in situ' means 'in the reaction'

$$NaNO_2 + HCl \rightarrow HNO_2 + NaCl$$

2) **Nitrous acid** reacts with **phenylamine** and **hydrochloric acid** to form **benzenediazonium chloride**. The temperature **must** be below **10 °C** to prevent a phenol forming instead.

$$C_6H_5NH_2 + HNO_2 + HCl \longrightarrow C_6H_5\overset{+}{N}\equiv N\ Cl^- + 2H_2O$$

Make the Azo Dye by Coupling the Diazonium Salt with a Phenol

1) First, the **phenol** has to be dissolved in **sodium hydroxide** solution to make **sodium phenoxide** solution.
2) It's then stood in **ice**, and chilled **benzenediazonium chloride** is added.
3) Here's the overall equation for the reaction:

$$C_6H_5\overset{+}{N}\equiv N\ Cl^- + C_6H_5OH + NaOH \longrightarrow C_6H_5-N{=}N-C_6H_4-OH + NaCl + H_2O$$

yellow-orange azo compound

4) The azo dye **precipitates** out of the solution immediately.
5) Phenol is a **coupling agent**. The lone pairs on its oxygen increase the **electron density** of the benzene ring, especially around carbons 2, 4 and 6 (see page 106). This gives the diazonium ion (a **weak electrophile**) something to attack.

Remember — electrophile means 'electron lover'.

Amines

Azo Dyes are used in *Food*, *Textiles* and *Paints*

1) Azo dyes produce **bright**, **vivid** colours, most of them in the **yellow** to **red spectrum**, though many other colours are possible too. Azo dyes make up about 70% of all dyes used in food and textiles.
2) Many azo dyes are used as **food colourings** (and have corresponding E numbers). Examples include tartrazine (E102), yellow 2G (E107), allura red (E129) and brilliant black BN (E151), but there are many, many more.
3) Because the molecules are very **stable**, azo dyes provide **lightfast** (i.e. strong light won't fade them), **permanent** colours for clothing.
4) They are added to materials like clay to produce **paint pigments**.

'Darn, I spilt my paint' on canvas, £2500

Some azo dyes have been linked to the condition "clown child".

5) Some azo dyes are used as **indicators**, e.g. methyl orange, because they change colour at different pHs.
6) In recent years there has been a lot of **concern** about the use of **artificial additives** in food. Some azo compounds that were previously used in foods have since been **banned** for health reasons — enzymes in the body can break some of them down to produce **toxic** or **carcinogenic** compounds. Others have been linked to **hyperactivity** in children.

Practice Questions

Q1 What are primary, secondary and tertiary amines? Draw and name an example of each.

Q2 Draw the structure of methyldiethylamine.

Q3 Why do amines act as bases?

Q4 Describe how you would make a diazonium salt.

Q5 What is formed when a diazonium salt reacts with a phenol?

Exam Questions

1 When 1-chloropropane is reacted with ammonia a mixture of different amines are produced.

a) Name two of the amines produced in the reaction. [2 marks]

b) How would you separate the mixture? [1 mark]

c) Outline the steps needed to reduce nitrobenzene to make phenylamine. [4 marks]

2 Sunset Yellow (E110) is a yellow dye used in orange squash and many foodstuffs.

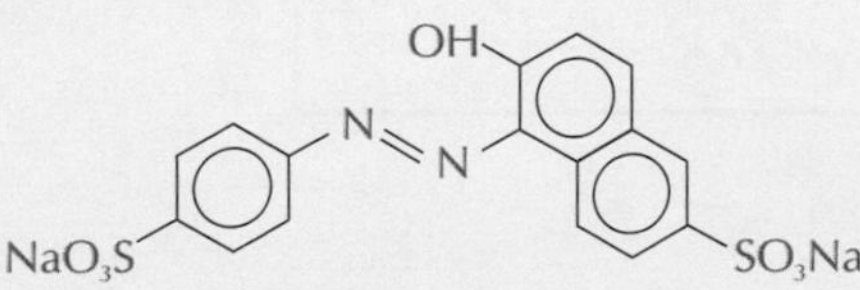

It has been withdrawn from use in many countries as it has been connected to hyperactivity in children.

a) Which class of dyes does it belong to? [1 mark]

b) Which part of its structure indicates this? [1 mark]

c) Draw the structures of two organic compounds that could be used to produce Sunset Yellow. [2 marks]

You've got to learn it — amine it might come up in your exam...

Rotting fish smells so bad because the flesh releases diamines as it decomposes. Is it fish that smells of amines or amines that smell of fish — it's one of those chicken or egg things that no one can answer. Well, enough philosophical pondering — we all know the answer to the meaning of life. It's A2 chemistry. Now make sure you can do all the questions above.

Polymers

And the organic chemistry joy continues... you're going to be eased nice and gently into this section with a double page on polymers. Nothing too scary here, so you shouldn't feel a thing. Well, maybe a slight pinprick.

There are Two Types of Polymerisation — Addition and Condensation

Alkenes Join Up to form Addition Polymers

1) **Polymers** are long chain molecules formed when lots of small molecules, called **monomers**, join together as if they're holding hands.
2) Alkenes will form polymers — the **double bonds** open up and join together to make long chains. The individual, small alkenes are the monomers.
3) This is called **addition polymerisation**. For example, **poly(ethene)** is made by the **addition polymerisation** of **ethene**.

monomer → polymer

The bit in brackets is the 'repeating unit'.
n is the number of repeating units.

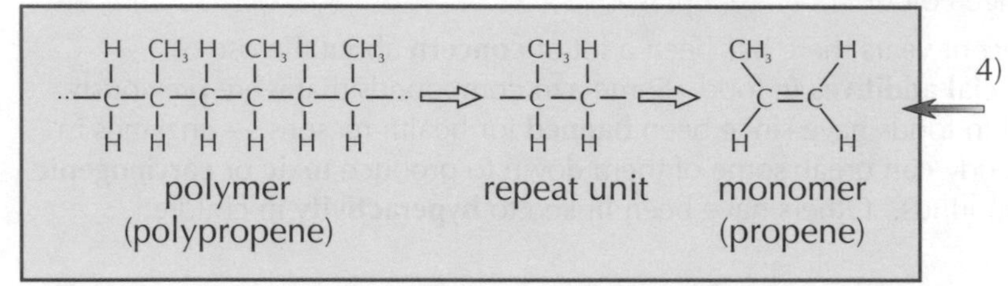

4) To find the **monomer** used to form an addition polymer, take the **repeating unit** and add a **double bond**.

5) Because of the loss of the double bond, poly(alkenes), like alkanes, are **unreactive**.

Condensation Polymers are formed as Water is Removed...

1) In **condensation** polymerisation, a small molecule, often **water**, is **lost** as the molecules link together.
2) Condensation polymers include **polyesters**, **polyamides** and **polypeptides**. Each of these is covered in more detail over the next few pages.
3) Each monomer in a condensation polymer has at least **two functional groups**. Each functional group reacts with a group on another monomer to form a **link**, creating the polymer **chain**.
4) In polyesters, an **ester link** (–COO–) is formed between the monomers.
5) In polyamides and polypeptides, **amide links** (–CONH–) are formed between the monomers. In polypeptides, these are usually called **peptide** bonds.

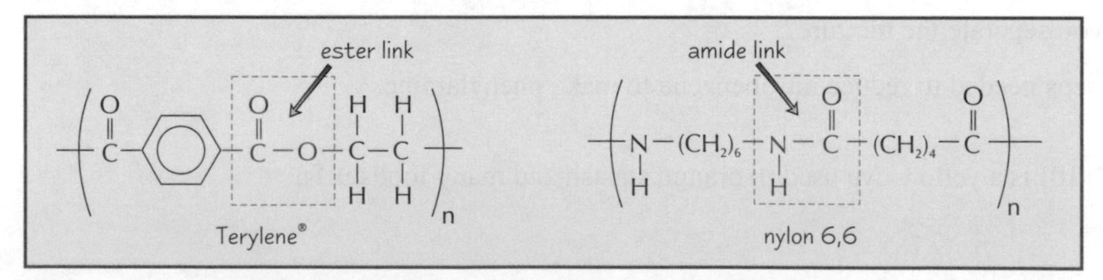

...and Broken Down by Adding It

1) The ester or amide link in polyesters and polyamides can be broken down by reaction with water in an acidic or alkaline solution. This is called **hydrolysis**. Even small amounts of acid will cause nylon to break down very quickly, so don't spill acids on your tights, boys...
2) The products of the hydrolysis are the **monomers** that were used to make the polymer — you're basically **reversing** the condensation reaction that was used to make them.

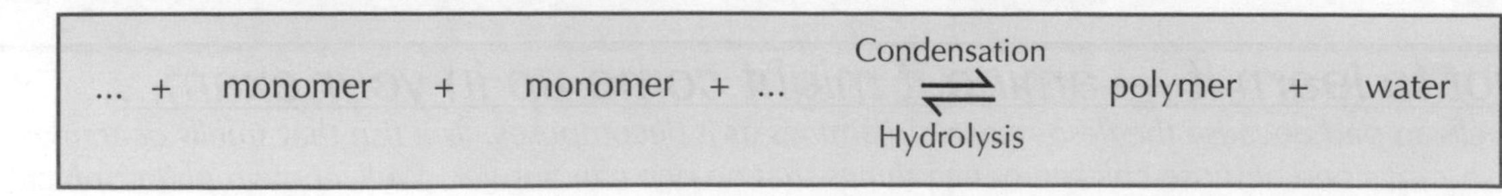

Polymers

Polyesters and Polyamides are **Biodegradable**... kind of

1) Because **amide links** in polyamides and **ester links** in polyesters can be easily **hydrolysed**, they're **biodegradable**. These links are found in nature, so there are **fungi** and **bacteria** that are able to degrade them.
2) It's not all hunky-dory though. It takes **absolutely ages** for synthetic polyamides and polyesters to decompose — e.g. nylon takes around 40 years.
3) Nylon and polyester can be **recycled** — you can buy polyester fleece jackets made from recycled PET (Polyethylene terephthalate) bottles.

Easily **Biodegradable** Polymers from **Renewable Sources** are the Aim

1) Although condensation polymers will break down by hydrolysis, the huge time this takes to happen in nature means that they still create **waste problems**. **Addition polymers** are even more of a problem — they are very **stable** molecules and won't be broken down by hydrolysis.
2) Also, most polymers we use are made from monomers derived from **crude oil** which is **not** a **renewable resource**.
3) To tackle these problems, chemists are trying to produce alternative polymers, which are easily **biodegradable** and **renewable**.

Example: The biodegradable polymer **poly(lactic acid)**

1) **Poly(lactic acid)**, or **PLA**, is a **polyester** made from **lactic acid**, which is produced by fermenting maize or sugar cane, which are both renewable crops — you can see it's structure on page 122.
2) PLA will **biodegrade** easily — it is hydrolysed by water if kept at a **high temperature** for **several days** in an industrial composter or more slowly at lower temperatures in **landfill** or home **compost heaps**.
3) PLA has many uses, including **rubbish bags**, food and electronic **packaging**, disposable **eating utensils** and internal **sutures** (stitches) that break down without having to open wounds to remove them.

Light can **Degrade** Some **Condensation Polymers**

Condensation polymers that contain **C=O (carbonyl) groups**, such as polyamides are **photodegradable** — they can be broken down by light as the C=O bond absorbs **ultra violet radiation**. This energy can cause bonds to break either side of the carbonyl group and the polymer breaks down into smaller units.

Practice Questions

Q1 What is the difference between addition and condensation polymerisation?

Q2 What type of bond do the monomers used to make addition polymers all have in their structure?

Q3 Give two environmental advantages of poly(lactic acid) over polythene.

Q4 What does photodegradable mean?

Exam Questions

1 Vinyl chloride is the monomer used in the manufacture of polyvinyl chloride (PVC), a polymer used to insulate electrical wires and in CD manufacture.

H₂C=CHCl — vinyl chloride (chloroethene)

a) What type of reaction is involved in the manufacture of PVC? [1 mark]

b) Draw the repeating unit in PVC. [1 mark]

c) Draw a three-unit section of the PVC molecule. [1 mark]

d) Why is PVC not likely to be biodegradable? [2 marks]

2 Dissolving stitches used in operations are made from hydroxycarboxylic acid polyesters.
Explain why the stitches dissolve in the human body.
Why is it not possible to use a polymer like poly(propene) for this purpose? [3 marks]

Wicked Witches are made of biodegradable polymers...

...and Dorothy, being something of a chemistry whizz, figured this out and famously used it to her advantage... A bucket of water, some alkaline cleaning fluid for good measure, and voila — witch is hydrolysed. Water is an excellent way to distinguish between good and bad witches. Just remember — only bad witches are biodegradable.

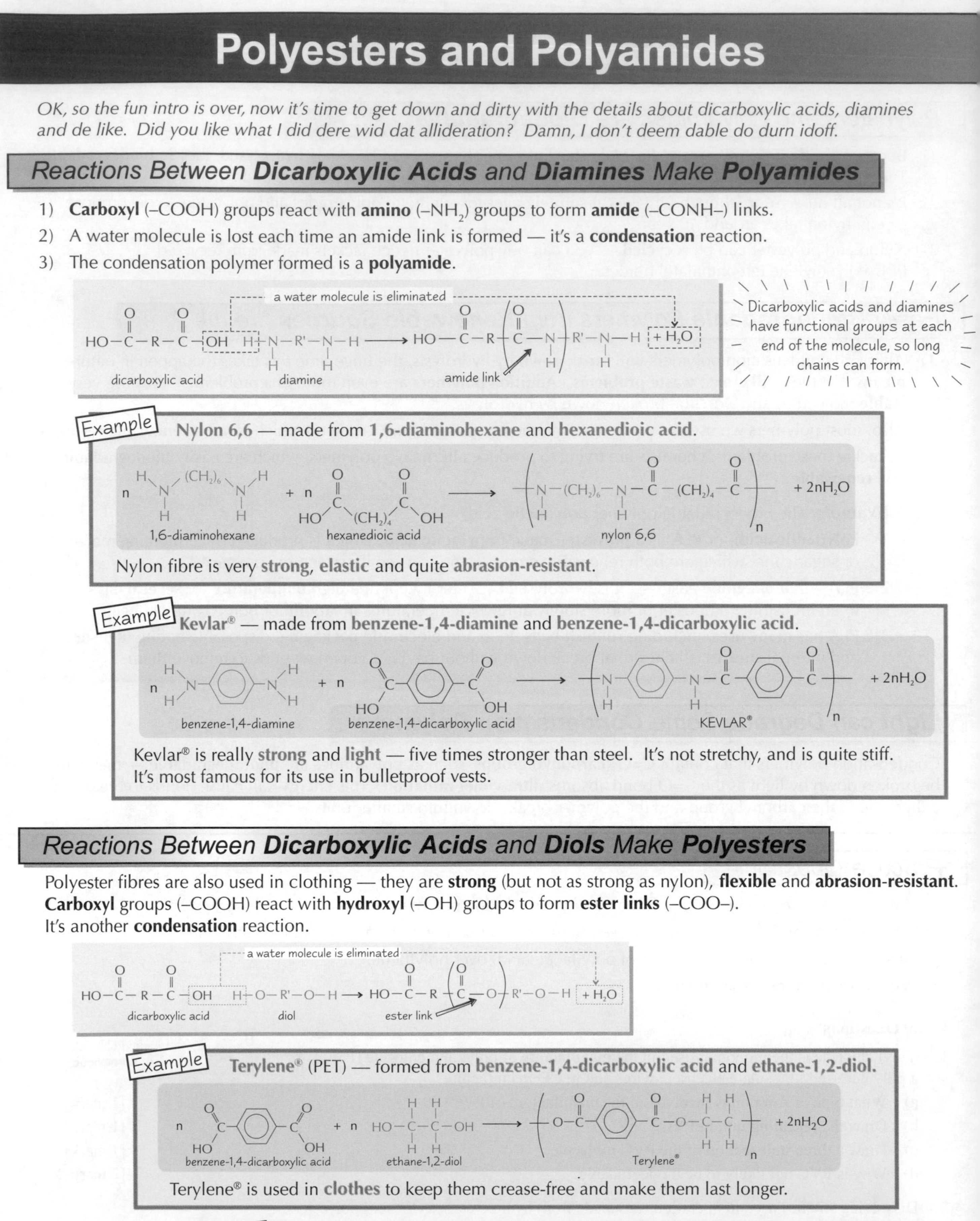

Polyesters and Polyamides

OK, so the fun intro is over, now it's time to get down and dirty with the details about dicarboxylic acids, diamines and de like. Did you like what I did dere wid dat allideration? Damn, I don't deem dable do durn idoff.

Reactions Between Dicarboxylic Acids *and* Diamines *Make* Polyamides

1) **Carboxyl** (–COOH) groups react with **amino** (–NH_2) groups to form **amide** (–CONH–) links.
2) A water molecule is lost each time an amide link is formed — it's a **condensation** reaction.
3) The condensation polymer formed is a **polyamide**.

Dicarboxylic acids and diamines have functional groups at each end of the molecule, so long chains can form.

Example **Nylon 6,6** — made from **1,6-diaminohexane** and **hexanedioic acid**.

Nylon fibre is very **strong**, **elastic** and quite **abrasion-resistant**.

Example **Kevlar®** — made from **benzene-1,4-diamine** and **benzene-1,4-dicarboxylic acid**.

Kevlar® is really **strong** and **light** — five times stronger than steel. It's not stretchy, and is quite stiff. It's most famous for its use in bulletproof vests.

Reactions Between Dicarboxylic Acids *and* Diols *Make* Polyesters

Polyester fibres are also used in clothing — they are **strong** (but not as strong as nylon), **flexible** and **abrasion-resistant**.
Carboxyl groups (–COOH) react with **hydroxyl** (–OH) groups to form **ester links** (–COO–).
It's another **condensation** reaction.

Example **Terylene®** (PET) — formed from **benzene-1,4-dicarboxylic acid** and **ethane-1,2-diol.**

Terylene® is used in **clothes** to keep them crease-free and make them last longer.

Example **Poly(lactic acid)** — a biodegradable and renewable polymer (see page 121).

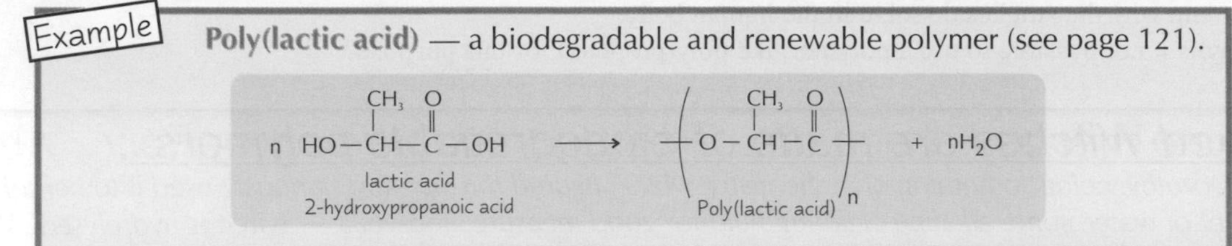

Polyesters and Polyamides

Hydrolysis Produces the Original **Monomers**

1) Condensation polymerisation can be reversed by **hydrolysis** — water molecules are added back in and the links are broken.
2) In practice, hydrolysis with just water is far too **slow**, so the reaction is done with an **acid** or **alkali.**
3) **Polyamides** will hydrolyse more easily with an **acid** than an alkali:

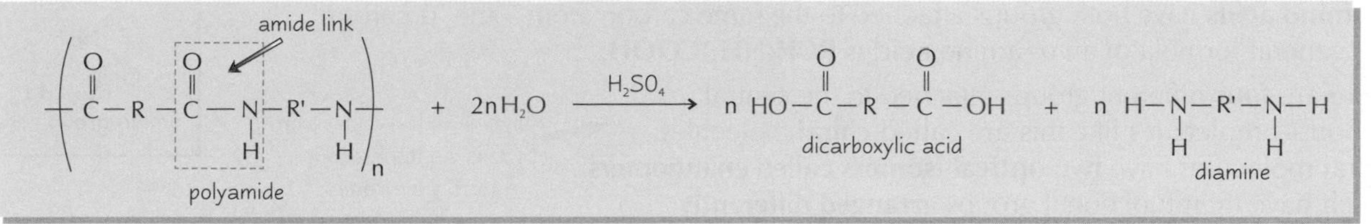

4) **Polyesters** will hydrolyse more easily with an **alkali**. A **metal salt** of the carboxylic acid is formed.

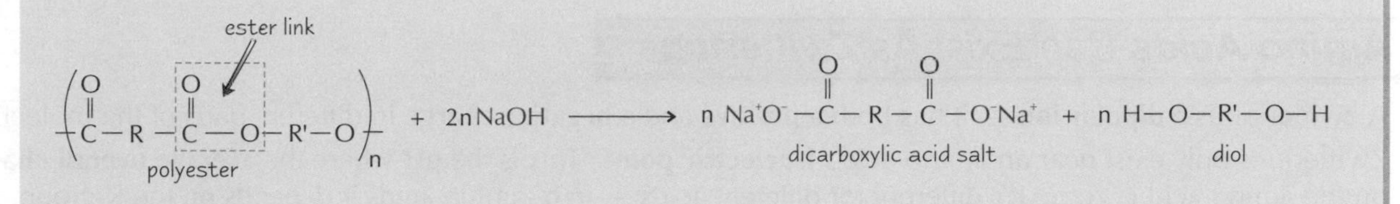

Practice Questions

Q1 What types of molecule can undergo condensation polymerisation?

Q2 Why is it called condensation polymerisation?

Q3 Which molecules are used to make a polyester?

Q4 What is a polyamide made from?

Exam Questions

1 a) Nylon 6,6 is the most commonly produced nylon. A section of the polymer is shown below:

$$-\underset{H}{N}-(CH_2)_6-\underset{H}{N}-\overset{O}{\overset{\|}{C}}-(CH_2)_4-\overset{O}{\overset{\|}{C}}-\underset{H}{N}\ (CH_2)_6-\underset{H}{N}-\overset{O}{\overset{\|}{C}}-(CH_2)_4-\overset{O}{\overset{\|}{C}}-\underset{H}{N}-(CH_2)_6-\underset{H}{N}-\overset{O}{\overset{\|}{C}}-(CH_2)_4-\overset{O}{\overset{\|}{C}}-$$

i) Draw the structural formulae of the monomers from which nylon 6,6 is formed. It is not necessary to draw the carbon chains out in full. [2 marks]

ii) Give a name for the type of linkage between the monomers in this polymer. [1 mark]

b) A polyester is formed by the reaction between the monomers hexane-1,6-dioic acid and hexane-1,6-diol.

i) Draw the repeating unit for the polyester. [1 mark]

ii) Explain why this is an example of condensation polymerisation. [1 mark]

2 Glycolic acid is produced from sugar cane, sugar beets and various fruit. It can be polymerised to produce a polymer used to make soluble sutures, which are used in surgery. The polymer is broken down by hydrolysis in the body and absorbed harmlessly.

$$HO-CH_2-\overset{O}{\overset{\|}{C}}-OH$$

Glycolic acid

a) Draw a section of this polymer made of three repeating units, and label the repeating unit. [2 marks]

b) Which group of polymers does this polymer belong to? [1 mark]

I have no friends? Yes I do... Poly... Ester... the other Poly... Ami... Des...

...yeah and we have loads of fun together — cinema trips, bowling, sometimes just a few drinks down the pub. What do you mean, 'they're not real friends'? They so are — they're my special chemistry friends. And they're much more interesting than your dumb friends. Bet your friends can't hydrolyse and condense back again. No, your friends suck. Losers. LOOZERRRRS

Amino Acids and Proteins

Time for a bit of biochemistry action now. We're still on polymers, but these are ones that do pretty important jobs in our bodies, without which we'd be nothing more than blobs of luminescent purple pulsating ectojelly.

Amino Acids have an **Amino Group** and a **Carboxyl** Group

An amino acid has a **basic amino group** (NH_2) and an **acidic carboxyl group** (COOH). This makes amino acids **amphoteric** — they've got both acidic and basic properties.

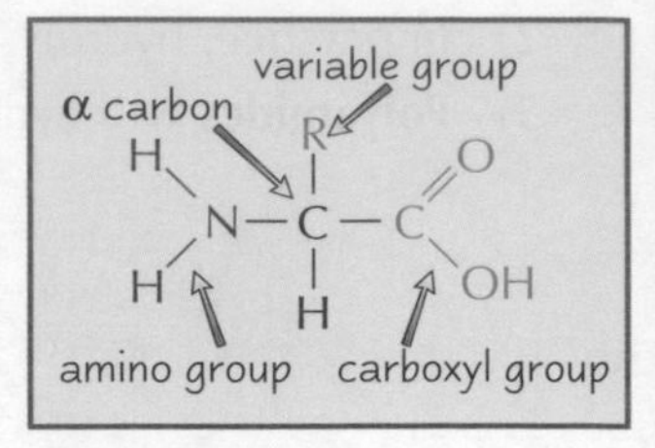

α–amino acids have both groups attached to the same carbon atom – the 'α carbon'. The general formula of an α–amino acid is **$RCH(NH_2)COOH$**.

There are **four** different groups attached to the central (α) carbon — molecules like this are called **chiral** molecules. Chiral molecules have two **optical isomers** called **enantiomers** which have their functional groups **arranged differently** — how this all works is explained on p130.

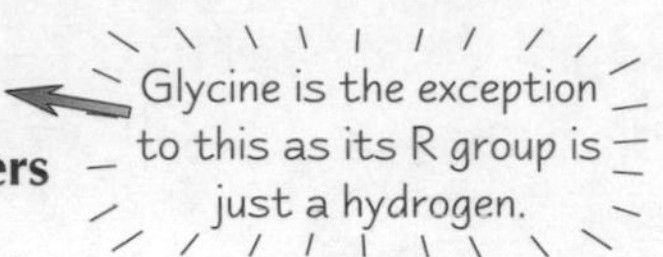

Amino Acids Can Exist As **Zwitterions**

A zwitterion is a **dipolar ion** — it has both a **positive** and a **negative charge** in different parts of the molecule. Zwitterions only exist near an amino acid's **isoelectric point**. This is the **pH** where the **average overall charge** on the amino acid is zero. It's different for different acids — in α–amino acids it depends on the R-group.

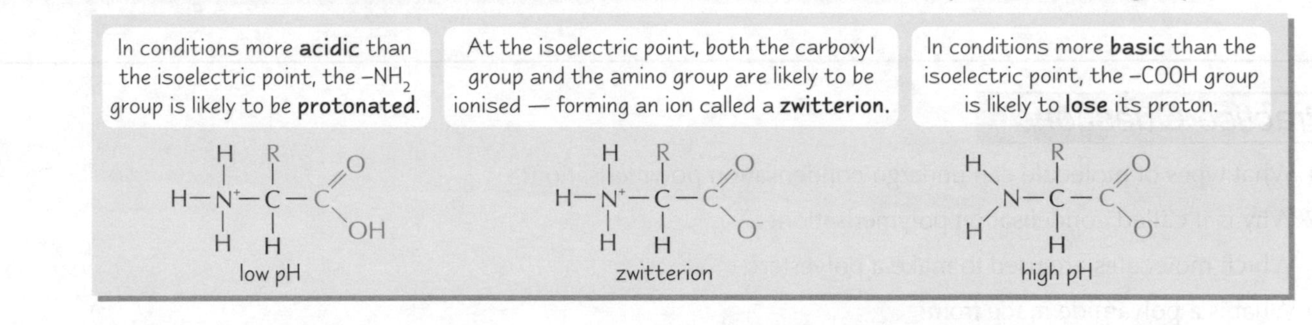

Polypeptides are **Condensation Polymers** of Amino Acids

Amino acids join together in a **condensation** reaction.
A **peptide bond** is made between the amino acids.
Here's how two amino acids join together to make a **dipeptide**:

Peptide bonds are exactly the same as amide bonds, but they're called 'peptide' when you're linking amino acids.

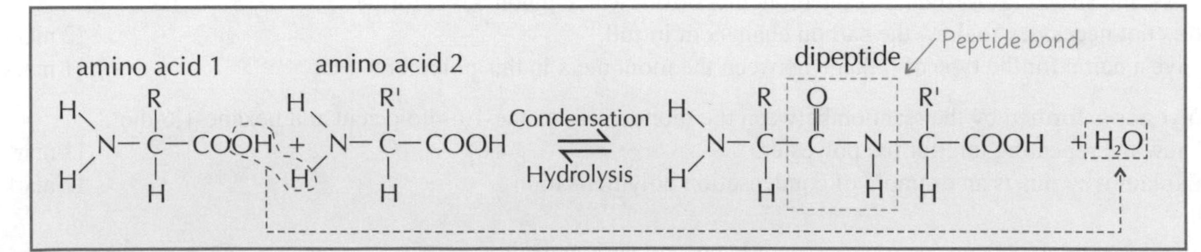

Polypeptides and **proteins** are made up of **lots** of amino acids joined together.
A **water molecule is lost** each time an amino acid joins on.

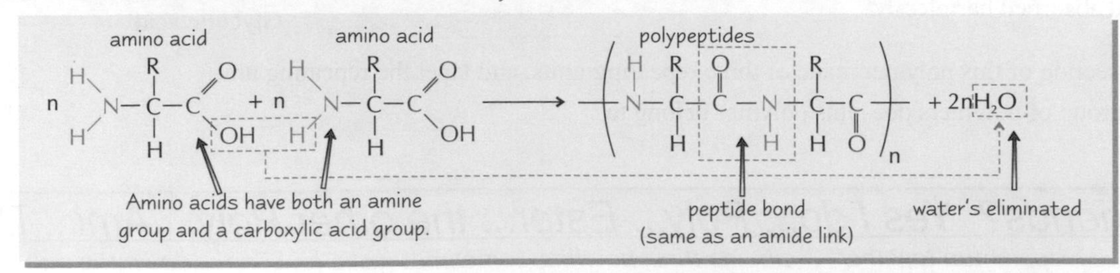

Amino Acids and Proteins

Proteins and ***Other Peptides*** *can be Taken Apart by* ***Hydrolysis***

When you eat proteins, **enzymes** in your digestive system break the proteins down to individual **amino acids** by **hydrolysis.** This process can be simulated in the lab by heating proteins with hydrochloric acid for 24 hours. A shorter reaction time will give a mixture of smaller peptides rather than individual acids. The reaction helps biochemists to work out the sequence of amino acids in a protein.

In the reaction, water molecules react with the **peptide links** and **break them apart**. The separate amino acids are then released.

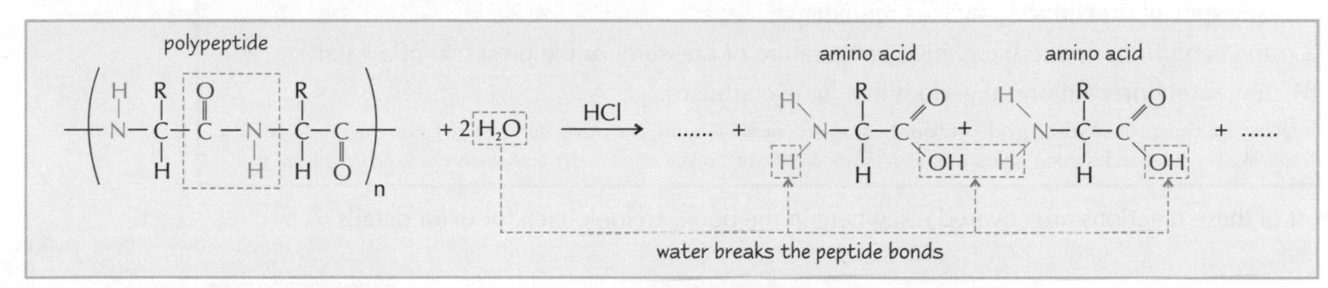

Hydrolysis can also be carried out using alkalis, but in this case the hydrogen in the –COOH group is replaced by a metal to form a **carboxylate salt** of the amino acid, e.g. **$RCH(NH_2)COO^-Na^+$**.

Practice Questions

Q1 Which two groups do all amino acids contain?

Q2 What is a zwitterion?

Q3 What reaction joins amino acids together?

Q4 What reaction takes proteins apart?

Exam Questions

1 Alanine is an α-amino acid found in many foods. It has the molecular formula $C_3H_7NO_2$.

a) What is meant by an α-amino acid? [1 mark]

b) Draw the structure of alanine. [1 mark]

c) Which part of this molecule will be different in other α-amino acids? [1 mark]

d) Alanine is a chiral molecule. What does this mean? [1 mark]

e) Draw the structure of the zwitterion produced by alanine. [1 mark]

2 A dipeptide X produces the amino acids glycine and phenylalanine when heated with hydrochloric acid overnight.

Glycine is: H_2N-CH_2-COOH (H₂N–C(H)(H)–COOH)

Phenylalanine is: $H_2N-CH(CH_2C_6H_5)-COOH$ (H₂N–C(H)(CH₂–benzene ring)–COOH)

a) What is meant by a dipeptide? [1 mark]

b) Draw one of the two possible structures of dipeptide X. [1 mark]

c) X is formed from glycine and phenylalanine in a condensation reaction.
Explain what this means. [1 mark]

d) What type of reaction occurs when X is heated with hydrochloric acid? [1 mark]

e) Only one of these two amino acids is chiral. Explain why. [1 mark]

Who killed the Zwittonians?

Zwitterions are actually the last physical remains of a race of highly advanced beings known as Zwittonians from the galaxy I Zwiky 19. It is thought that they lived approximately 9.75 billion years ago before evolving beyond the need for physical bodies, exploding in a ball of energy and emitting zwitterion particles throughout the cosmos. Fascinating.... Also untrue.

Organic Synthesis

In your exam you may be asked to suggest a pathway for the synthesis of a particular molecule. These pages contain a summary of some of the reactions you should know.

Chemists use **Synthesis Routes** to Get from One Compound to Another

Chemists have got to be able to make one compound from another. It's vital for things like **designing medicines**. It's also good for making imitations of **useful natural substances** when the real things are hard to extract.

If you're asked how to make one compound from another in the exam, make sure you include:

1) any **special procedures**, such as refluxing.
2) the **conditions** needed, e.g. high temperature or pressure, or the presence of a catalyst.
3) any **safety** precautions, e.g. do it in a fume cupboard.

If there are things like hydrogen chloride or hydrogen cyanide around, you really don't want to go breathing them in. Stuff like bromine and strong acids and alkalis are corrosive, so you don't want to splash them on your bare skin.

Most of these reactions are covered elsewhere in the book, so look back for extra details.

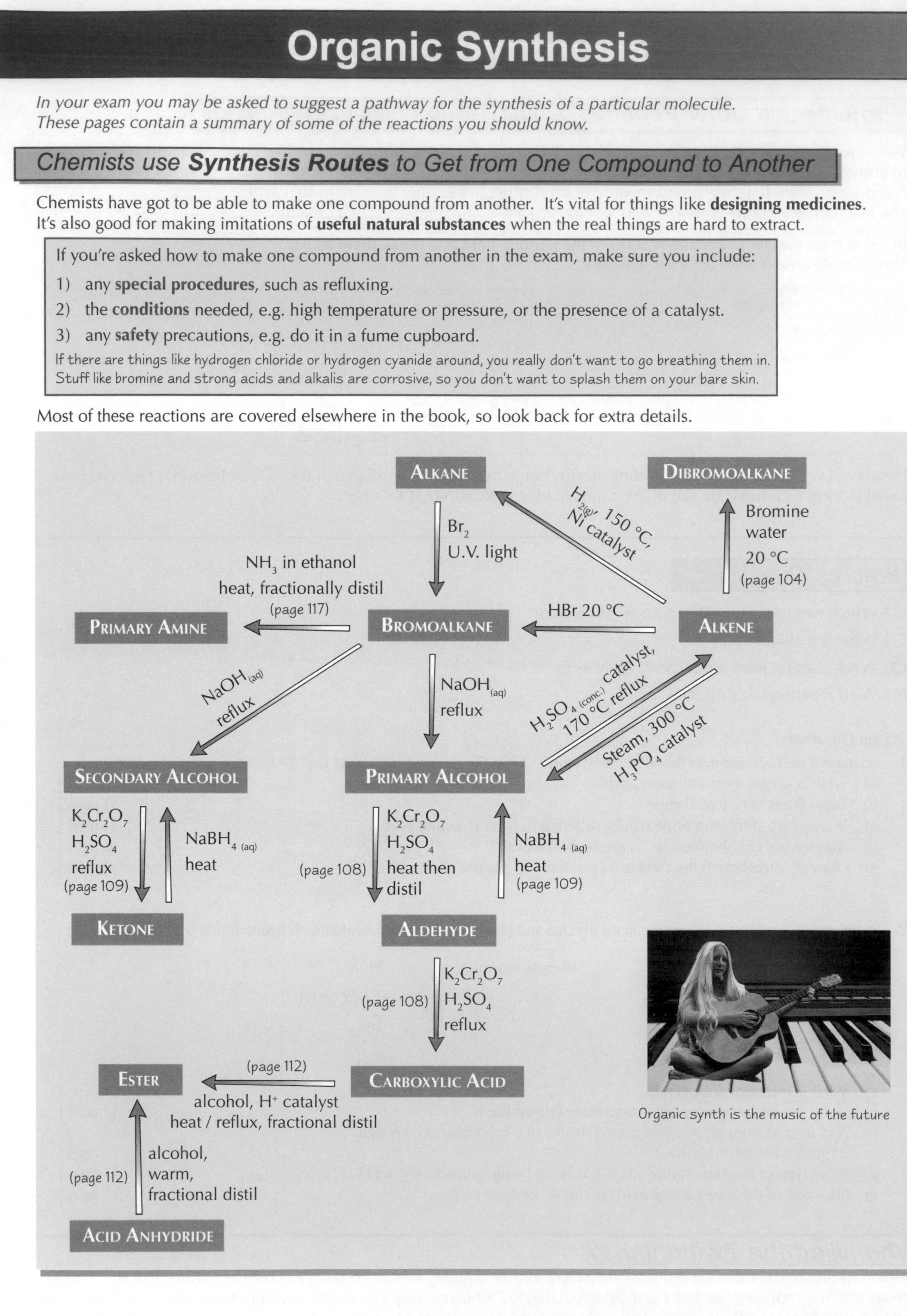

Organic synth is the music of the future

Organic Synthesis

Synthesis Route for Making **Aromatic Compounds**

There aren't so many of these reactions to learn — so make sure you know all the itty-bitty details.
If you can't remember any of the reactions, look back to the relevant pages and take a quick peek over them.

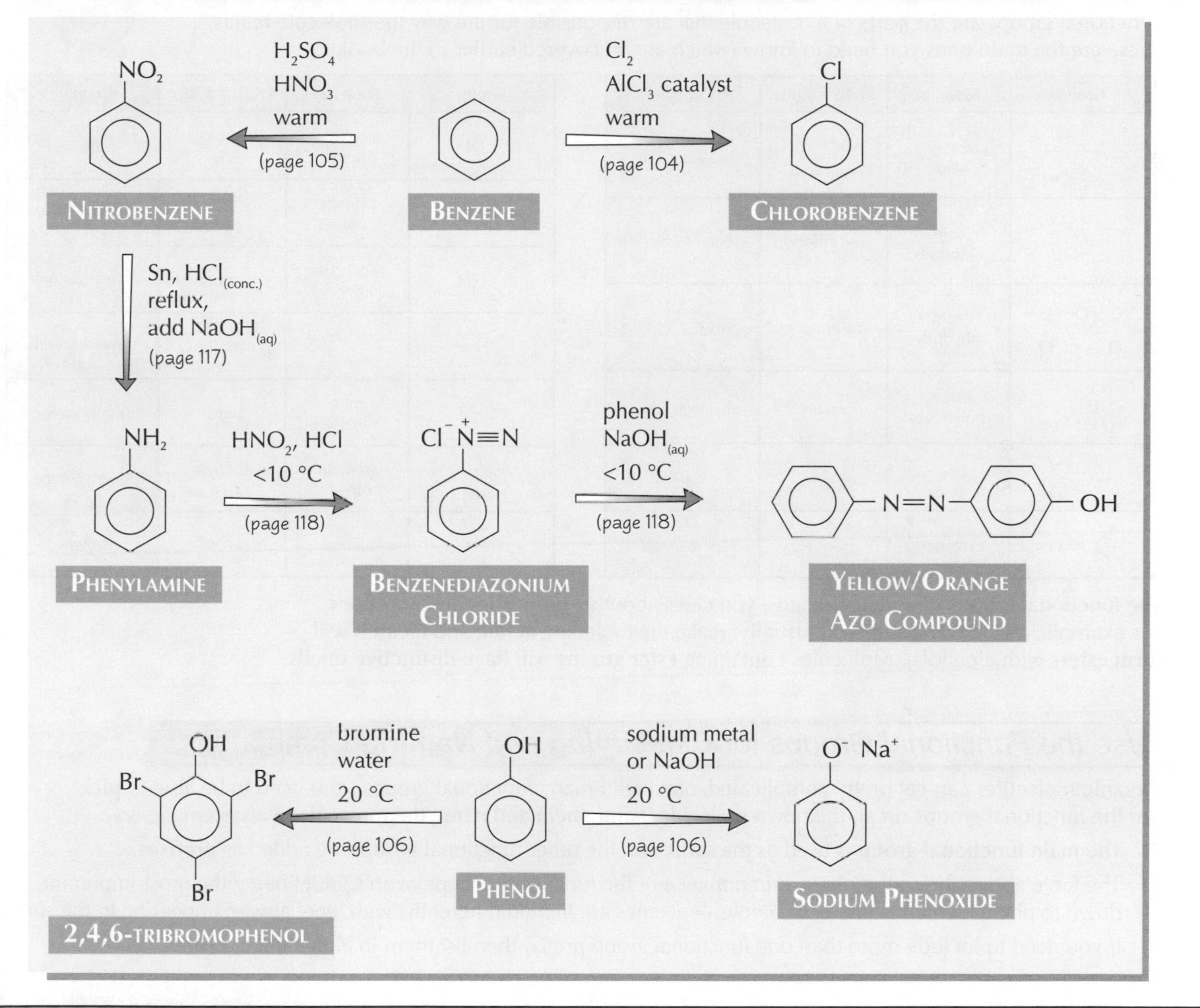

Practice Questions

Q1 How do you convert an alkane to a primary amine?

Q2 How do you make an alkene from an aldehyde?

Q3 How do you make phenylamine from benzene?

Exam Questions

1 Ethyl methanoate is one of the compounds responsible for the smell of raspberries.
Outline, with reaction conditions, how it could be synthesised in the laboratory from methanol. [7 marks]

2 How would you synthesise propanol starting with propane? State the reaction conditions and reagents needed for each step and any particular safety considerations. [7 marks]

I saw a farmer turn a tractor into a field once — now that's impressive...

There's loads of information here. Tons and tons of it. But you've covered pretty much all of it before, so it shouldn't be too hard to make sure it's firmly embedded in your head. If it's not, you know what to do — go back over it again. Then cover the diagrams up, and try to draw them out from memory. Keep going until you can do it perfectly.

Functional Groups

It's been a bit of a hard slog of equations and reaction conditions and dubious puns to get this far. But with a bit of luck it should all be starting to come together now. If not, don't panic — there's always revision bunny.

Functional Groups are the Most Important Parts of a Molecule

Functional groups are the parts of a molecule that are responsible for the way the molecule reacts. These are the main ones you need to know (which are all covered earlier in the book)...

Group	Found in	Prefix / Suffix	Example
–C(=O)OH	carboxylic acids	carboxy– –oic acid	ethanoic acid
–C(=O)Cl	acid chlorides	–oyl chloride	ethanoyl chloride
–C(=O)–O–C(=O)–	acid anhydrides	–oic anhydide	ethanoic anhydride
–C(=O)–O–	esters, polyesters	–oate	ethyl methanoate
–C(=O)H	aldehydes	–al	propanal
>C=O	ketones	–one	propanone
–OH	alcohols, phenols	hydroxy– –ol	propanol
$-NH_2$	primary amines	amino– –amine	methylamine
>NH	secondary amines	–amine	dimethylamine
>N–	tertiary amines	–amine	trimethylamine
$-NO_2$	nitro benzenes	nitro-	nitrobenzene
(benzene ring)	aromatic compounds	phenyl– –benzene	phenylamine
>C=C<	alkenes	-ene	butene

The functional groups in a molecule give you clues about its **properties** and **reactions**. For example, a **–COOH group** will (usually) make the molecule **acidic** and mean it will **form esters** with alcohols. Molecules containing **ester groups** will have **distinctive smells**.

Use the Functional Groups for Classifying and Naming Compounds

Organic molecules can get pretty complicated, often with many functional groups. You need to be able to **pick out** the functional groups on an unknown molecule, **name them** and **name the molecule** in a systematic way.

1) The **main functional group** is used as the **suffix** and the other functional groups are added as **prefixes**.
2) The table above shows the order of importance of the functional groups, with COOH being the most important, down to phenyl which is the least. (Note — alkenes are treated differently, with 'ene' always appearing in the suffix.)
3) If you need to include more than one functional group prefix, then list them in alphabetical order.

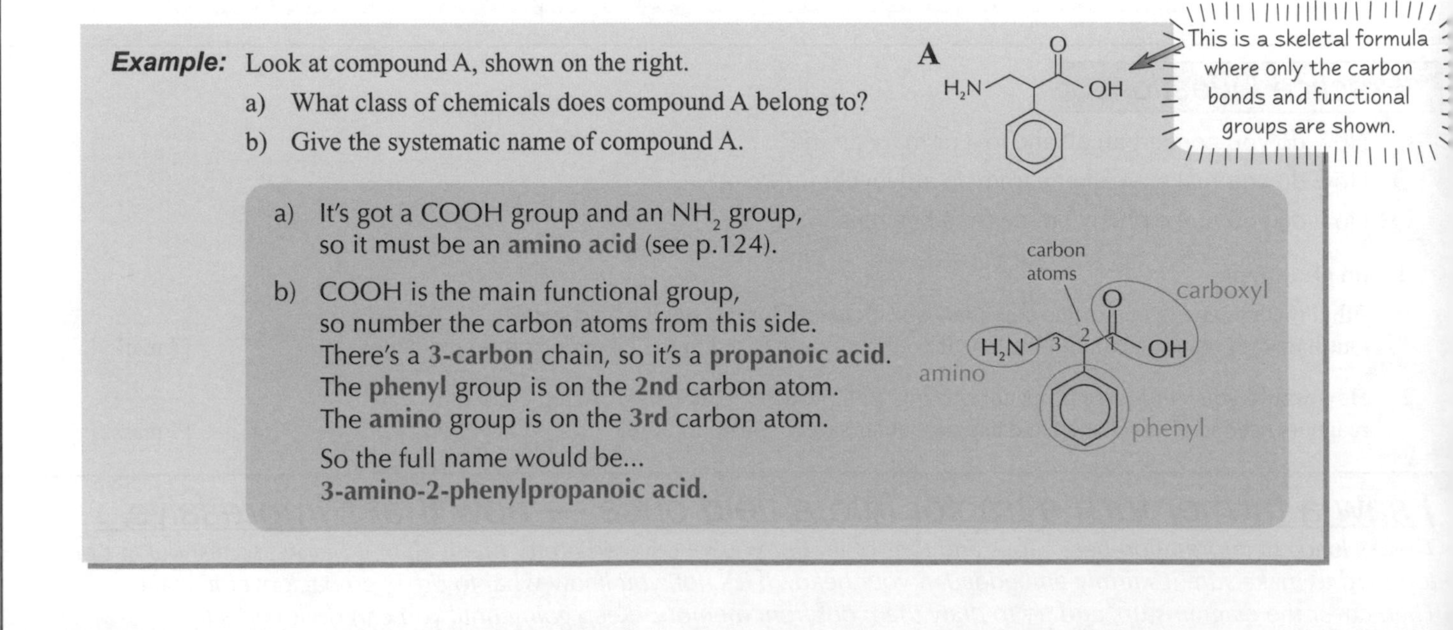

Example: Look at compound A, shown on the right.

a) What class of chemicals does compound A belong to?

b) Give the systematic name of compound A.

a) It's got a COOH group and an NH_2 group, so it must be an **amino acid** (see p.124).

b) COOH is the main functional group, so number the carbon atoms from this side. There's a **3-carbon** chain, so it's a **propanoic acid**. The **phenyl** group is on the **2nd** carbon atom. The **amino** group is on the **3rd** carbon atom.

So the full name would be...
3-amino-2-phenylpropanoic acid.

Functional Groups

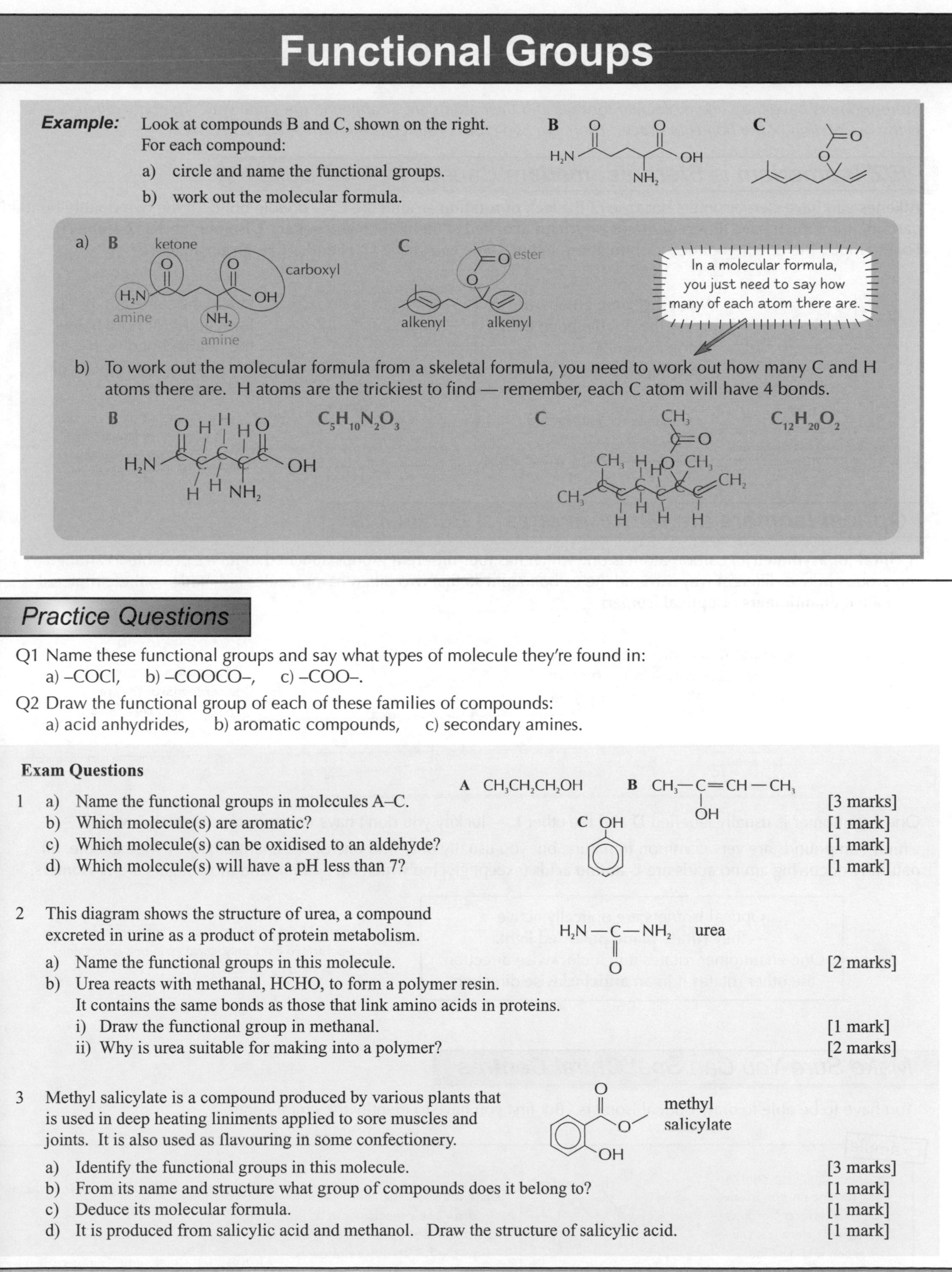

Example: Look at compounds B and C, shown on the right.
For each compound:

a) circle and name the functional groups.

b) work out the molecular formula.

a) **B** — ketone, carboxyl, amine (H_2N), amine (NH_2)

C — ester, alkenyl, alkenyl

In a molecular formula, you just need to say how many of each atom there are.

b) To work out the molecular formula from a skeletal formula, you need to work out how many C and H atoms there are. H atoms are the trickiest to find — remember, each C atom will have 4 bonds.

B $C_5H_{10}N_2O_3$

C $C_{12}H_{20}O_2$

Practice Questions

Q1 Name these functional groups and say what types of molecule they're found in:
a) –COCl, b) –COOCO–, c) –COO–.

Q2 Draw the functional group of each of these families of compounds:
a) acid anhydrides, b) aromatic compounds, c) secondary amines.

Exam Questions

A $CH_3CH_2CH_2OH$ **B** $CH_3-C(OH)=CH-CH_3$ **C** phenol (OH on benzene ring)

1 a) Name the functional groups in molecules A–C. [3 marks]
b) Which molecule(s) are aromatic? [1 mark]
c) Which molecule(s) can be oxidised to an aldehyde? [1 mark]
d) Which molecule(s) will have a pH less than 7? [1 mark]

2 This diagram shows the structure of urea, a compound excreted in urine as a product of protein metabolism.

$H_2N-C(=O)-NH_2$ urea

a) Name the functional groups in this molecule. [2 marks]
b) Urea reacts with methanal, HCHO, to form a polymer resin.
It contains the same bonds as those that link amino acids in proteins.
i) Draw the functional group in methanal. [1 mark]
ii) Why is urea suitable for making into a polymer? [2 marks]

3 Methyl salicylate is a compound produced by various plants that is used in deep heating liniments applied to sore muscles and joints. It is also used as flavouring in some confectionery.

methyl salicylate

a) Identify the functional groups in this molecule. [3 marks]
b) From its name and structure what group of compounds does it belong to? [1 mark]
c) Deduce its molecular formula. [1 mark]
d) It is produced from salicylic acid and methanol. Draw the structure of salicylic acid. [1 mark]

I used to be in a band — we played 2,4,6-tritechno hiphopnoic acid jazz...

As well as recognising functional groups, this page gives you practice of a few other useful skills you'll need for your exam, e.g. interpreting different types of formula. If you're rusty on the difference between structural, molecular and displayed formulae, have a look back at the AS stuff – it's fundamental stuff that could trip you up in the exam if you don't know it.

Stereoisomerism and Chirality

Stereoisomers have the same molecular formula and their atoms are arranged in the same way. The only difference is the orientation of the bonds in space. There are two types of stereoisomerism — E/Z and optical.

***E/Z Isomerism** is **Stereoisomerism** Caused by **C=C** Bonds*

Alkenes can have stereoisomers because of the **lack of rotation** around the C=C double bond. If the two double-bonded carbon atoms each have **different atoms** or **groups** attached to them, then you get an '**E-isomer**' and a '**Z-isomer**'. For example, the double-bonded carbon atoms in but-2-ene each have an **H** and a $\mathbf{CH_3}$ group attached.

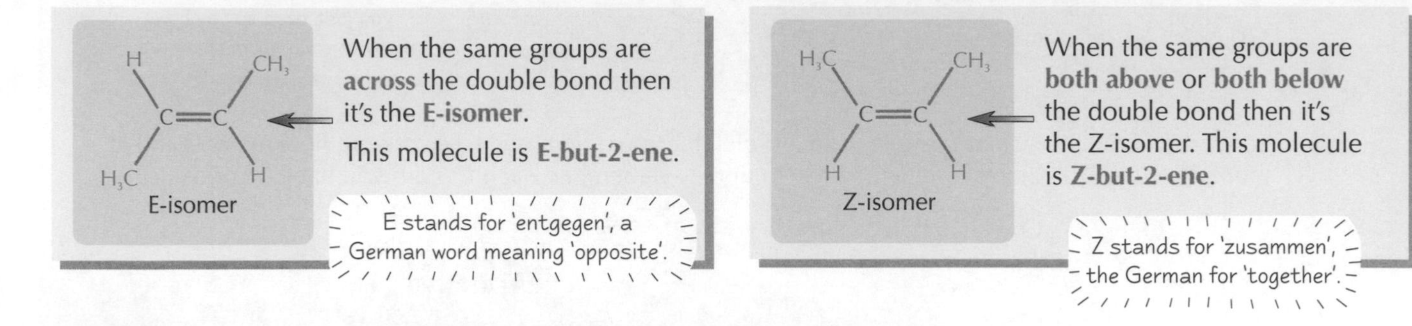

When the same groups are **across** the double bond then it's the **E-isomer**.

This molecule is **E-but-2-ene**.

E stands for 'entgegen', a German word meaning 'opposite'.

When the same groups are **both above** or **both below** the double bond then it's the Z-isomer. This molecule is **Z-but-2-ene**.

Z stands for 'zusammen', the German for 'together'.

***Optical Isomers** are **Mirror Images** of Each Other*

A **chiral** (or **asymmetric**) carbon atom is one which has **four different** groups attached to it. It's possible to arrange the groups in two different ways around the carbon atom so that two different molecules are made — these molecules are called **enantiomers** or **optical isomers**.

The enantiomers are **mirror images** and no matter which way you turn them, they can't be **superimposed.**

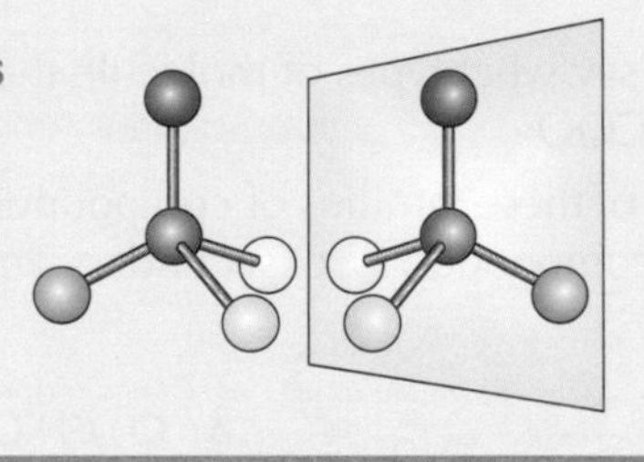

If the molecules can be superimposed, they're achiral — and there's no optical isomerism.

One enantiomer is usually labelled **D** and the other **L** — luckily you don't have to worry about which is which.

Chiral compounds are very common in nature, but you usually only find **one** of the enantiomers — for example, all naturally occurring amino acids are **L–amino acids** (except glycine which isn't chiral) and most sugars are **D-isomers**.

Optical isomers are optically active — they **rotate plane-polarised light**. One enantiomer rotates it in a **clockwise** direction, the other rotates it in an **anticlockwise** direction.

Normal light vibrates in all directions, but plane-polarised light only vibrates in one direction.

*Make Sure You Can Spot **Chiral Centres***

You have to be able to draw optical isomers. But first you have to identify the chiral centre...

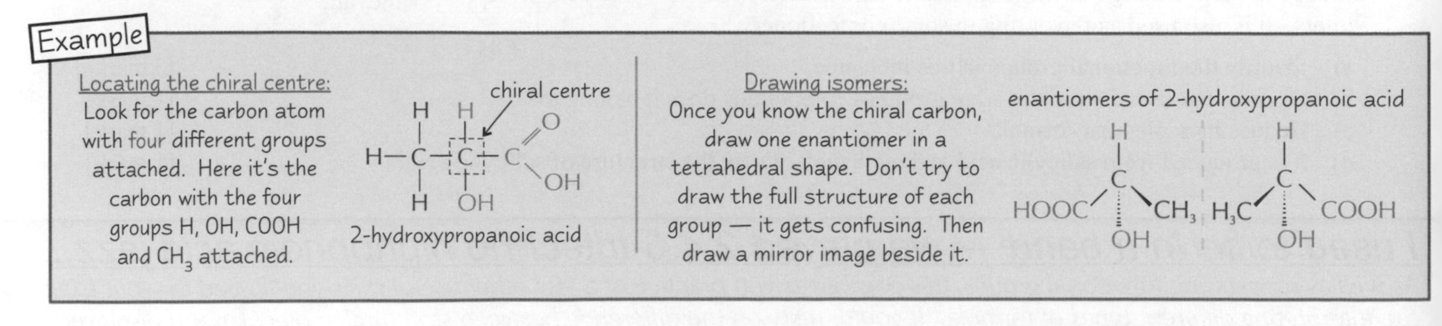

Locating the chiral centre:
Look for the carbon atom with four different groups attached. Here it's the carbon with the four groups H, OH, COOH and CH_3 attached.

Drawing isomers:
Once you know the chiral carbon, draw one enantiomer in a tetrahedral shape. Don't try to draw the full structure of each group — it gets confusing. Then draw a mirror image beside it.

Stereoisomerism and Chirality

Pharmaceutical Drugs Must Only Contain One Optical Isomer

1) Unlike naturally-occurring chiral molecules, those prepared in a lab tend to contain an **equal split** of enantiomers — this is called a **racemic** mixture. This creates problems when producing **pharmaceutical drugs**.
2) Drugs work by binding to **active sites** on **enzymes** or other **receptor molecules** in the body and **changing** chemical reactions. The drug must be the right **shape** to fit the active site — only **one enantiomer** will do. The other enantiomer might fit another active site, and may have **no effect** at all, or cause **harmful side-effects**.
3) So, usually, synthetic chiral drugs have to be made so that they only contain **one enantiomer**. This has the **benefit** that only half the dose is needed. It also reduces the risk of the drug companies being sued over side effects.
4) The problem is that optical isomers are very **tricky to separate** — producing single-enantiomer drugs is **expensive**.

Ethambutol, a drug used to treat TB, is produced as a single enantiomer because the other causes blindness.

The painkiller Ibuprofen is sold as a racemic mixture — the inactive enantiomer is harmless and the cost of separating the mixture is very high.

Methods for producing single-enantiomer drugs include (often in combination):

1) Using natural enzymes or bacteria in the process which tend to produce only one isomer.
2) Using naturally-occuring single optical isomer compounds as starting materials, e.g. sugars, amino acids.
3) Using chemical chiral synthesis — this basically involves using carefully chosen reagents and conditions which will ensure only one isomer is produced.

Chemical chiral synthesis methods usually rely on chemically modifying the reagent molecule in a way that physically blocks most approaches to it, so that it can only be 'attacked' from one side. For example, you could turn your reagent into a cyclic molecule, or bond your reagent molecules to a polymer support and let the other reactants flow over them.

4) Using chiral catalysts — these basically do the same job of only producing only one isomer, but have the advantage that only a small amount is needed because they're reused in the reaction. (Getting large quantities of single-enantiomer compounds for these reactions is expensive.)

Practice Questions

Q1 Explain how molecules with C=C bonds can produce stereoisomers — what are the two types called?

Q2 What is a chiral carbon atom? Draw a diagram of a molecule that contains a chiral carbon atom.

Q3 Give four methods that could be used to produce a single optical isomer for use as a pharmaceutical drug.

Exam Questions

1 a) Identify the chiral carbon atom in each of the three molecules shown on the right. [3 marks]

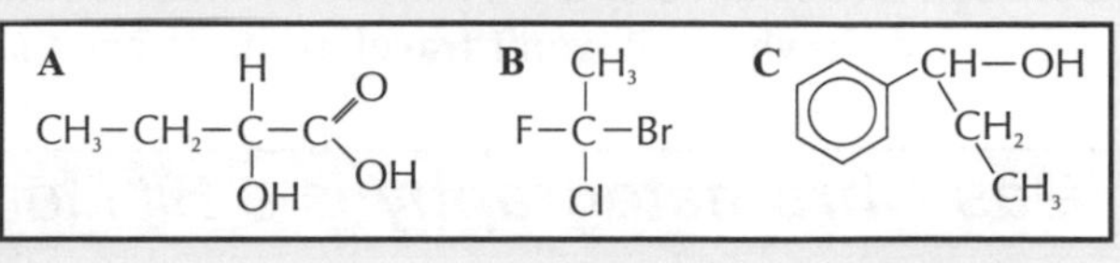

b) Draw the two optical isomers of molecule B. [2 marks]

c) Explain why these two isomers are said to be optically active. [1 mark]

2 Parkinson's disease involves a deficiency of dopamine. It is treated by giving patients L-DOPA (dihydroxyphenylalanine), a naturally occurring amino acid, which is converted to dopamine in the brain.

a) DOPA is a chiral molecule. Its structure is shown above on the right. Mark the structure's chiral centre. [1 mark]

b) A D,L-DOPA racemic mixture was synthesised in 1911, but today natural L-DOPA is isolated from fava beans for use as a pharmaceutical.

i) Suggest two reasons why L-DOPA is used in preference to the D,L-DOPA mixture. [2 marks]

ii) Explain the meaning of the term 'racemic mixture'. [1 mark]

iii) Why is it potentially dangerous to use the racemic mixture as a pharmaceutical drug? [1 mark]

I isolated it from some fava beans and a nice Chianti...

This isomer stuff's not all bad — you get to draw little pretty pictures of molecules. If you're having difficulty picturing them as 3D shapes, you could always make models with Blu-tack® and those matchsticks that're propping your eyelids open. Blu-tack®'s very therapeutic anyway and squishing it about'll help relieve all that revision stress. It's great stuff.

Chromatography

You've probably tried chromatography with a spot of ink on a piece of filter paper — it's a classic experiment.

Chromatography is Good for **Separating** and **Identifying** Things

Chromatography is used to **separate** stuff in a mixture — once it's separated out, you can often **identify** the components. There are quite a few different types of chromatography — you might have tried paper chromatography — but the ones you need to know about are **thin-layer chromatography** (TLC) and **gas chromatography** (GC).
They both have two phases:

1) A **mobile phase** — where the molecules can move. This is always a liquid or a gas.
2) A **stationary phase** — where the molecules can't move. This must be a solid, or a liquid on a solid support.

The mobile phase **moves through** the stationary phase. As this happens, the components in the mixture **separate out** between the phases.

Thin-Layer Chromatography Separates Components by ***Adsorption***

In **thin-layer chromatography** (TLC):

1) The **mobile phase** is a **solvent**, such as ethanol, which passes over the stationary phase.
2) The **stationary phase** is a **thin layer of solid** (0.1-0.3 mm), e.g. silica gel or alumina powder, on a **glass or plastic plate**.

Here's the method:

1) Draw a **pencil line** near the bottom of the plate and put a **spot** of the mixture to be separated on the line.
2) Dip the bottom of the plate (not the spot) into a **solvent**.
3) As the solvent spreads up the plate, the different substances in the mixture move with it, but at **different rates** — so they separate out.
4) When the solvent's **nearly** reached the top, take the plate out and **mark** the distance that the solvent has moved (**solvent front**) in pencil.
5) You can work out what was in the mixture by calculating an R_f **value** for each spot and looking them up in a **table of known values**.

solvent front
chromatography plate
spot of unknown substance
point of origin
solvent

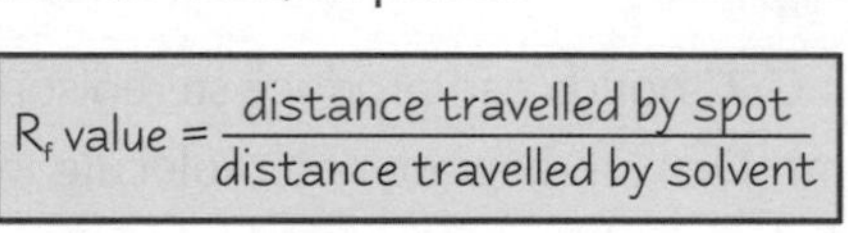
$$R_f \text{ value} = \frac{\text{distance travelled by spot}}{\text{distance travelled by solvent}}$$

The stationary phase and solvent used will affect the R_f value.

How far each part of the mixture travels depends on **how strongly** it's **attracted** to the stationary phase. The **attraction** between a substance and the surface of the stationary phase is called **adsorption**. A substance that is **strongly adsorbed** will move **slowly**, so it **won't travel as far** as one that's only **weakly adsorbed**. This means it will have a **different** R_f **value**.

Gas Chromatography is a Bit More ***High-Tech***

1) In **gas chromatography** (GC) the stationary phase is a **viscous liquid**, such as an oil, or a **solid**, which coats the inside of a long tube. The tube's **coiled** to save space, and built into an oven.
2) The mobile phase is an **unreactive carrier gas** such as nitrogen or helium.
3) The **sample** to be analysed is **injected** into the stream of carrier gas, which carries it through the **tube** and over the stationary phase.

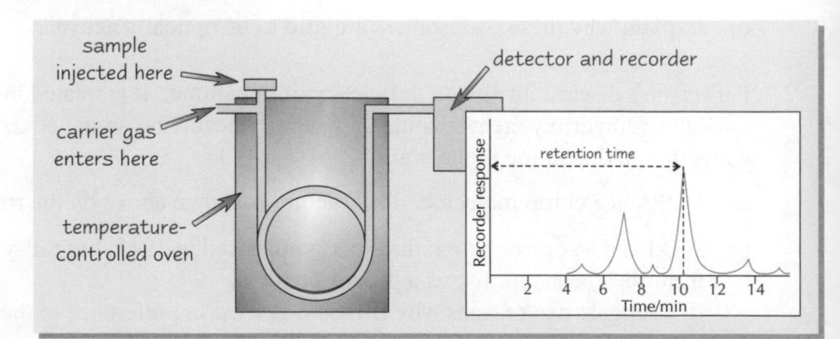

4) The components of the mixture constantly **dissolve in the stationary phase**, **evaporate** into the mobile phase and then **redissolve** as they travel through the tube.
5) The **solubility** of each component of the mixture determines **how long** it spends **dissolved in the stationary phase** and how long it spends **moving along** the tube in the **mobile phase**. A substance with a high solubility will spend more time dissolved, so will take longer to travel through the tube to the detector than one with a lower solubility. The time taken to reach the detector is called the **retention time**. It can be used to help **identify** the substances.

Chromatography

GC Chromatograms Show the Proportions of the Components in a Mixture

A **gas chromatogram** shows a **series of peaks** at the times when the **detector** senses something other than the carrier gas **leaving the tube**. They can be used to **identify the substances** within a sample and their **relative proportions**.

1) Each **peak** on a chromatogram corresponds to a substance with a particular **retention time**.
2) **Retention times** are measured from **zero** to the **centre** of each peak, and can be looked up in a **reference table** to **identify** the **substances** present.
3) The **area** under each peak is proportional to the relative **amount of each substance** in the original mixture. Remember, it's **area**, not height, that's important — the **tallest** peak on the chromatogram **won't always** represent the **most abundant substance**.

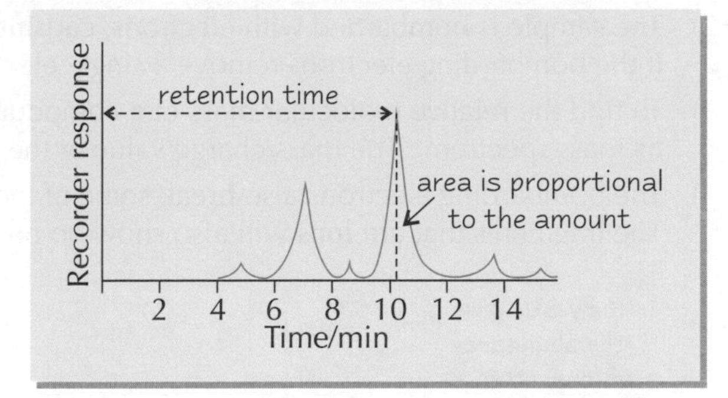

Gas Chromatography has Limitations

Although a very useful and widely used technique, GC does have **limitations** when it comes to identifying chemicals.

1) **Compounds** which are **similar** often have **very similar retention times**, so they're difficult to identify accurately. A mixture of **two similar substances** may only produce **one peak** so you **can't tell how much** of each one there is.
   Handily, you can combine GC with mass spectrometry to make a much more powerful identification tool — there's more on this on page 135.
2) You can only use GC to identify substances that you already have **reliable reference retention times** for. (That means someone must have run a sample of the same pure substance under exactly the same conditions before.)

Practice Questions

Q1 Explain the terms 'stationary phase' and 'mobile phase' in the context of chromatography.

Q2 What is the stationary phase in TLC?

Q3 What is the mobile phase in GC?

Q4 Describe how you would calculate the R_f value of a substance on a TLC plate.

Exam Questions

1 Look at this diagram of a chromatogram produced using TLC on a mixture of substances A and B.

Solvent front; A; B; Initial spot; 8 cm; 7 cm

a) Calculate the R_f value of spot A. [2 marks]

b) Explain why substance A has moved further up the plate than substance B. [3 marks]

2 A scientist has a mixture of several organic chemicals. He wants to know if it contains any hexene. He runs a sample of pure hexene through a GC machine and finds that its retention time is 5 minutes. Then he runs a sample of his mixture through the same machine, under the same conditions, and produces the chromatogram shown on the right.

Recorder response; 2 4 6 8 10 12 14; Time/min

a) What feature of the chromatogram suggests that the sample contains hexene? [1 mark]

b) Give a reason why the researcher may still not be absolutely certain that his mixture contains hexene. [1 mark]

3 A mixture of 25% ethanol and 75% benzene is run through a GC apparatus.

a) Describe what happens to the mixture in the apparatus. [4 marks]

b) Explain why the substances separate. [2 marks]

c) How will the resulting chromatogram show the proportions of ethanol and benzene present in the mixture? [1 mark]

A little bit of TLC is what you need...

If you only remember one thing about chromatography, remember that it's really good at separating mixtures, but not so reliable at identifying the substances that make up the mixture. Or does that count as two things? Hmm... well it's probably not the best idea to only learn one thing from each page anyway. Learn lots of stuff, that's my advice.

Mass Spectrometry and Chromatography

Mass spectrometry is an analysis technique that can be used with chromatography to positively identify compounds.

Mass Spectrometry Can Help to Identify Compounds

1) A mass spectrum is produced when a sample of a **gaseous compound** is analysed in a mass spectrometer.
2) The sample is bombarded with electrons, causing other electrons to break off from the molecules. If the bombarding electrons remove a single electron from a molecule, the **molecular ion**, $M^+_{(g)}$, is formed.
3) To find the relative molecular mass of a compound you can look at the **molecular ion peak** (the **M peak**) on its mass spectrum. The mass/charge value of the molecular ion peak is the **molecular mass** of the compound.
4) The bombarding electrons also break some of the molecules up into **fragments**. The fragments that are **ions** will also show up on the mass spectrum, giving a **fragmentation pattern**.

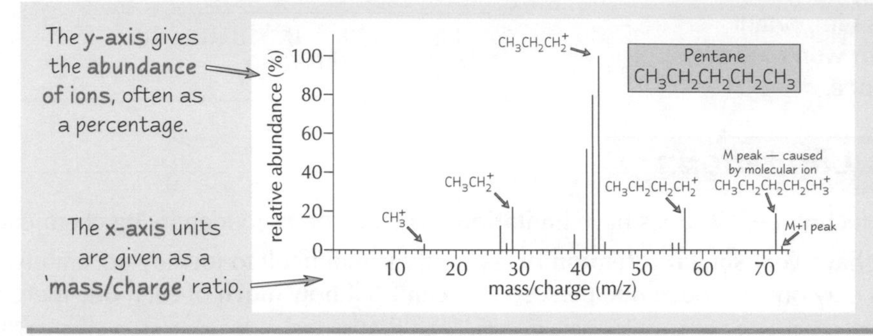

Here's the mass spectrum of **pentane**.

Each **peak** on the mass spectrum is caused by a different **fragment** ion. The **M peak** is at 72 — so the compound's M_r must be 72.

For most **organic compounds** the M peak is the one with the second highest mass/charge ratio. (The smaller peak to the right of the M peak is the **M+1 peak** — it's caused by the presence of the carbon isotope ^{13}C.)

You Can Join the Fragments Together to Find the Molecule's Structure

Fragmentation patterns are really useful — you can use them to identify **molecules** and work out their **structures**.

For **propane**, the molecular ion is $CH_3CH_2CH_3^+$. The fragments it breaks into include $\mathbf{CH_3^+}$ (M_r = 15) and $\mathbf{CH_3CH_2^+}$ (M_r = 29).

Only the **ions** will show up on the mass spectrum — the **free radicals** are 'lost'.

$CH_3CH_2CH_3^+ \rightarrow CH_3CH_2\bullet$ (free radical) $+ CH_3^+$ (ion)

$CH_3CH_2CH_3^+ \rightarrow CH_3CH_2^+$ (ion) $+ \bullet CH_3$ (free radical)

To work out the structural formula, you've got to work out what **ion** could have made each peak from its **m/z value**. (You assume that the m/z value of a peak matches the **mass** of the ion that made it.)

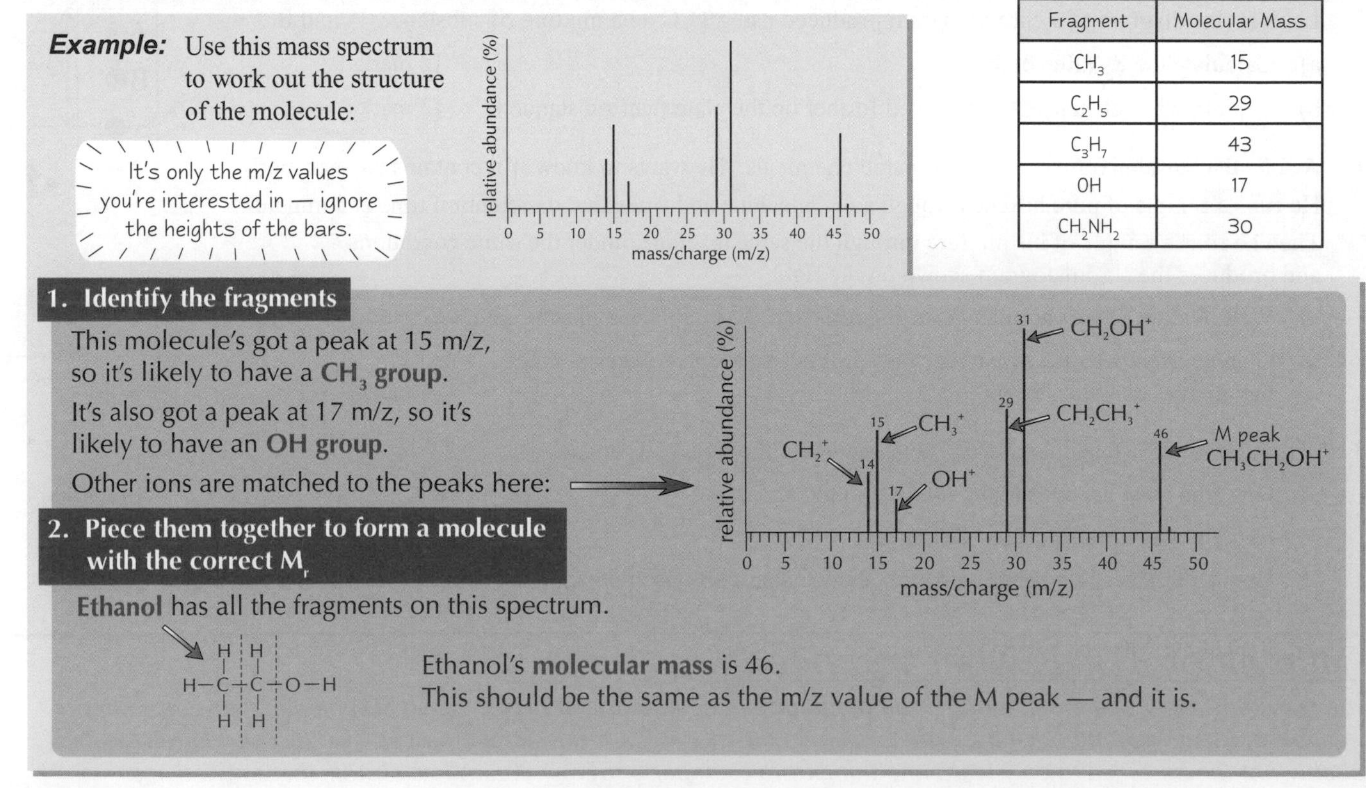

Example: Use this mass spectrum to work out the structure of the molecule:

It's only the m/z values you're interested in — ignore the heights of the bars.

Fragment	Molecular Mass
CH_3	15
C_2H_5	29
C_3H_7	43
OH	17
CH_2NH_2	30

1. Identify the fragments

This molecule's got a peak at 15 m/z, so it's likely to have a **CH_3 group**.

It's also got a peak at 17 m/z, so it's likely to have an **OH group**.

Other ions are matched to the peaks here:

2. Piece them together to form a molecule with the correct M_r

Ethanol has all the fragments on this spectrum.

Ethanol's **molecular mass** is 46.
This should be the same as the m/z value of the M peak — and it is.

Mass Spectrometry and Chromatography

Mass Spectrometry is Used to Differentiate Between Similar Molecules

Even if two **different compounds** contain **the same atoms**, you can still tell them apart with mass spectrometry because they won't produce exactly the same set of fragments.

Example: The mass spectra of **propanal** and **propanone**:

1) **Propanal** and **propanone** have the same empirical formula — C_3H_6O. So they also have the **same** M_r **(58)**.
2) But they have different molecular **structures** — one is an **aldehyde**, and the other's a **ketone**.

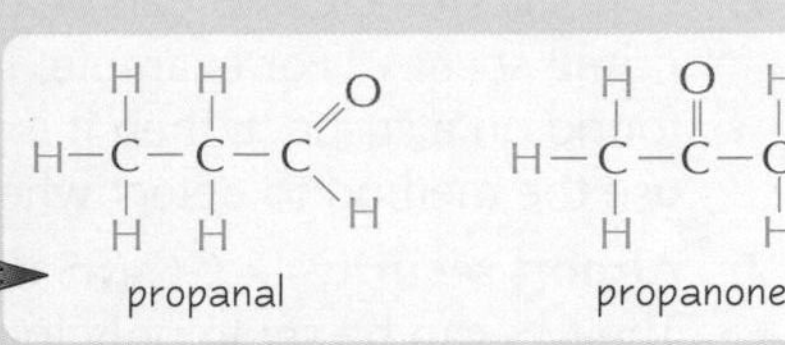

3) Because their structures are different, they'll break up into **different fragments** in a mass spectrometer. So the **mass spectrum** of propanal is different from the mass spectrum of propanone.

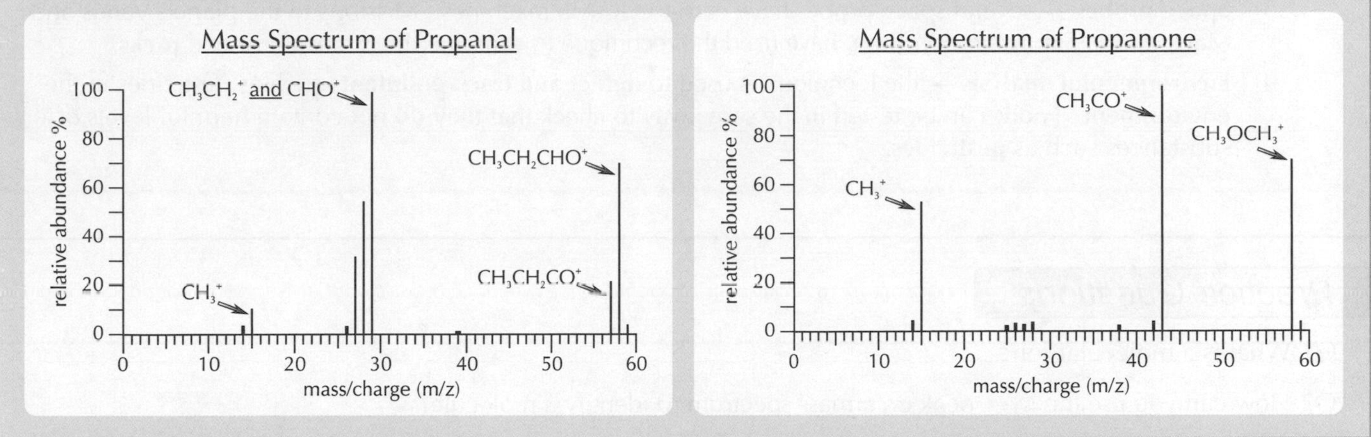

Every compound produces a different mass spectrum — so the spectrum is like a **fingerprint** for that compound. Large computer **databases** of mass spectra can be used to identify a sample of a compound from its spectrum.

Mass Spectrometry can be Combined with Gas Chromatography

Gas chromatography (see pages 132-133) is very good at **separating** a mixture into its individual components, but not so good at identifying those components. **Mass spectrometry**, on the other hand, is very good at **identifying** unknown compounds, but would give confusing results from a mixture of substances.
If you put these **two techniques together**, you get an **extremely useful** analytical tool.

Gas chromatography-mass spectrometry (or GC-MS for short) **combines the benefits** of gas chromatography and mass spectrometry to make a super analysis tool.
The sample is **separated** using **gas chromatography**, but instead of going to a detector, the separated components are fed into a **mass spectrometer**.
The spectrometer produces a **mass spectrum** for each **component**, which can be used to **identify** each one and show what the original **sample** consisted of.

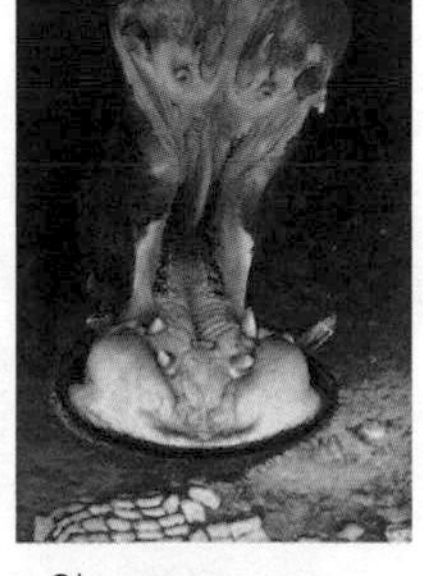
Oh, excuse me — chemistry always sends me to sleep, I'm afraid

The **advantage** of this method over normal GC is that the components separated out by the chromatography can be **positively identified**, which can be impossible from a chromatogram alone.
Computers can be used to match up the **mass spectrum** for each component of the mixture against a **database**, so the whole process can be **automated**.

You can also combine **high pressure liquid chromatography**, or **HPLC**, with **mass spectrometry** to get **HPLC-MS**.

1) In **HPLC**, the **stationary phase** is a **solid** that is packed into a glass **column**, like tiny silica beads.
2) The **mobile phase** (a solvent) and the **mixture** are **pushed** through the column under **high pressure**. This allows the separation to happen much **faster** than if the solvent just dripped through.
3) As with GC, HPLC is more useful for **separating** mixtures of substances than **identifying** them — **combining it** with **mass spectrometry** gives a better **identification** tool than either method alone.

Mass Spectrometry and Chromatography

GC-MS is used in Forensics and Security

GC-MS is a really **important analytical tool**, and not just in chemistry labs — check out the four uses below.

1) **Forensics** — GC-MS can be used to **identify unknown substances** found on **victims** or **suspects** or at **crime scenes**. For example, if GC-MS shows that a substance found at a crime scene is **identical** to one found on a suspect, then it is evidence that the suspect was at the crime scene. Or **fire investigators** can use the method to detect whether fires were started **deliberately** using substances such as petrol or paraffin.
2) **Airport security** — GC-MS can be used to look for **specific substances** — e.g. **explosives** or **illegal drugs**. The MS can be set to only look at a substance produced at a particular retention time on the GC to find out if it is present or not. The whole process is quick — it takes just a few minutes — and is accurate enough to be used in **court** as evidence.
3) **Space probes** — several space probes have carried GC-MS machines. Missions to the planets Venus and Mars, and to Saturn's moon Titan, have used the technique to examine the **atmosphere** and **rocks**.
4) **Environmental analysis** — the technique is used to **detect and track pollutants** such as pesticides in the environment. **Foods** can be tested in the same way to check that they do not contain harmful levels of substances such as **pesticides**.

Practice Questions

Q1 What is a molecular ion?

Q2 How can you use the M+1 peak on a mass spectrum to identify a molecule?

Q3 What pattern of peaks on a mass spectrum tell you that chlorine is present?

Q4 Give two uses of GC-MS.

Exam Questions

1 Below is the mass spectrum of a carboxylic acid. Use the spectrum to answer this question.

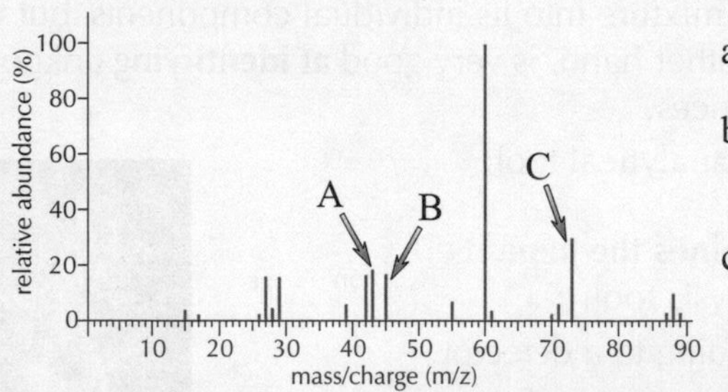

a) What is the molecular mass of this acid? [1 mark]

b) Suggest the formulae of the fragment ions that are responsible for the peaks labelled A, B and C. [3 marks]

c) Use your answers from parts (a) and (b) to draw the structure of the acid, and give its name. [2 marks]

2 The Huygens probe sent to Titan, the giant moon of Saturn, carried a GC-MS machine to examine the atmosphere. The atmosphere surrounding Titan consists of 98.4% nitrogen with the rest mainly methane, and trace amounts of many other gases including several hydrocarbons.

a) One of the gases found in the atmosphere is ethane.
Why is a GC-MS machine necessary to verify the presence of ethane, rather than just GC? [2 marks]

b) Give the formula and the mass of the molecular ion that ethane produces. [2 marks]

c) The mass spectrum of ethane also has peaks at m/z = 29 and m/z = 15.
i) Which fragment ion produces the peak at m/z = 29? [1 mark]
ii) Which fragment ion produces the peak at m/z = 15? [1 mark]

Mass spectrometry — weight watching for molecules...

So mass spectrometry's a bit like weighing yourself, then taking bits off your body, weighing them separately, then trying to work out how they all fit together. Luckily you won't get anything as complicated as a body, and you won't need to cut yourself up either. Good news all round then. Just learn this page and watch out for the M peak in the exam.

NMR Spectroscopy

NMR isn't the easiest of things, so ingest this information one piece at a time — a bit like eating a bar of chocolate.

NMR Gives You Information about the Structure of Molecules

Nuclear magnetic resonance (**NMR**) **spectroscopy** is an analysis technique that you can use to work out the **structure** of an organic molecule. The way that NMR works is pretty **complicated**, but you only need to know the **basics**:

1) A sample of a compound is placed in a **strong magnetic field** and exposed to a range of different **frequencies** of **low-energy radio waves**.
2) The **nuclei** of certain atoms within the molecule **absorb energy** from the radio waves.
3) The amount of energy that a nucleus absorbs at each frequency will depend on the **environment** that it's in — there's more about this further down the page.
4) The **pattern** of these absorptions gives you information about the **positions** of certain atoms within the molecule, and about **how many** atoms of that type the molecule contains.
5) You can piece these bits of information together to work out the **structure of the molecule**.

The two types of NMR spectroscopy you need to know about are **carbon-13** (or **^{13}C**) **NMR** and **high resolution proton NMR**.

Carbon-13 NMR gives you information about the **number of carbon atoms** that are in a molecule, and the **environments** that they are in.

High resolution proton NMR gives you information about the **number of hydrogen atoms** that are in a molecule, and the **environments** that they're in.

Nuclei in Different Environments Absorb Different Amounts of Energy

1) A nucleus is partly **shielded** from the effects of external magnetic fields by its **surrounding electrons**.
2) Any **other atoms** and **groups of atoms** that are around a nucleus will also affect its amount of electron shielding.
   E.g. If a carbon atom bonds to a more electronegative atom (like oxygen) the amount of electron shielding around its nucleus will decrease.
3) This means that the nuclei in a molecule feel different magnetic fields depending on their **environments**. This means that they will absorb **different amounts** of energy at **different frequencies**.
4) It's these **differences in absorption** of energy between environments that you're looking for in **NMR spectroscopy**.
5) An atom's **environment** depends on **all** the groups that it's connected to, going **right along the molecule** — not just the atoms it's actually bonded to. To be in the **same environment**, two atoms must be joined to **exactly the same things**.

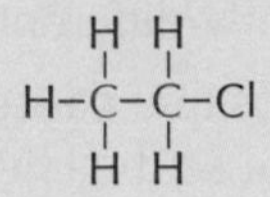

```
 H H
 | |
H-C-C-Cl
 | |
 H H
```

Chloroethane has 2 carbon environments — its carbons are bonded to different atoms.

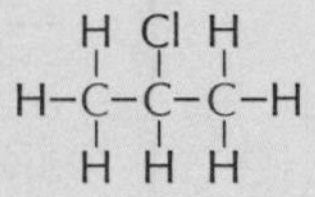

```
 H Cl H
 |  | |
H-C-C-C-H
 |  | |
 H  H H
```

2-chloropropane has 2 carbon environments:
- 1 C in a CHCl group, bonded to $(CH_3)_2$
- 2 Cs in CH_3 groups, bonded to $CHCl(CH_3)$

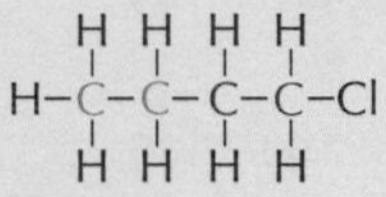

```
 H H H H
 | | | |
H-C-C-C-C-Cl
 | | | |
 H H H H
```

1-chlorobutane has 4 carbon environments. (The two carbons in CH_2 groups are different distances from the electronegative Cl atom — so their environments are different.)

Tetramethylsilane is Used as a Standard

The diagram below shows a typical **carbon-13 NMR spectrum**. The **peaks** show the **frequencies** at which **energy was absorbed** by the carbon nuclei. **Each peak** represents one **carbon environment** — so this molecule has two.

1) The **differences in absorption** are measured relative to a **standard substance** — **tetramethylsilane** (**TMS**).
2) TMS produces a **single absorption peak** in both types of NMR because all its carbon and hydrogen nuclei are in the **same environment.**
3) It's chosen as a standard because the **absorption peak** is at a **lower frequency** than just about everything else.
4) This peak is given a value of **0** and all the peaks in other substances are measured as **chemical shifts** relative to this.

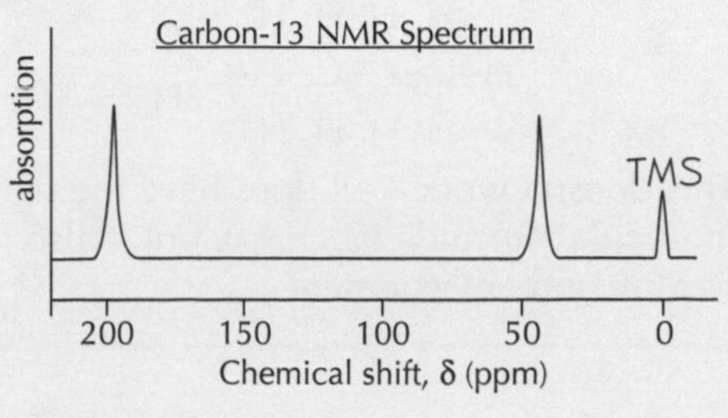

Carbon-13 NMR Spectrum

absorption
TMS
200 150 100 50 0
Chemical shift, δ (ppm)

Chemical shift is the **difference in the radio frequency** absorbed by the nuclei (hydrogen or carbon) in the molecule being analysed and that absorbed by the same nuclei in **TMS**. They're given the symbol **δ** and are measured in **parts per million**, or **ppm**. A small amount of TMS is often added to samples to give a **reference peak** on the spectrum.

NMR Spectroscopy

^{13}C NMR Spectra Tell You About Carbon Environments

It's very likely that you'll be given an **NMR spectrum** to **interpret** in your exam. It might be a **carbon-13 NMR spectrum** or a **proton NMR spectrum** (you might even see both if you're really lucky), so you need to have both kinds sussed. First, here's a **step-by-step guide** to interpreting carbon-13 spectra.

1) Count the Number of Carbon Environments

First, count the **number of peaks** in the spectrum — this is the **number of carbon environments** in the molecule. If there's a peak at **δ = 0**, **don't count it** — it's the reference peak from **TMS**.

The spectrum on the right has **three peaks** — so the molecule must have **three different carbon environments**. This **doesn't** necessarily mean it only has **three carbons**, as it could have **more than one** in the **same environment**. In fact the molecular formula of this molecule is $\mathbf{C_5H_{10}O}$, so it must have **several carbons** in the **same environment**.

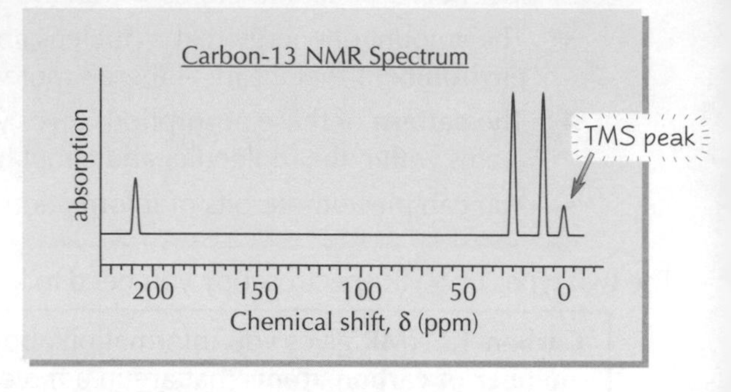

2) Look Up the Chemical Shifts in a Data Table

^{13}C NMR Chemical Shifts Relative to TMS	
Chemical shift, δ (ppm)	**Type of Carbon**
5 – 55	C – C
30 – 70	C – Cl or C – Br
35 – 60	C – N (amines)
50 – 70	C – O
115 – 140	C = C (alkenes)
110 – 165	aromatic
160 – 185	carbonyl (ester, amide, or carboxylic acid)
190 – 220	carbonyl (ketone or aldehyde)

In your exam you'll get a **data sheet** that will include a **table** like this one. The table shows the **chemical shifts** experienced by **carbon nuclei** in **different environments**.

You need to **match up** the **peaks** in the spectrum with the **chemical shifts** in the table to work out which **carbon environments** they could represent.

For example, the peak at **δ ≈ 10** in the spectrum above represents a **C–C** bond. The peak at **δ ≈ 25** is also due to a **C–C** bond. The carbons causing this peak have a different chemical shift to those causing the first peak — so they must be in a slightly different environment.

Matching peaks to the groups that cause them isn't always straightforward, because the chemical shifts can **overlap**. For example, a peak at **δ ≈ 30** might be caused by **C–C**, **C–Cl** or **C–Br**.

The peak at **δ ≈ 210**, is due to a **C=O** group in an **aldehyde** or a **ketone** — but you **don't** know which.

3) Try Out Possible Structures

An **aldehyde** with 5 carbons:

H–C–C–C–C–CHO (each of the four C's bonded to two H's)

This doesn't work — it does have the right **molecular formula** ($C_5H_{10}O$), but it also has **five carbon environments**.

A **ketone** with five carbons:

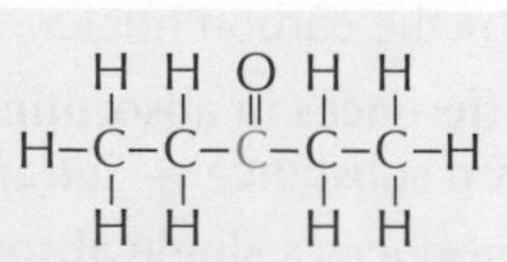

This works. **Pentan-3-one** has **three** carbon environments — two $\mathbf{CH_3}$ carbons, each bonded to $CH_2COCH_2CH_3$, two $\mathbf{CH_2}$ carbons, each bonded to CH_3 and $COCH_2CH_3$, and one **CO** carbon bonded to $(CH_2CH_3)_2$. It has the right **molecular formula** ($C_5H_{10}O$) too.

So, the molecule analysed was **pentan-3-one**.

It can't be pentan-2-one — that has 5 carbon environments.

NMR Spectroscopy

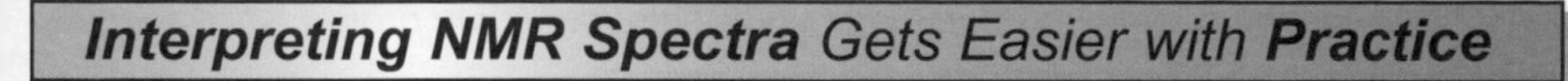

Interpreting NMR Spectra Gets Easier with *Practice*

EXAMPLE

The diagram shows the carbon-13 NMR spectrum of an alcohol with the molecular formula $C_4H_{10}O$. Analyse and interpret the spectrum to identify the structure of the alcohol.

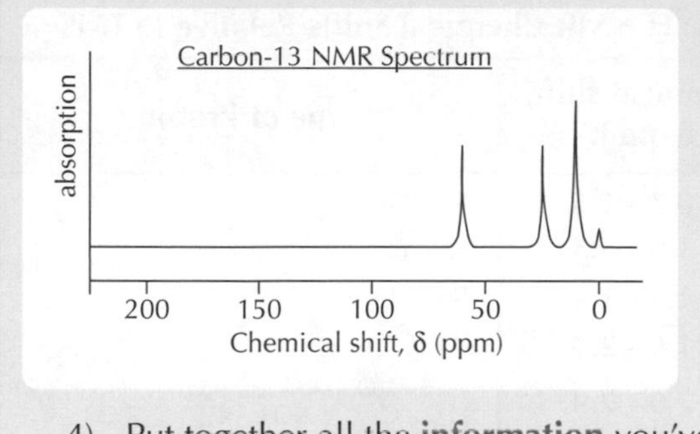

1) Looking at the **table** on the **previous page**, the peak with a **chemical shift** of **δ ≈ 65** is likely to be due to a **C–O** bond.
2) The two peaks around **δ ≈ 20** probably both represent carbons in **C–C** bonds, but with slightly different environments. Remember the alcohol doesn't contain any **chlorine**, **bromine** or **nitrogen** so you can **ignore** those entries in the table.
3) The spectrum has **three peaks**, so the alcohol must have three **carbon environments**. There are **four carbons** in the alcohol, so two of the carbons must be in the **same environment**.
4) Put together all the **information** you've got so far, and try out some **structures**:

This has a C–O bond and some C–C bonds, which is right. But all four carbons are in different environments.

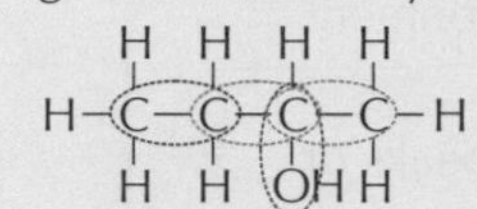

Again, this one has the right types of bond. But, the carbons are still all in different environments.

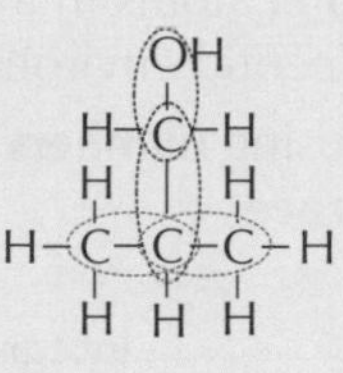

This molecule has C–O bonds and C–C bonds and two of the carbons are in the same environment. So this must be the correct structure.

MRI Scanners Use *NMR Technology*

Magnetic resonance imaging (**MRI**) is a **scanning** technique that's used in **hospitals** to study the **internal structures** of the body. MRI uses the **same technology** as NMR spectroscopy — the patient is placed inside a very **large magnet** and **radio waves** are directed at the area of the body being investigated. **Hydrogen nuclei** in **water molecules** in the body **absorb** energy from the radio waves at certain frequencies. The **frequency** depends on the kind of **tissue** that the water is in, so an **image** of the different tissues can be built up.

The **benefit** of MRI is that it **doesn't** use **damaging radiation** like **X-rays** or gamma rays, but does give **high quality images** of soft tissue like the **brain**. The technique is used to diagnose and monitor **cancerous tumours**, examine **bones** and **joints** for signs of injury, and to study the **brain** and **cardiovascular system**.

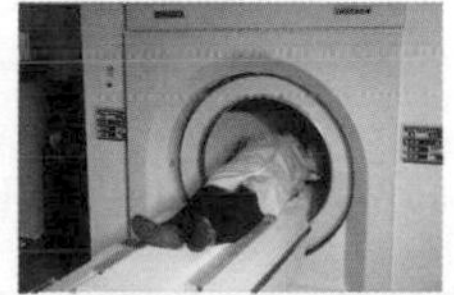

Never stick your head into a giant washing machine.

Practice Questions

Q1 What part of the electromagnetic spectrum does NMR spectroscopy use?

Q2 What is meant by chemical shift? What compound is used as a reference for chemical shifts?

Q3 How can you tell from a carbon-13 NMR spectrum how many carbon environments a molecule contains?

Q4 What is the medical scanning technique that uses the same technology as NMR?

Exam Question

1 The carbon-13 NMR spectrum shown on the right was produced by a compound with the molecular formula C_3H_9N.

a) Explain why there is a peak at δ = 0. [1 mark]

b) The compound does not have the formula $CH_3CH_2CH_2NH_2$. Explain how the spectrum shows this. [2 marks]

c) Suggest and explain a possible structure for the compound. [4 marks]

Carbon-13 NMR Spectrum
absorption
200 150 100 50 0
Chemical shift, δ (ppm)

Why did the carbon peak? Because it saw the radio wave...

The ideas behind NMR are difficult, but don't worry too much if you don't really understand them. The important thing is to make sure you know how to interpret a spectrum — that's what will get you marks in the exam. If you're having trouble, go over the examples and practice questions a few more times. You should have the "ahh... I get it" moment sooner or later.

More NMR Spectroscopy

So, you know how to interpret carbon-13 NMR spectra — now it's time to get your teeth into some proton NMR spectra.

^{1}H NMR Spectra Tell You About Hydrogen Environments

Interpreting **proton NMR spectra** is similar to interpreting carbon-13 NMR spectra:

1) Each peak represents one **hydrogen environment**.

For example, 1-chloropropane has 3 hydrogen environments.

2) Look up the **chemical shifts** in a **data table** to identify possible environments. They're different from ^{13}C NMR, so make sure you're looking at the **correct data table**.
3) In ^{1}H NMR, the **relative area** under each peak tells you the relative number of H atoms in each environment. For example, if the area under two peaks is in the **ratio** 2:1, there will be **two** H atoms in the first environment for every **one** in the second environment.
4) Areas can be shown using **numbers** above the peaks or with an **integration trace**:

Relative area under peaks — Integration trace — Absorption — Chemical shift, δ (ppm)

The integration trace is the red line.

The height increases are proportional to the area under each peak.

^{1}H NMR Chemical Shifts Relative to TMS

Chemical shift, δ (ppm)	Type of Proton		
0.7 – 1.6	$R-CH_3$		
1.0 – 5.5	N – H	R – OH	
1.2 – 1.4	$R-CH_2-R$		
1.6 – 2.0	R_3CH		
2.0 – 2.9	$H_3C-C(=O)$	$RCH_2-C(=O)$	$R_2CH-C(=O)$
2.3 – 2.7	$C_6H_5-CH_3$	$C_6H_5-CH_2R$	$C_6H_5-CHR_2$
2.3 – 2.9	$N-CH_3$	$N-CH_2R$	$N-CHR_2$
3.3 – 4.3	$O-CH_3$	$O-CH_2R$	$O-CHR_2$
3.0 – 4.2	$Br/Cl-CH_3$	$Br/Cl-CH_2R$	$Br/Cl-CHR_2$
4.5 – 10.0	C_6H_5-OH		
4.5 – 6.0	$-CH=CH-$		
5.0 – 12.0	$-C(=O)NH_2$	$-C(=O)NH-$	
6.5 – 8.0	C_6H_5-H		
9.0 – 10.0	$-C(=O)H$		
11.0 – 12.0	$-C(=O)O-H$		

The big difference between carbon-13 NMR and proton NMR spectra is that the peaks in a spectrum **split** according to how the **hydrogen environments are arranged**. Putting all this info together should let you work out the structure.

Spin-Spin Coupling Splits the Peaks in a Proton NMR Spectrum

In a proton NMR spectrum, a peak that represents a hydrogen environment can be **split**. The splitting is caused by the influence of hydrogen atoms that are bonded to **neighbouring carbons**. This effect is called **spin-spin coupling**. Only hydrogen nuclei on **adjacent** carbon atoms affect each other.

These **split peaks** are called **multiplets**. They always split into one more than the number of hydrogens on the neighbouring carbon atoms — it's called the **n + 1 rule**. For example, if there are **2 hydrogens** on the adjacent carbon atoms, the peak will be split into 2 + 1 = 3.

You can work out the **number** of **neighbouring hydrogens** by looking at how many the peak splits into:
- If a peak's split into **two** (a **doublet**) then there's **one hydrogen** on the neighbouring carbon atoms.
- If a peak's split into **three** (a **triplet**) then there are **two hydrogens** on the neighbouring carbon atoms.
- If a peak's split into **four** (a **quartet**) then there are **three hydrogens** on the neighbouring carbon atoms.

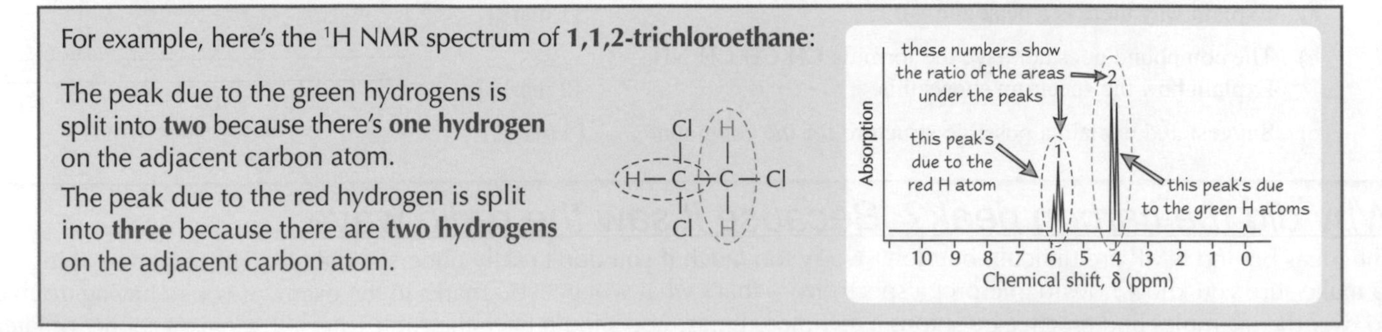
For example, here's the ^{1}H NMR spectrum of **1,1,2-trichloroethane**:

The peak due to the green hydrogens is split into **two** because there's **one hydrogen** on the adjacent carbon atom.

The peak due to the red hydrogen is split into **three** because there are **two hydrogens** on the adjacent carbon atom.

More NMR Spectroscopy

Deuterated Solvents are used in Proton NMR Spectroscopy

NMR spectra are recorded with the molecule that is being analysed in **solution**. But if you used a ordinary solvent like water or ethanol, the **hydrogen nuclei** in the solvent would **add peaks** to the spectrum and confuse things. To overcome this, the **hydrogen nuclei** in the solvent are **replaced** with **deuterium** (D) — an **isotope** of hydrogen with **one proton** and **one neutron**. Deuterium nuclei don't absorb the radio wave energy, so they don't add peaks to the spectrum. A commonly used example of a '**deuterated solvent**' is **deuterated chloroform, $CDCl_3$**.

OH and NH Protons can be Identified by Proton Exchange Using D_2O

The **chemical shift** due to protons attached to oxygen (OH) or nitrogen (NH) is very **variable** — check out the huge **ranges** given in the **table** on the previous page. They make quite a **broad** peak that isn't usually split.

Don't panic, though, as there's a clever little trick chemists use to identify OH and NH protons:

1) Run **two** spectra of the molecule — one with a little **deuterium oxide**, D_2O, added.
2) If an OH or NH proton is present it'll swap with deuterium and, hey presto, the peak will **disappear**. (This is because deuterium doesn't absorb the radio wave energy).

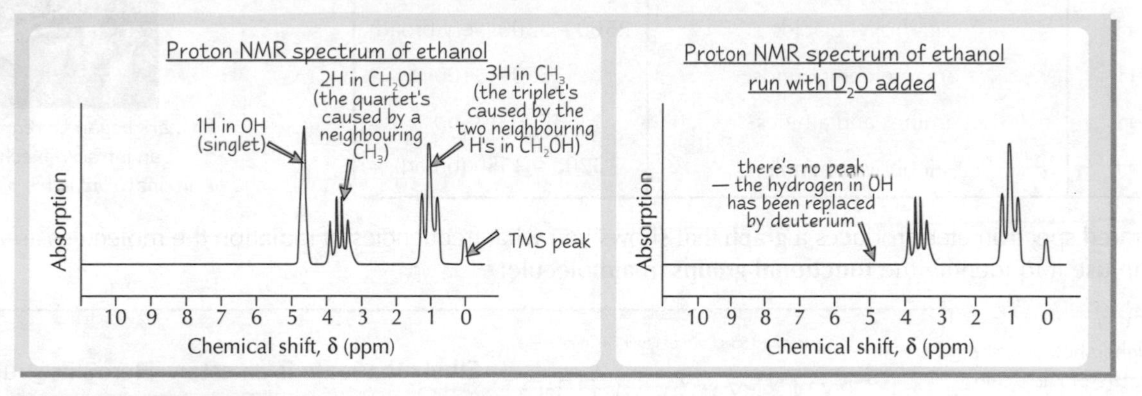

Practice Questions

Q1 What causes the peaks on a high resolution proton NMR spectrum to split?

Q2 What causes a triplet of peaks on a high resolution proton NMR spectrum?

Q3 What are deuterated solvents? Why are they needed?

Q4 How can you get rid of a peak caused by an OH group?

Exam Question

1 The proton NMR spectrum below is for an alkyl halide. Use the table of chemical shifts on page 42 to answer this question.

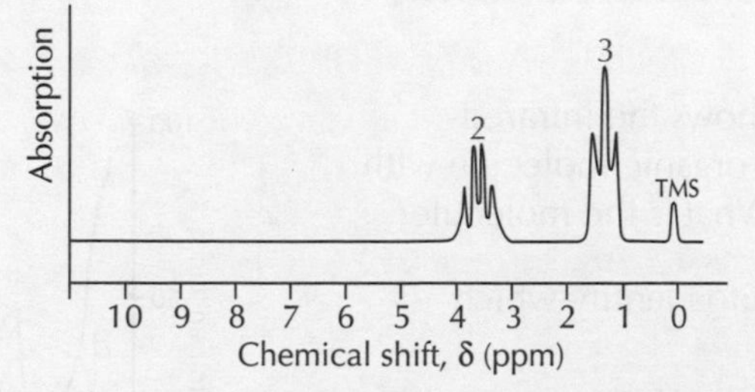

a) What is the likely environment of the two protons with a shift of 3.6 p.p.m.? [1 mark]

b) What is the likely environment of the three protons with a shift of 1.3 p.p.m.? [1 mark]

c) The molecular mass of the molecule is 64. Suggest a possible structure and explain your suggestion. [2 marks]

d) Explain the shapes of the two peaks. [4 marks]

Never mind splitting peaks — this stuff's likely to cause splitting headaches...

Is your head spinning yet? I know mine is. Round and round like a merry-go-round. It's a hard life when you're tied to a desk trying to get NMR spectroscopy firmly fixed in your head. You must be looking quite peaky by now... so go on, learn this stuff, take the dog around the block, then come back and see if you can still remember it all.

Infrared Spectroscopy

Eeek... more spectroscopy. Infrared (IR to its friends) radiation has less energy than visible light, and a longer wavelength.

Infrared Spectroscopy Lets You Identify Organic Molecules

1) In infrared (IR) spectroscopy, a beam of **IR radiation** is passed through a sample of a chemical.
2) The IR radiation is absorbed by the **covalent bonds** in the molecules, increasing their **vibrational** energy.
3) **Bonds between different atoms** absorb **different frequencies** of IR radiation. Bonds in different **places** in a molecule absorb different frequencies too — so the O–H group in an **alcohol** and the O–H in a **carboxylic acid** absorb different frequencies.

   This table shows what **frequencies** different bonds absorb:

Bond	Where it's found	Frequency/ Wavenumber (cm^{-1})
C–O	alcohols, carboxylic acids and esters	1000 – 1300
C=O	aldehydes, ketones, carboxylic acids, esters and amides	1640 – 1750
O–H	carboxylic acids	2500 – 3300 (very broad)
C–H	organic compounds	2850 – 3100
N–H	amines and amides	3200 – 3500
O–H	alcohols, phenols	3200 – 3550 (broad)

This data will be on the data sheet in the exam, so you don't need to learn it. BUT you do need to understand how to use it.

Clark began to regret having an infrared mechanism installed in his glasses.

4) An infrared spectrometer produces a **graph** that shows you what frequencies of radiation the molecules are absorbing. You can use it to identify the **functional groups** in a molecule:

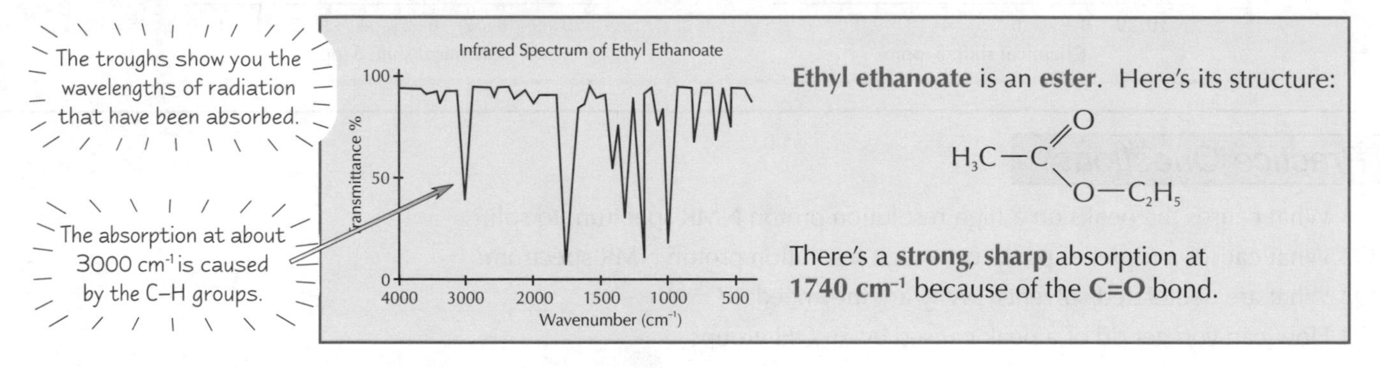

First Identify the Type of Molecule...

Example: The diagram on the right shows the infrared absorption spectrum of an organic molecule with a molecular mass of 46. What is the molecule?

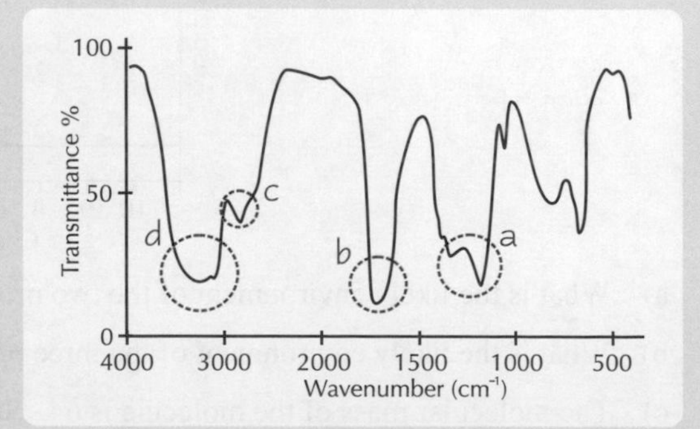

Start off by looking at the troughs to try to identify which **functional groups** the molecule has:

a) A trough at around **1200 cm^{-1}**. This could be due to the **C–O bond** in an **alcohol**, **carboxylic acid** or **ester**.
b) A trough at around **1700 cm^{-1}**, which is the characteristic place for a **C=O bond** in an aldehyde, ketone, carboxylic acid, ester or amide.
c) A small trough just below **3000 cm^{-1}**, which is characteristic of **C–H bonds** in an organic molecule.
d) A very broad trough at around **3100 cm^{-1}**. This is due to the **O–H bond** in a **carboxylic acid**.

So the only type of molecule that could create this pattern of absorptions is a **carboxylic acid**.

Infrared Spectroscopy

...Then Use the **Molecular Mass** to **Work Out Which One It Is**

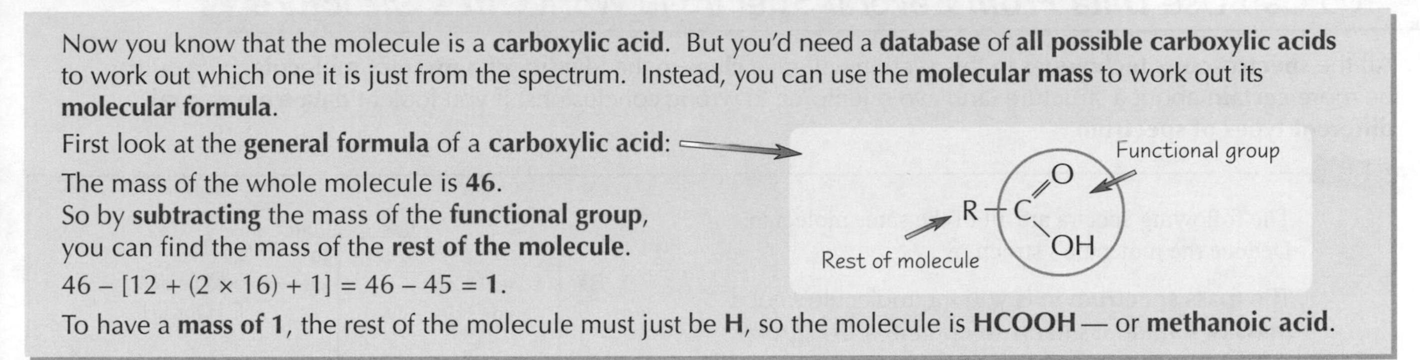

Now you know that the molecule is a **carboxylic acid**. But you'd need a **database** of **all possible carboxylic acids** to work out which one it is just from the spectrum. Instead, you can use the **molecular mass** to work out its **molecular formula**.

First look at the **general formula** of a **carboxylic acid**:

The mass of the whole molecule is **46**.

So by **subtracting** the mass of the **functional group**, you can find the mass of the **rest of the molecule**.

46 – [12 + (2 × 16) + 1] = 46 – 45 = **1**.

To have a **mass of 1**, the rest of the molecule must just be **H**, so the molecule is **HCOOH** — or **methanoic acid.**

Practice Questions

Q1 What happens when bonds absorb infrared radiation?

Q2 What do the troughs on a spectrum show?

Q3 Which bond absorbs strongly at 1700 cm^{-1}?

Q4 What extra information can help you identify a molecule?

Exam Questions

1 A molecule with a molecular mass of 74 gives the following IR spectrum.

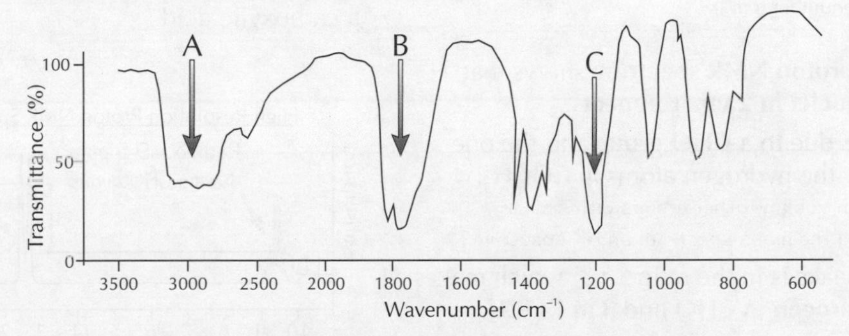

a) What type of bonds are likely to have produced the troughs labelled A, B and C? [3 marks]

b) Suggest a molecular formula and name for this molecule. Explain your suggestion. [3 marks]

2 Substance A gives this IR spectrum. It has a molecular mass of 46.

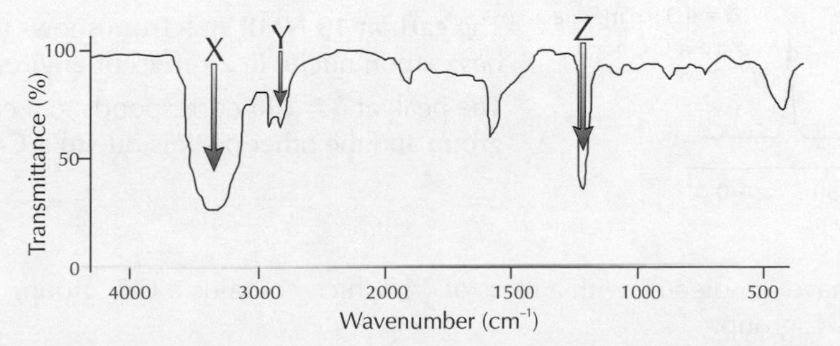

a) Suggest what type of bonds could have caused the troughs labelled X, Y and Z. [3 marks]

b) Suggest a molecular formula and explain your reasoning. [3 marks]

Ooooh — I'm picking up some good vibrations...

Now, I've warned you — infrared glasses are not for fun. They're highly advanced pieces of technology which if placed in the wrong hands could cause havoc and destruction across the universe. There's not much to learn on these pages — so make sure you can apply it. You'll be given a data table, so you don't have to bother learning all the wavenumber ranges.

More on Spectra

Yes, I know, it's yet another page on spectra — but it's the last one (alright, two) I promise.

You Can Use Data From Several Spectra to Work Out a Structure

All the **spectroscopy techniques** in this section will **give clues** to the **identity of a mystery molecule**, but you can be more **certain** about a structure (and avoid jumping to wrong conclusions) if you look at **data from several different types of spectrum**.

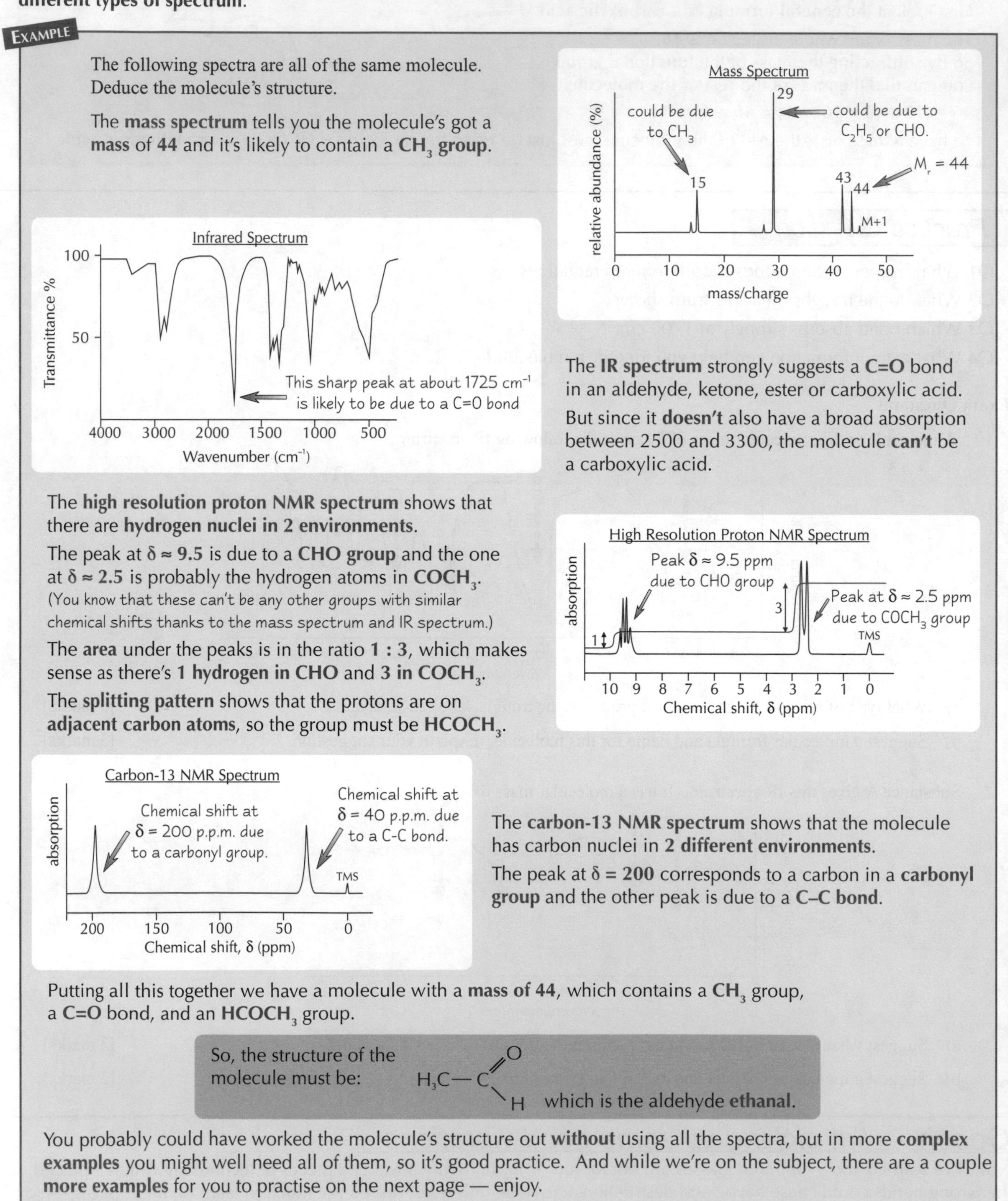

The following spectra are all of the same molecule. Deduce the molecule's structure.

The **mass spectrum** tells you the molecule's got a **mass of 44** and it's likely to contain a **CH_3 group**.

The **IR spectrum** strongly suggests a **C=O** bond in an aldehyde, ketone, ester or carboxylic acid.

But since it **doesn't** also have a broad absorption between 2500 and 3300, the molecule **can't** be a carboxylic acid.

The **high resolution proton NMR spectrum** shows that there are **hydrogen nuclei in 2 environments**.

The peak at **δ ≈ 9.5** is due to a **CHO group** and the one at **δ ≈ 2.5** is probably the hydrogen atoms in **$COCH_3$**.

(You know that these can't be any other groups with similar chemical shifts thanks to the mass spectrum and IR spectrum.)

The **area** under the peaks is in the ratio **1 : 3**, which makes sense as there's **1 hydrogen in CHO** and **3 in $COCH_3$**.

The **splitting pattern** shows that the protons are on **adjacent carbon atoms**, so the group must be **$HCOCH_3$**.

The **carbon-13 NMR spectrum** shows that the molecule has carbon nuclei in **2 different environments**.

The peak at **δ = 200** corresponds to a carbon in a **carbonyl group** and the other peak is due to a **C–C bond**.

Putting all this together we have a molecule with a **mass of 44**, which contains a **CH_3** group, a **C=O** bond, and an **$HCOCH_3$** group.

So, the structure of the molecule must be: $H_3C-C(=O)-H$ which is the aldehyde **ethanal**.

You probably could have worked the molecule's structure out **without** using all the spectra, but in more **complex examples** you might well need all of them, so it's good practice. And while we're on the subject, there are a couple **more examples** for you to practise on the next page — enjoy.

More on Spectra

Practice Questions

Q1 Which type of spectrum gives you the mass of a molecule?

Q2 Which spectrum can tell you how many carbon environments are in a molecule?

Q3 Which spectrum can tell you how many different hydrogen environments there are in a molecule?

Q4 Which spectrum involves radio wave radiation?

Exam Questions

1 The four spectra below were produced by running different tests on samples of the same pure organic compound.

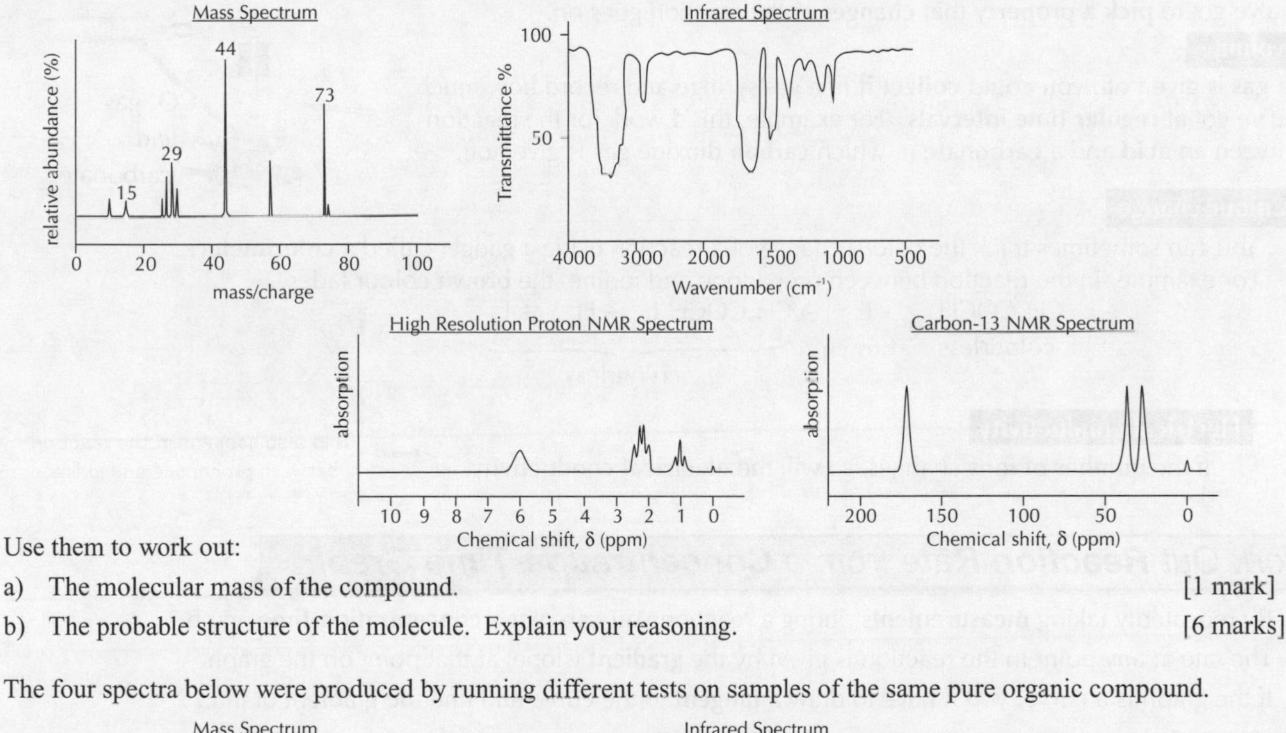

Use them to work out:

a) The molecular mass of the compound. [1 mark]

b) The probable structure of the molecule. Explain your reasoning. [6 marks]

2 The four spectra below were produced by running different tests on samples of the same pure organic compound.

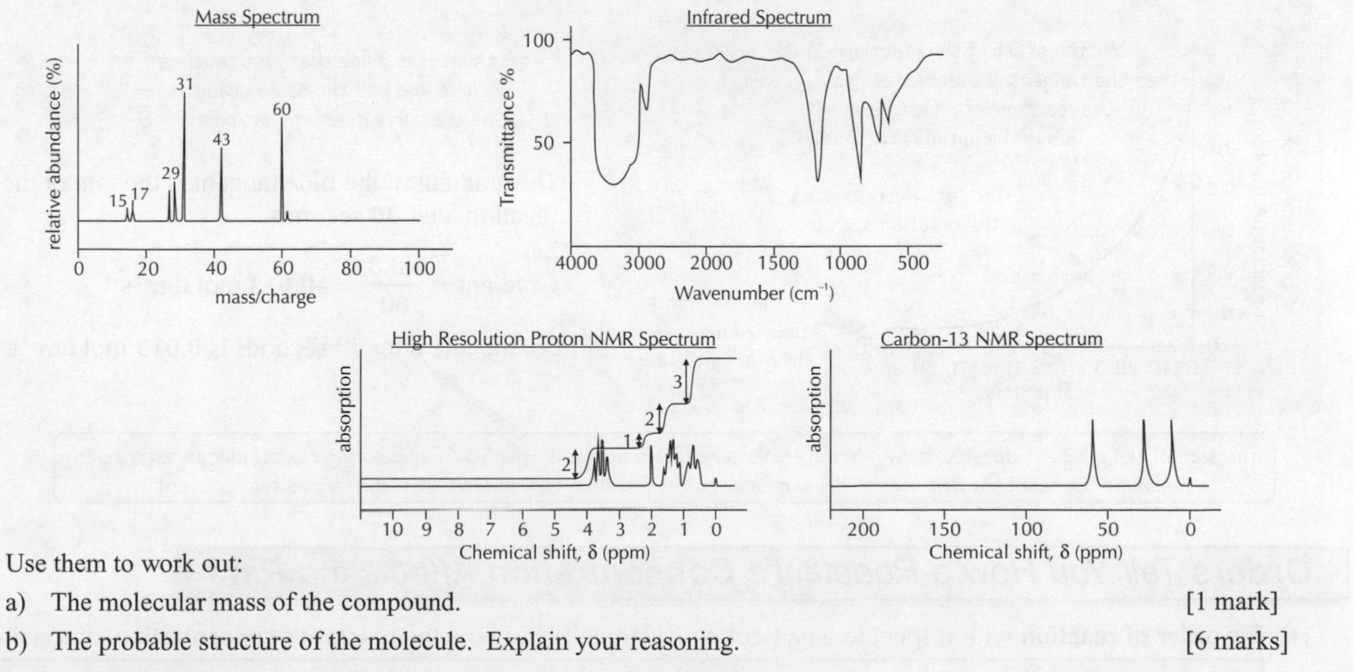

Use them to work out:

a) The molecular mass of the compound. [1 mark]

b) The probable structure of the molecule. Explain your reasoning. [6 marks]

Spectral analysis — psychology for ghosts...

So that's analysis done and dusted, you'll be pleased to hear. But before you rush off to learn about rates, take a moment to check that you really know how to interpret all the different spectra. You might want to go back and have a look at page 128 too if you're having trouble remembering what all the different functional groups look like.

Rate Graphs and Orders

This section's a whole lot of fun. Well, it is if you like learning about speed of reactions anyway, and who doesn't...

The Reaction Rate tells you How Fast Reactants are Converted to Products

The **reaction rate** is the **change in the amount** of reactants or products **per unit time** (normally per second).

If the reactants are in **solution**, the rate'll be **change in concentration per second** and the units will be **mol dm^{-3} s^{-1}**.

There are Loads of Ways to Follow the Rate of a Reaction

Although there are quite a few ways to follow reactions, not every method works for every reaction. You've got to **pick a property** that **changes** as the reaction goes on.

Gas volume

If a **gas** is given off, you could **collect it** in a gas syringe and record how much you've got at **regular time intervals**. For example, this'd work for the reaction between an **acid** and a **carbonate** in which **carbon dioxide gas** is given off.

Colour change

You can sometimes track the colour change of a reaction using a gadget called a **colorimeter**. For example, in the reaction between propanone and iodine, the **brown** colour fades.

$$CH_3COCH_{3(aq)} + I_{2(aq)} \rightarrow CH_3COCH_2I_{(aq)} + H^+_{(aq)} + I^-_{(aq)}$$

colourless (propanone), brown (I_2), colourless (products)

Electrical conductivity

If the **number of ions** changes, so will the **electrical conductivity**.

This also happens in the reaction between propanone and iodine.

Work Out Reaction Rate from a Concentration-Time Graph

1) By repeatedly taking **measurements** during a reaction you can plot a **concentration-time** graph.
2) The rate at any point in the reaction is given by the **gradient** (slope) at that point on the graph.
3) If the graph is a curve, you'll have to draw a **tangent** to the curve and find the gradient of that.

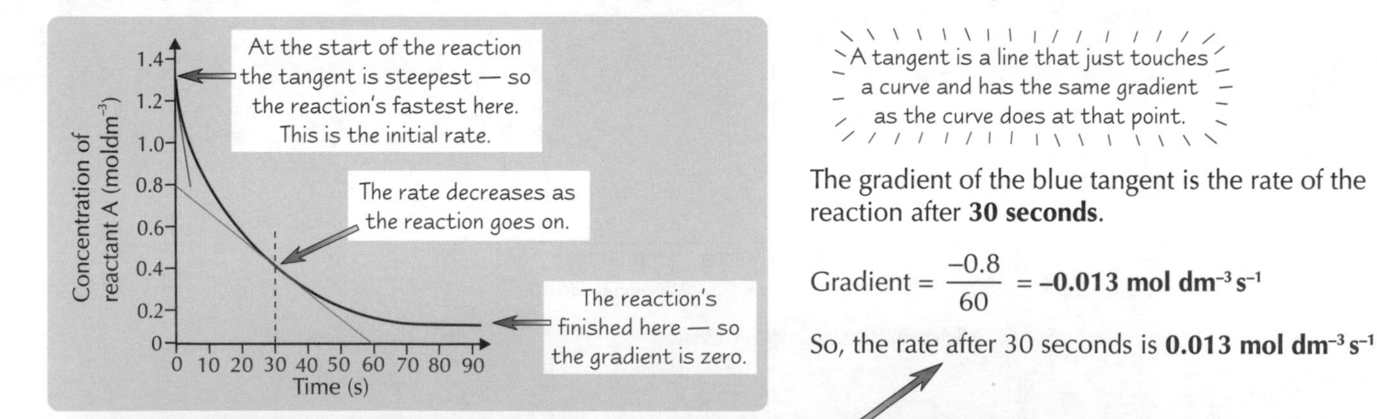

A tangent is a line that just touches a curve and has the same gradient as the curve does at that point.

The gradient of the blue tangent is the rate of the reaction after **30 seconds**.

$$\text{Gradient} = \frac{-0.8}{60} = \mathbf{-0.013\ mol\ dm^{-3}\ s^{-1}}$$

So, the rate after 30 seconds is **0.013 mol dm^{-3} s^{-1}**

The sign of the gradient doesn't really matter — it's a negative gradient when you're measuring reactant concentration because the reactant decreases. If you measured the product concentration, it'd be a positive gradient.

Orders Tell You How a Reactant's Concentration Affects the Rate

1) The **order of reaction** with respect to a particular reactant tells you how the **reactant's concentration** affects the **rate**.

   If you double the reactant's concentration and the rate **stays the same**, the order with respect to that reactant is **0**.
   If you double the reactant's concentration and the rate **also doubles**, the order with respect to that reactant is **1**.
   If you double the reactant's concentration and the rate **quadruples**, the order with respect to that reactant is **2**.

2) You can only find **orders of reaction** from **experiments**. You **can't** work them out from chemical equations.

Rate Graphs and Orders

The *Shape* of a *Rate-Concentration Graph* Tells You the *Order*

You can use your concentration-time graph to construct a **rate-concentration graph**, which you can then use to work out the order of the reaction. Here's how:

1) Find the **gradient** (which is the rate, remember) at various points along the concentration-time graph. This gives you a **set of points** for the rate-concentration graph.
2) Just **plot the points** and then **join them up** with a line or smooth curve, and you're done. The **shape** of the new graph tells you the **order**...

The notation [X] means 'the concentration of reactant X'.

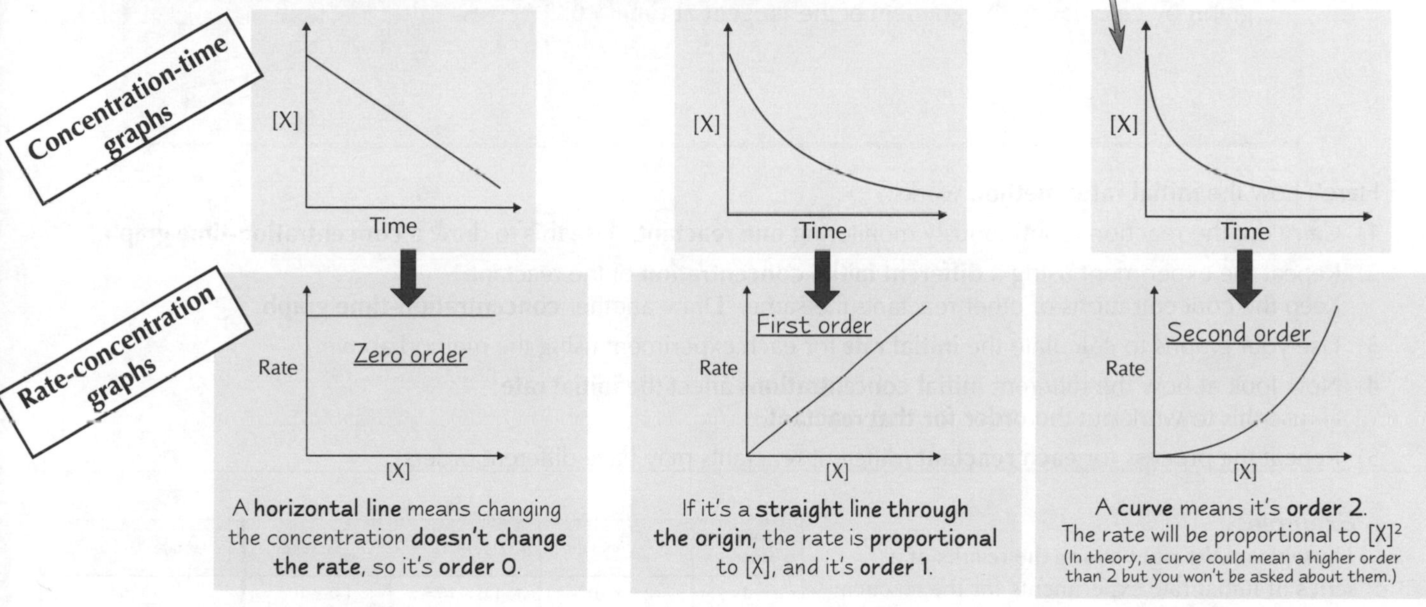

A **horizontal line** means changing the concentration **doesn't change the rate**, so it's **order 0**.

If it's a **straight line through the origin**, the rate is **proportional** to [X], and it's **order 1**.

A **curve** means it's **order 2**. The rate will be proportional to $[X]^2$ (In theory, a curve could mean a higher order than 2 but you won't be asked about them.)

Practice Questions

Q1 Give two things that could be measured to follow the rate of a reaction.

Q2 What does the gradient of a concentration-time graph measure?

Q3 If you double the concentration of a reactant and the rate doubles, what is the order of reaction with respect to that reactant?

Q4 Sketch a typical rate-concentration graph for a second order reaction.

Exam Questions

1 It takes 200 seconds to completely dissolve a 0.4 g piece of magnesium in 25 ml of dilute hydrochloric acid. It takes 100 seconds if the concentration of the acid is doubled.

a) What is the order of the reaction with respect to the concentration of the acid? [1 mark]

b) Sketch a graph to show the relationship between the concentration of the acid and the overall rate of the reaction. [2 marks]

c) What could be measured to follow the rate of this reaction in more detail? [2 marks]

2 The rate of decomposition of hydrogen peroxide was followed by monitoring the concentration of hydrogen peroxide.

$$2H_2O_{2(aq)} \rightarrow 2H_2O_{(l)} + O_{2(g)}$$

Time (minutes)	0	20	40	60	80	100
$[H_2O_2]$ (mol dm^{-3})	2.00	1.00	0.50	0.25	0.125	0.0625

a) Suggest an alternative method that could have been used to follow the rate of this reaction. [2 marks]

b) Using the data above, plot a graph and determine the rate of the reaction after 30 minutes. [6 marks]

Mmmm, look at those seductive curves...

...sorry, chemistry gets me a bit over-excited sometimes. I think I'm OK now. Remember — the [X]-time graphs on this page slope downwards, i.e. have negative gradients, because they're showing the concentration of reactants. If you measure concentration of products instead the graph would be flipped the other way up — it'd slope upwards instead, but still level off in the same way.

Initial Rates and Half-Life

This is where it starts getting a bit mathsy. But don't panic, just take a deep breath and dive in... And don't bash your head on the bottom. Oh, and don't start sentences with 'And' or 'But' — English teachers hate it. 'Butt' is fine though.

The **Initial Rates Method** can be used to work out **Orders** too

On the previous page, reaction order was found by turning a concentration-time graph into a rate-concentration graph. Another way to find order is by looking at **initial reaction rates**.

The **initial rate of a reaction** is the rate right at the **start** of the reaction. You can find this from a **concentration-time** graph by calculating the **gradient** of the **tangent** at **time = 0**.

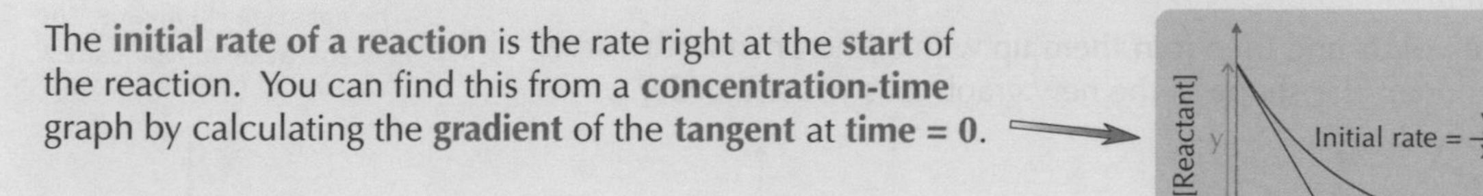

Here's how the **initial rates method** works:

1) Carry out the reaction, continuously monitoring **one reactant**. Use this to draw a **concentration-time graph**.
2) Repeat the experiment using a **different initial concentration** of the reactant. Keep the concentrations of other reactants the same. Draw another **concentration-time graph**.
3) Use your graphs to calculate the **initial rate** for each experiment using the method above.
4) Now look at how the different **initial concentrations** affect the **initial rate** — use this to work out the **order for that reactant**.
5) Repeat the process for **each reactant** (different reactants may have different orders).

Example:
The table on the right shows the results of a series of initial rate experiments for the reaction:

$NO_{(g)} + CO_{(g)} + O_{2(g)} \rightarrow NO_{2(g)} + CO_{2(g)}$

Write down the order with respect to each reactant.

Experiment number	$[NO_{(g)}]$ (mol dm^{-3})	$[CO_{(g)}]$ (mol dm^{-3})	$[O_{2(g)}]$ (mol dm^{-3})	Initial rate (mol dm^{-3} s^{-1})
1	2.0×10^{-2}	1.0×10^{-2}	1.0×10^{-2}	0.176
2	4.0×10^{-2}	1.0×10^{-2}	1.0×10^{-2}	0.704
3	2.0×10^{-2}	2.0×10^{-2}	1.0×10^{-2}	0.176
4	2.0×10^{-2}	1.0×10^{-2}	2.0×10^{-2}	0.176

1) Look at experiments 1 and 2 — when **$[NO_{(g)}]$** doubles (but all the other concentrations stay constant), the rate **quadruples**. So the reaction is **second order** with respect to NO.
2) Look at experiments 1 and 3 — when **$[CO_{(g)}]$** doubles (but all the other concentrations stay constant), the rate **stays the same**. So the reaction is **zero order** with respect to CO.
3) Look at experiments 1 and 4 — when **$[O_{2(g)}]$** doubles (but all the other concentrations stay constant), the rate **stays the same**. So the reaction is **zero order** with respect to O_2.

Clock Reactions can be used to **Simplify** the Initial Rate Method

The method described above is a bit faffy — lots of measuring and drawing graphs.
In clock reactions, the initial rate can be **easily estimated**.

1) In a **clock reaction**, you can easily **measure the time** it takes for a given amount of product to form — usually there's a sudden colour change. The **shorter** the time, the faster the **initial rate**.
2) It's a much easier way to find **initial rates** than drawing lots of **concentration-time graphs**.

The most famous clock reaction is the **iodine-clock** reaction.

- sodium thiosulfate solution and starch are added to hydrogen peroxide and iodide ions in acid solution.
- the important product is **iodine** — after a certain amount of time, the solution **suddenly** turns dark blue.
- varying iodide or hydrogen peroxide concentration while keeping the others constant will give **different times** for the colour change. These can be used to work out the **reaction order**.

Initial Rates and Half-Life

Half-life is the Time for Half the Reactant to Disappear

The **half-life** of a reaction is the time it takes for **half of the reactant** to be used up.
The **half-life** of a first order reaction is **independent of the concentration.**

Example:

This graph shows the decomposition of hydrogen peroxide, H_2O_2.

- Use the graph to measure the half-life at various points:
  $[H_2O_2]$ from **4** to **2** mol dm^{-3} = **200 s,**
  $[H_2O_2]$ from **2** to **1** mol dm^{-3} = **200 s,**
  $[H_2O_2]$ from **1** to **0.5** mol dm^{-3} = **200 s.**
- The half-life is constant, regardless of the concentration, so it's a **first order reaction** with respect to $[H_2O_2]$.

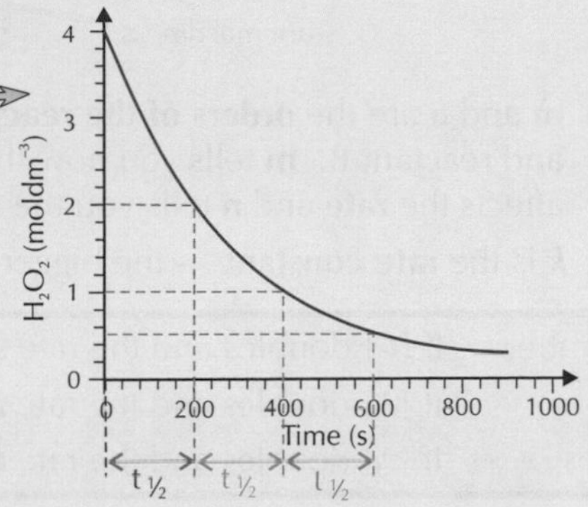

H_2O_2 (moldm^{-3})
4
3
2
1
0
0 200 400 600 800 1000
Time (s)
t ½ t ½ t ½

Practice Questions

Q1 What is meant by the initial rate of a reaction?

Q2 Describe how to calculate the initial rate of a reaction.

Q3 What is a clock reaction?

Q4 What does the term 'half-life' mean?

Exam Questions

1 The table below shows the results of a series of initial rate experiments for the reaction between substances D and E.

Experiment	[D] (mol dm^{-3})	[E] (mol dm^{-3})	Initial rate × 10^{-3} (mol dm^{-3} s^{-1})
1	0.2	0.2	1.30
2	0.4	0.2	5.19
3	0.2	0.4	2.61

Find the order of the reaction with respect to reactants D and E. Explain your reasoning. [4 marks]

2 The table shows the results of an experiment on the decomposition of nitrogen(V) oxide at constant temperature.

$$2N_2O_5 \rightarrow 4NO_2 + O_2$$

Time (s)	0	50	100	150	200	250	300
$[N_2O_5]$ (mol dm^{-3})	2.50	1.66	1.14	0.76	0.50	0.32	0.22

a) Plot a graph of these results. [3 marks]

b) From the graph, find the times for the concentration of N_2O_5 to decrease:
  i) to half its original concentration. [2 marks]
  ii) from 2.0 mol dm^{-3} to 1.0 mol dm^{-3}. [2 marks]

c) Giving a reason, deduce the order of this reaction. [2 marks]

If you thought this was fun, just wait till you get a load of...

...page 150. That's even better. It's got a proper maths equation. It's also got shape-shifters, dogs and industrial men. And speaking of other things that are fun, can I recommend... flying kites, peeling bananas, making models of your friends out of apples, the literary works of Virginia Woolf, counting spots on the carpet, eating all the pies and cuddling with a boy.

Rate Equations

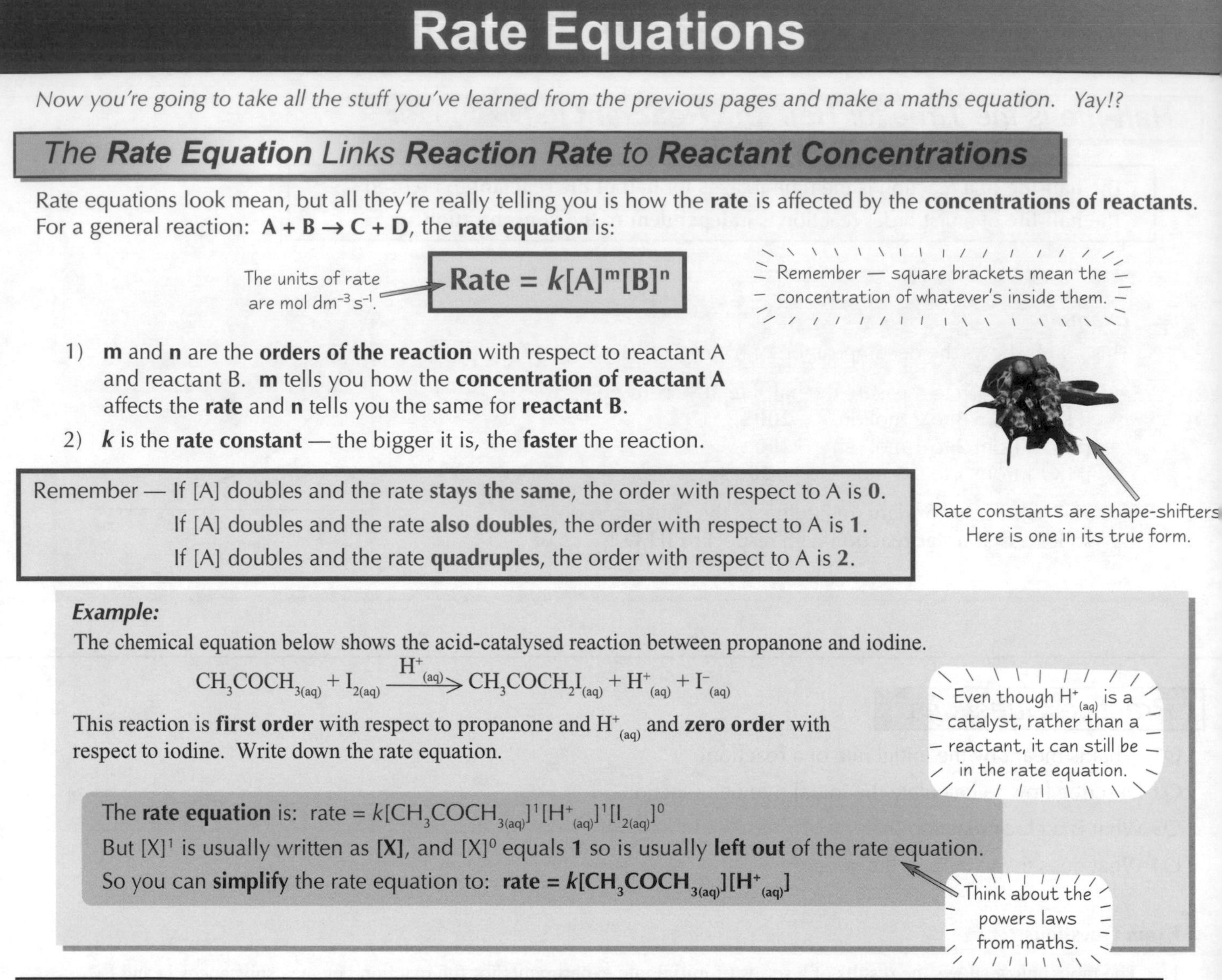

Now you're going to take all the stuff you've learned from the previous pages and make a maths equation. Yay!?

The Rate Equation Links Reaction Rate to Reactant Concentrations

Rate equations look mean, but all they're really telling you is how the **rate** is affected by the **concentrations of reactants**. For a general reaction: **A + B → C + D**, the **rate equation** is:

$$\text{Rate} = k[A]^m[B]^n$$

The units of rate are mol dm^{-3} s^{-1}.

Remember — square brackets mean the concentration of whatever's inside them.

1) **m** and **n** are the **orders of the reaction** with respect to reactant A and reactant B. **m** tells you how the **concentration of reactant A** affects the **rate** and **n** tells you the same for **reactant B**.
2) ***k*** is the **rate constant** — the bigger it is, the **faster** the reaction.

Rate constants are shape-shifters. Here is one in its true form.

Remember — If [A] doubles and the rate **stays the same**, the order with respect to A is **0**.
If [A] doubles and the rate **also doubles**, the order with respect to A is **1**.
If [A] doubles and the rate **quadruples**, the order with respect to A is **2**.

Example:

The chemical equation below shows the acid-catalysed reaction between propanone and iodine.

$$CH_3COCH_{3(aq)} + I_{2(aq)} \xrightarrow{H^+_{(aq)}} CH_3COCH_2I_{(aq)} + H^+_{(aq)} + I^-_{(aq)}$$

This reaction is **first order** with respect to propanone and $H^+_{(aq)}$ and **zero order** with respect to iodine. Write down the rate equation.

Even though $H^+_{(aq)}$ is a catalyst, rather than a reactant, it can still be in the rate equation.

The **rate equation** is: rate = $k[CH_3COCH_{3(aq)}]^1[H^+_{(aq)}]^1[I_{2(aq)}]^0$

But $[X]^1$ is usually written as **[X]**, and $[X]^0$ equals **1** so is usually **left out** of the rate equation.

So you can **simplify** the rate equation to: **rate =** $\mathbf{k[CH_3COCH_{3(aq)}][H^+_{(aq)}]}$

Think about the powers laws from maths.

You can Calculate the Rate Constant from the Orders and Rate of Reaction

Once the rate and the orders of the reaction have been found by experiment, you can work out the **rate constant, *k***. The **units** of the rate constant vary, so you have to **work them out**.

Example:

The reaction below was found to be second order with respect to NO and zero order with respect to CO and O_2. The rate is 1.76×10^{-3} mol dm^{-3} s^{-1}, when $[NO_{(g)}] = [CO_{(g)}] = [O_{2(g)}] = 2.00 \times 10^{-3}$ mol dm^{-3}.

$$NO_{(g)} + CO_{(g)} + O_{2(g)} \rightarrow NO_{2(g)} + CO_{2(g)}$$

Find the value of the rate constant.

First write out the **rate equation**: Rate = $k[NO_{(g)}]^2[CO_{(g)}]^0[O_{2(g)}]^0 = k[NO_{(g)}]^2$

Next insert the **concentration** and the **rate**. **Rearrange** the equation and calculate the value of ***k***:

$$\text{Rate} = k[NO_{(g)}]^2, \text{ so, } 1.76 \times 10^{-3} = k \times (2.00 \times 10^{-3})^2 \Rightarrow k = \frac{1.76 \times 10^{-3}}{(2.00 \times 10^{-3})^2} = 440$$

Find the **units for *k*** by putting the other units in the rate equation:

$$\text{Rate} = k[NO_{(g)}]^2, \text{ so mol dm}^{-3}\text{ s}^{-1} = k \times (\text{mol dm}^{-3})^2 \Rightarrow k = \frac{\text{mol dm}^{-3}\text{ s}^{-1}}{(\text{mol dm}^{-3})^2} = \frac{\text{s}^{-1}}{\text{mol dm}^{-3}} = \text{dm}^3\text{ mol}^{-1}\text{ s}^{-1}$$

So the answer is: $\mathbf{k = 440\ dm^3\ mol^{-1}\ s^{-1}}$

Rate Equations

Temperature Changes Affect the Rate Constant

1) Reactions happen when the reactant particles **collide** and have enough energy to **break** the existing bonds.
2) Increasing the temperature **speeds up** the reactant particles, so that they collide **more often**. It also increases the chances of the particles reacting when they do hit each other, as they have more energy.
3) In other words, **increasing temperature** increases the **reaction rate**.
4) According to the rate equation, reaction rate depends **only** on the rate constant and reactant concentrations. Since temperature **does** increase the reaction rate, it must **change the rate constant**.

The **rate constant** applies to a **particular reaction** at a **certain temperature**.
At a **higher** temperature, the reaction will have a **higher** rate constant.

Practice Questions

Q1 Write down a general rate equation for a reaction with reactants A, B and C.

Q2 How do you work out the units of k?

Q3 How does temperature affect the value of k?

Exam Questions

1 The following reaction is second order with respect to NO and first order with respect to H_2.

$$2NO_{(g)} + 2H_{2(g)} \rightarrow 2H_2O_{(g)} + N_{2(g)}$$

a) Write a rate equation for the reaction. [2 marks]

b) The rate of the reaction at 800 °C was determined to be 0.00267 mol dm^{-3} s^{-1} when $[H_{2(g)}]$ = 0.0020 mol dm^{-3} and $[NO_{(g)}]$ = 0.0040 mol dm^{-3}.

i) Calculate a value for the rate constant at 800 °C, including units. [3 marks]

ii) Predict the effect on the rate constant of decreasing the temperature of the reaction to 600 °C. [1 mark]

2 The ester ethyl ethanoate, $CH_3COOC_2H_5$, is hydrolysed by heating with dilute acid to give ethanol and ethanoic acid. The reaction is first order with respect to the concentration of H^+ and the ester.

a) Write the rate equation for the reaction. [1 mark]

b) When the initial concentration of the acid is 2.0 mol dm^{-3} and the ester is 0.25 mol dm^{-3}, the initial rate is 2.2×10^{-3} mol dm^{-3} s^{-1}. Calculate a value for the rate constant at this temperature and give its units. [3 marks]

c) The temperature is kept constant and more solvent is added to the initial mixture so that the volume doubles. Calculate the new initial rate. [2 marks]

3 A reaction between substances X and Y, which is first order with respect to X and Y, has an initial rate of 1.6×10^{-3} mol dm^{-3} s^{-1} at 300 K.
The reaction rate is doubled when the temperature rises by 10 K (a 3.33% increase).

Which of the following changes has a greater effect on the reaction rate?

Increasing the temperature by 10 K

OR Increasing the concentration of X from 2.00 mol dm^{-3} to 2.06 mol dm^{-3} (a 3.33% increase)? [4 marks]

Rate constants are masters of disguise...

Here's some of their most common disguises. If you think you've spotted one, heating it will transform it back.

Rates and Reaction Mechanisms

It's a cold, miserable grey day outside, but on the plus side I just had a really nice slice of carrot and ginger cake. Anyway, this page is about the connection between rate equations and reaction mechanisms.

The Rate-Determining Step is the Slowest Step in a Multi-Step Reaction

Mechanisms can have **one step** or a **series of steps**. In a series of steps, each step can have a **different rate**. The **overall rate** is decided by the step with the **slowest** rate — the **rate-determining step**.

Otherwise known as the rate-limiting step.

Reactants in the Rate Equation Affect the Rate

The rate equation is handy for helping you work out the **mechanism** of a chemical reaction.
You need to be able to pick out which reactants from the chemical equation are involved in the **rate-determining step**. Here are the **rules** for doing this:

> If a reactant appears in the **rate equation**, it must be affecting the **rate**.
> So this reactant, or something derived from it, must be in the **rate-determining step**.
>
> If a reactant **doesn't** appear in the **rate equation**, then it **won't** be involved in the **rate-determining step** (and neither will anything derived from it).

Catalysts can appear in rate equations, so they can be in rate-determining steps too.

Some **important points** to remember about rate-determining steps and mechanisms are:

1) The rate-determining step **doesn't** have to be the first step in a mechanism.
2) The reaction mechanism **can't** usually be predicted from **just** the chemical equation.

You Can Predict the Rate Equation from the Rate-Determining Step...

> The **order of a reaction** with respect to a reactant shows the **number of molecules** of that reactant which are involved in the **rate-determining step**.

So, if a reaction's second order with respect to X, there'll be two molecules of X in the rate-determining step.

For example, the mechanism for the reaction between **chlorine free radicals** and **ozone**, O_3, consists of **two steps**:

$$Cl\bullet_{(g)} + O_{3(g)} \rightarrow ClO\bullet_{(g)} + O_{2(g)}$$ — **slow (rate-determining step)**

$$ClO\bullet_{(g)} + O\bullet_{(g)} \rightarrow Cl\bullet_{(g)} + O_{2(g)}$$ — **fast**

$Cl\bullet$ and O_3 must both be in the rate equation, so the rate equation is of the form: **rate =** $k[Cl\bullet]^m[O_3]^n$.
There's only **one** $Cl\bullet$ radical and **one** O_3 molecule in the rate-determining step, so the **orders**, m and n, are both **1**.
So the rate equation is **rate =** $k[Cl\bullet][O_3]$.

...And You Can Predict the Mechanism from the Rate Equation

Knowing exactly which reactants are in the **rate-determining step** gives you an idea of the reaction **mechanism**.

For example, here are two possible mechanisms for the reaction $(CH_3)_3CBr + OH^- \rightarrow (CH_3)_3COH + Br^-$

$$(CH_3)_3C\text{–}Br + OH^- \rightarrow (CH_3)_3C\text{–}OH + Br^-$$

or

$$(CH_3)_3C\text{–}Br \rightarrow (CH_3)_3C^+ + Br^-$$ — **slow (rate-determining step)**

$$(CH_3)_3C^+ + OH^- \rightarrow (CH_3)_3C\text{–}OH$$ — **fast**

The actual **rate equation** was worked out by rate experiments: **rate =** $k[(CH_3)_3CBr]$
OH^- isn't in the **rate equation**, so it **can't** be involved in the rate-determining step.
The **second mechanism** is most likely to be correct because OH^- **isn't** in the rate-determining step.

Rates and Reaction Mechanisms

You have to *Take Care* when Suggesting a *Mechanism*

If you're suggesting a mechanism, **watch out** — things might not always be what they seem.
For example, when nitrogen(V) oxide, N_2O_5, decomposes, it forms nitrogen(IV) oxide and oxygen:

$$2N_2O_{5(g)} \rightarrow 4NO_{2(g)} + O_{2(g)}$$

From the chemical equation, it looks like **two** N_2O_5 molecules react with each other. So you might predict that the reaction is **second order** with respect to N_2O_5 ... but you'd be wrong.

Experimentally, it's been found that the reaction is **first order** with respect to N_2O_5 — the rate equation is: **rate =** $k[N_2O_5]$. This shows that there's only one molecule of N_2O_5 in the rate-determining step.

One **possible mechanism** that fits the rate equation is:

Only one molecule of N_2O_5 is in the rate-determining step, fitting in with the rate equation.

$N_2O_{5(g)} \rightarrow NO_{2(g)} + NO_{3(g)}$ — **slow (rate-determining step)**

$NO_{3(g)} + N_2O_{5(g)} \rightarrow 3NO_{2(g)} + O_{2(g)}$ — **fast**

The two steps add up to the overall chemical equation. You can cancel the $NO_{3(g)}$ as it appears on both sides.

Practice Questions

Q1 What is meant by the rate-determining step?

Q2 Is the rate-determining step the first step in the reaction?

Q3 What is the connection between the rate equation and the rate-determining step?

Q4 How can the rate-determining step help you to understand the mechanism?

Exam Questions

1 The following reaction is first order with respect to H_2 and first order with respect to ICl.

$$H_{2(g)} + 2ICl_{(g)} \rightarrow I_{2(g)} + 2HCl_{(g)}$$

a) Write the rate equation for this reaction. [1 mark]

b) The mechanism for this reaction consists of two steps.

i) Identify the molecules that are in the rate-determining step. Justify your answer. [2 marks]

ii) A chemist suggested the following mechanism for the reaction.

$2ICl_{(g)} \rightarrow I_{2(g)} + Cl_{2(g)}$ slow

$H_{2(g)} + Cl_{2(g)} \rightarrow 2HCl_{(g)}$ fast

Suggest, with reasons, whether this mechanism is likely to be correct. [2 marks]

2 The reaction between HBr and oxygen gas occurs rapidly at 700 K.

It can be represented by the equation $4HBr_{(g)} + O_{2(g)} \rightarrow 2H_2O_{(g)} + 2Br_{2(g)}$

The rate equation found by experiment is Rate = $k[HBr][O_2]$

a) Explain why the reaction cannot be a one-step reaction. [3 marks]

b) Each of the 4 steps of this reaction involves the reaction of 1 molecule of HBr.
Two of the steps are the same. The rate-determining step is the first one and results in the formation of $HBrO_2$. Write equations for the full set of 4 reactions. [4 marks]

I found rate-determining step aerobics a bit on the slow side...

These pages show you how rate equations, orders of reaction and reaction mechanisms all tie together and how each actually means something in the grand scheme of A2 Chemistry. It's all very profound. So get it all learnt and answer the questions and then you'll have plenty of time to practise the quickstep for your Strictly Come Dancing routine.

The Equilibrium Constant

'Oh no, not another page on equilibria', I hear you cry. Well actually it's the first one, so quit moaning.

*At **Equilibrium** the Amounts of Reactants and Products **Stay the Same***

1) Lots of changes are **reversible** — they can go **both ways**. To show a change is reversible, you stick in a $\rightleftharpoons$.
2) As the **reactants** get used up, the **forward** reaction **slows down** — and as more **product** is formed, the **reverse** reaction **speeds up**. After a while, the forward reaction will be going at exactly the **same rate** as the backward reaction.
   The amounts of reactants and products **won't be changing** any more, so it'll seem like **nothing's happening**. It's a bit like you're **digging a hole** while someone else is **filling it in** at exactly the **same speed**. This is called a **dynamic equilibrium**.
3) Equilibria can be set up in **physical** systems, e.g.:

   When **liquid bromine** is shaken in a closed flask, some of it changes to orange **bromine gas**. After a while, **equilibrium** is reached — bromine liquid is **still** changing to bromine gas and bromine gas is still changing to bromine liquid, but they are changing at the **same rate**.

   $Br_{2(l)} \rightleftharpoons Br_{2(g)}$

   ...and **chemical** systems, e.g.:

   If **hydrogen gas** and **iodine gas** are mixed together in a closed flask, **hydrogen iodide** is formed.

   $H_{2(g)} + I_{2(g)} \rightleftharpoons 2HI_{(g)}$

   Imagine that **1.0 mole** of hydrogen gas is mixed with **1.0 mole** of iodine gas at a constant temperature of **640 K**. When this mixture reaches equilibrium, there will be **1.6 moles** of hydrogen iodide and **0.2 moles** of both hydrogen gas and iodine gas. No matter how long you leave them at this temperature, the **equilibrium** amounts **never change**. As with the physical system, it's all a matter of the forward and backward rates **being equal**.
4) A **dynamic equilibrium** can only happen in a **closed system** at a **constant temperature**.

A closed system just means nothing can get in or out.

*K_c is the **Equilibrium Constant***

If you know the **molar concentration** of each substance at equilibrium, you can work out the **equilibrium constant**, K_c. Your value of K_c will only be true for that particular **temperature**.

Before you can calculate K_c, you have to write an **expression** for it. Here's how:

For the general reaction $aA + bB \rightleftharpoons dD + eE$, $K_c = \frac{[D]^d [E]^e}{[A]^a [B]^b}$

The products go on the top line. The square brackets, [ ], mean concentration in mol dm^{-3}.

The lower-case letters a, b, d and e are the number of moles of each substance.

So for the reaction $H_{2(g)} + I_{2(g)} \rightleftharpoons 2HI_{(g)}$, $K_c = \frac{[HI]^2}{[H_2]^1 [I_2]^1}$. This simplifies to $K_c = \frac{[HI]^2}{[H_2][I_2]}$.

*Calculate K_c by **Sticking Numbers** into the Expression*

If you know the **equilibrium concentrations**, just bung them in your expression. Then, using a handy calculator, you can work out the **value** of K_c. The **units** are trickier though — they **vary**, so you have to work them out after each calculation.

Example: If the volume of the closed flask in the hydrogen iodide example above is 2.0 dm^3, what is the equilibrium constant for the reaction at 640 K? The equilibrium concentrations are:

$[HI] = 0.8$ mol dm^{-3}, $[H_2] = 0.1$ mol dm^{-3}, and $[I_2] = 0.1$ mol dm^{-3}.

Just stick the concentrations into the **expression** for K_c: $K_c = \frac{[HI]^2}{[H_2][I_2]} = \frac{0.8^2}{0.1 \times 0.1} = 64$ ← This is the value of K_c.

To work out the **units** of K_c put the units in the expression instead of the numbers:

$K_c = \frac{(\text{mol dm}^{-3})^2}{(\text{mol dm}^{-3})(\text{mol dm}^{-3})} = 0$, so there are **no units** for K_c because the concentration units cancel.

So K_c is just **64**.

The Equilibrium Constant

You Might Need to *Work Out* the *Equilibrium Concentrations*

You might have to figure out some of the **equilibrium concentrations** before you can find K_c:

Example: 0.20 moles of phosphorus(V) chloride decomposes at 600 K in a vessel of 5.00 dm³. The equilibrium mixture is found to contain 0.08 moles of chlorine. Write the expression for K_c and calculate its value, including units.

$$PCl_{5(g)} \rightleftharpoons PCl_{3(g)} + Cl_{2(g)}$$

First find out how many moles of PCl_5 and PCl_3 there are at equilibrium:

The **equation** tells you that when **1 mole of PCl_5** decomposes, **1 mole of PCl_3** and **1 mole of Cl_2** are formed.
So if 0.08 moles of chlorine are produced at equilibrium, then there will be **0.08 moles** of PCl_3 as well.
0.08 mol of PCl_5 must have decomposed, so there will be **0.12 moles** left (0.2 – 0.08).

Divide each number of moles by the volume of the flask to give the molar concentrations:

$[PCl_3] = [Cl_2] = 0.08 \div 5.00 = \mathbf{0.016\ mol\ dm^{-3}}$ $\quad [PCl_5] = 0.12 \div 5.00 = \mathbf{0.024\ mol\ dm^{-3}}$

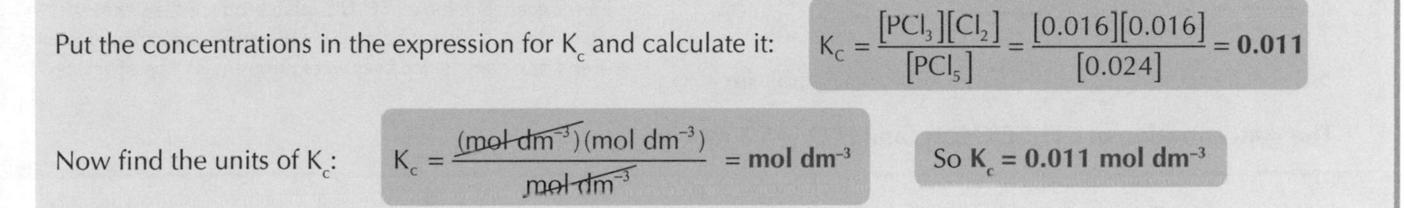

Put the concentrations in the expression for K_c and calculate it: $K_c = \frac{[PCl_3][Cl_2]}{[PCl_5]} = \frac{[0.016][0.016]}{[0.024]} = \mathbf{0.011}$

Now find the units of K_c: $K_c = \frac{(\text{mol dm}^{-3})(\text{mol dm}^{-3})}{\text{mol dm}^{-3}} = \text{mol dm}^{-3}$ So $K_c = \mathbf{0.011\ mol\ dm^{-3}}$

Practice Questions

Q1 Describe what a dynamic equilibrium is.

Q2 Write the expression for K_c for the following reaction: $2NO_{(g)} + O_{2(g)} \rightleftharpoons 2NO_{2(g)}$

Q3 Explain why you need to work out the units for K_c.

Q4 What are the units for K_c in the reaction in question 2?

Exam Questions

1 In the Haber process to make ammonia, nitrogen and hydrogen are reacted over an iron catalyst.

$$N_{2\,(g)} + 3H_{2\,(g)} \rightleftharpoons 2NH_{3\,(g)}$$

At 900 K, the equilibrium concentrations are $[N_2] = 1.06$ mol dm⁻³, $[H_2] = 1.41$ mol dm⁻³ and $[NH_3] = 0.150$ mol dm⁻³

a) Write an expression for the equilibrium constant for this reaction. [2 marks]

b) Calculate a value for K_c at this temperature and give its units. [3 marks]

2 Nitrogen dioxide dissociates according to the equation $2NO_{2(g)} \rightleftharpoons 2NO_{(g)} + O_{2(g)}$.

When 42.5 g of nitrogen dioxide were heated in a vessel of volume 22.8 dm³ at 500 °C, 14.1 g of oxygen were found in the equilibrium mixture.

a) Calculate i) the number of moles of nitrogen dioxide originally. [1 mark]
ii) the number of moles of each gas in the equilibrium mixture. [3 marks]

b) Write an expression for K_c for this reaction. Calculate the value for K_c at 500 °C and give its units. [5 marks]

A big K_c means heaps of product...

Most organic reactions and plenty of inorganic reactions are reversible. Sometimes the backwards reaction's about as speedy as a dead snail though, so some reactions might be thought of as only going one way. It's like if you're walking forwards, continental drift could be moving you backwards at the same time, just reeeeally slowly.

More on the Equilibrium Constant

You didn't think it was going to be that easy, did you? No, there's lots more to learn about the equilibrium constant... Did you know his friends call him Reggie, he's a fan of hardcore hip hop and he's a sworn enemy of the rate constant?

K_c can be used to Find Concentrations in an Equilibrium Mixture

Example: When ethanoic acid was allowed to reach equilibrium with ethanol at 25 °C, it was found that the equilibrium mixture contained 2.0 mol dm^{-3} ethanoic acid and 3.5 mol dm^{-3} ethanol. The K_c of the equilibrium is 4.0 at 25 °C. What are the concentrations of the other components?

$$CH_3COOH_{(l)} + C_2H_5OH_{(l)} \rightleftharpoons CH_3COOC_2H_{5(l)} + H_2O_{(l)}$$

Put all the values you know in the K_c expression: $K_c = \frac{[CH_3COOC_2H_5][H_2O]}{[CH_3COOH][C_2H_5OH]} \Rightarrow 4.0 = \frac{[CH_3COOC_2H_5][H_2O]}{2.0 \times 3.5}$

Rearranging this gives: $[CH_3COOC_2H_5][H_2O] = 4.0 \times 2.0 \times 3.5 = 28.0$

But from the equation, $[CH_3COOC_2H_5] = [H_2O]$.

This step might take a while to get your head around — the equation tells you that for every mole of $CH_3COOC_2H_5$ produced, one mole of H_2O is also produced, so their concentrations will always be equal. (The reactant concentrations aren't the same since they were different at the start).

So: $[CH_3COOC_2H_5] = [H_2O] = \sqrt{28} = 5.3$ mol dm^{-3}

The concentration of $CH_3COOC_2H_5$ and H_2O is 5.3 mol dm^{-3}

If Conditions Change *the* Position of Equilibrium *Will Move*

If you **change** the **concentration**, **pressure** or **temperature** of a reversible reaction, you're going to **alter** the **position of equilibrium**. This just means you'll end up with **different amounts** of reactants and products at equilibrium.

If the position of equilibrium moves to the **left**, you'll get more **reactants**. $\mathbf{H_{2(g)} + I_{2(g)}} \rightleftharpoons 2HI_{(g)}$

If the position of equilibrium moves to the **right**, you'll get more **products**. $H_{2(g)} + I_{2(g)} \rightleftharpoons \mathbf{2HI_{(g)}}$

There's a rule that lets you predict how the **position of equilibrium** will change if a **condition changes**. Here it is:

If there's a change in **concentration**, **pressure** or **temperature**, the equilibrium will move to help **counteract** the change.

So, basically, if you **raise the temperature**, the position of equilibrium will shift to try to **cool things down**.
And if you **raise the pressure or concentration**, the position of equilibrium will shift to try to **reduce it again**.

Temperature Changes Alter K_c

TEMPERATURE

1) If you **increase** the temperature of a reaction, the equilibrium will shift in the **endothermic direction** to absorb the extra heat.
2) **Decreasing** the reaction temperature will shift the equilibrium in the **exothermic direction** to replace the lost heat.
3) If the change means **more product** is formed, K_c will **increase**. If it means **less product** is formed, then K_c will **decrease**.

If the forward direction of a reversible reaction is **endothermic**, the reverse direction will be **exothermic**, and vice versa.

An **exothermic** reaction **releases heat** and has a **negative** ΔH.
An **endothermic** reaction **absorbs heat** and has a **positive** ΔH.

The reaction below is exothermic in the forward direction. If you increase the temperature, the equilibrium shifts to the left to absorb the extra heat. This means that less product is formed.

Exothermic ⟹

$$2SO_{2(g)} + O_{2(g)} \rightleftharpoons 2SO_{3(g)} \quad \Delta H = -197 \text{ kJ mol}^{-1}$$

⟸ Endothermic

$$K_c = \frac{[SO_3]^2}{[SO_2]^2[O_2]}$$

There's less product, so K_c decreases.

More on the Equilibrium Constant

Concentration and Pressure Changes Don't Affect K_c

CONCENTRATION

The value of the **equilibrium constant**, K_c, is **fixed** at a given temperature. So if the concentration of one thing in the equilibrium mixture **changes** then the concentrations of the others must change to keep the value of K_c the same.

$$CH_3COOH_{(l)} + C_2H_5OH_{(l)} \rightleftharpoons CH_3COOC_2H_{5(l)} + H_2O_{(l)}$$

If you increase the concentration of CH_3COOH then the equilibrium will move to the right to get rid of the extra CH_3COOH — so more $CH_3COOC_2H_5$ and H_2O are produced. This keeps the equilibrium constant the same.

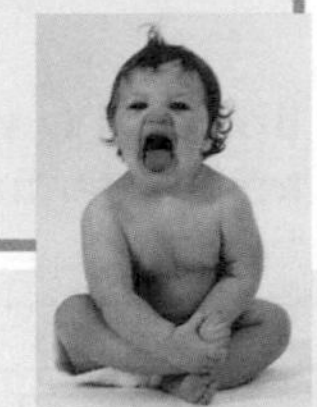

The removal of his dummy was a change that Maxwell always opposed.

PRESSURE (changing this only really affects **equilibria involving gases**)

Increasing the pressure shifts the equilibrium to the side with **fewer** gas molecules — this **reduces** the pressure. **Decreasing** the pressure shifts the equilibrium to the side with **more** gas molecules. This **raises** the pressure again. K_c stays the **same**, no matter what you do to the pressure.

There are 3 moles on the left, but only 2 on the right. So an increase in pressure would shift the equilibrium to the right. ⟹ $2SO_{2(g)} + O_{2(g)} \rightleftharpoons 2SO_{3(g)}$

Catalysts have **NO EFFECT** on the **position of equilibrium.**
They **can't** increase **yield** — but they **do** mean equilibrium is approached **faster**.

Practice Questions

Q1 If you raise the temperature of a reversible reaction, in which direction will the reaction move?

Q2 Does temperature change affect K_c?

Q3 Why does concentration not affect K_c?

Q4 What effect do catalysts have on the equilibrium of a reaction?

Exam Questions

1 The following equilibrium was established at temperature T_1:

$$2SO_{2(g)} + O_{2(g)} \rightleftharpoons 2SO_{3(g)} \quad \Delta H = -196 \text{ kJmol}^{-1}.$$

K_c at T_1 was found to be 0.67 $\text{mol}^{-1}\text{ dm}^3$.

a) When equilibrium was established at a different temperature, T_2, the value of K_c was found to have increased. State which of T_1 or T_2 is the lower temperature and explain why. [3 marks]

b) The experiment was repeated exactly the same in all respects at T_1, except a flask of smaller volume was used. How would this change affect the yield of sulfur trioxide and the value of K_c? [2 marks]

2 The reaction between methane and steam is used to produce hydrogen. The forward reaction is endothermic.

$$CH_{4\,(g)} + H_2O_{\,(g)} \rightleftharpoons CO_{\,(g)} + 3H_{2\,(g)}$$

a) Write an equation for K_c for this reaction. [2 marks]

b) How will the value of K_c be affected by:
i) increasing the temperature,
ii) using a catalyst. [2 marks]

c) How will the composition of the equilibrium mixture be affected by increasing the pressure? [2 marks]

Shift to the left, and then jump to the right...

Hmm, sounds like there's a song in there somewhere. I'm getting an image now of chemists in lab coats dancing at a Xmas party... Let's not go there. Instead just make sure you really get your head round this concept of changing conditions and the equilibrium shifting to compensate. Reread until you've definitely got it — it makes this topic much easier to learn.

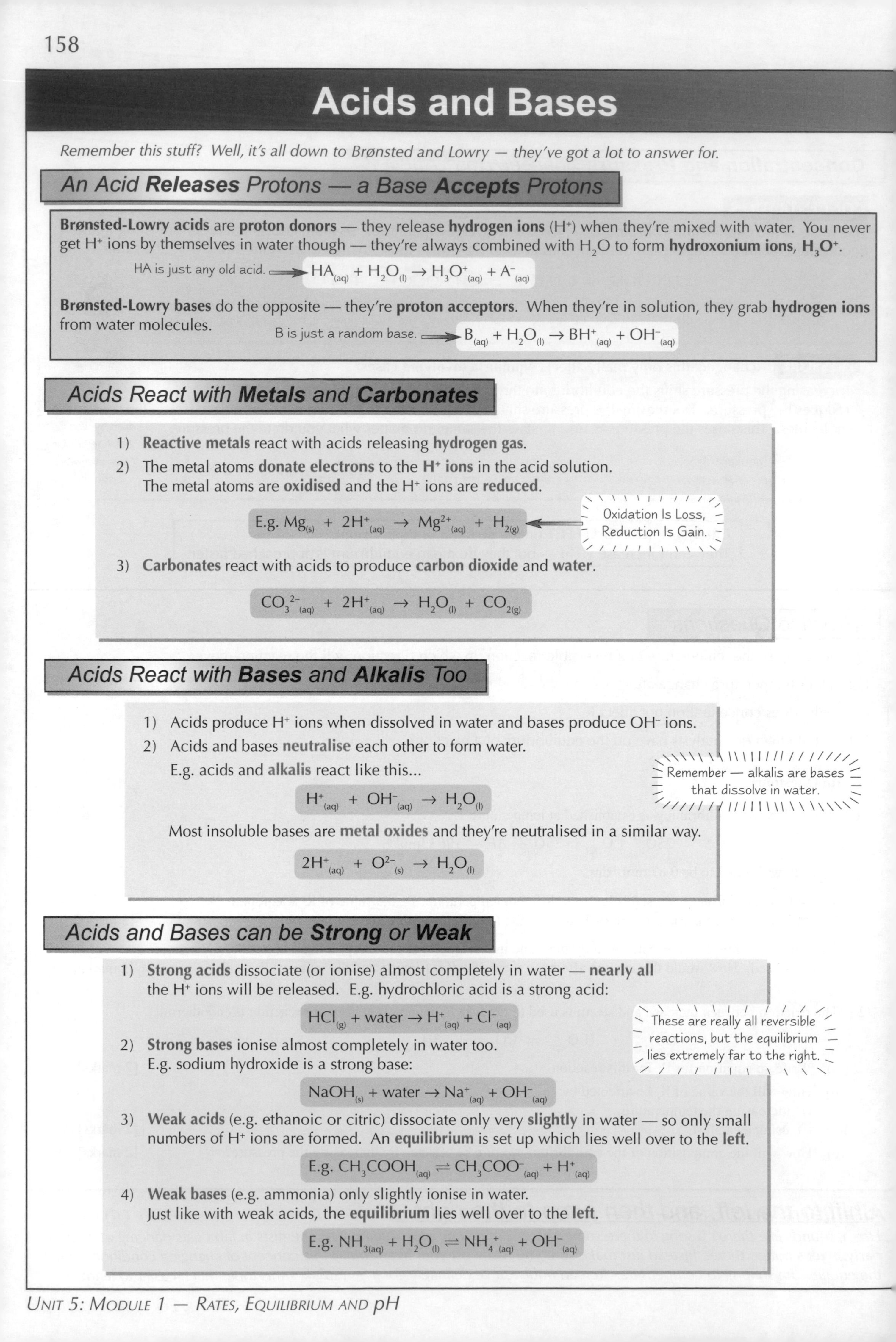

Acids and Bases

Remember this stuff? Well, it's all down to Brønsted and Lowry — they've got a lot to answer for.

An Acid **Releases** Protons — a Base **Accepts** Protons

Brønsted-Lowry acids are **proton donors** — they release **hydrogen ions** (H^+) when they're mixed with water. You never get H^+ ions by themselves in water though — they're always combined with H_2O to form **hydroxonium ions, H_3O^+**.

HA is just any old acid. → $HA_{(aq)} + H_2O_{(l)} \rightarrow H_3O^+_{(aq)} + A^-_{(aq)}$

Brønsted-Lowry bases do the opposite — they're **proton acceptors**. When they're in solution, they grab **hydrogen ions** from water molecules.

B is just a random base. → $B_{(aq)} + H_2O_{(l)} \rightarrow BH^+_{(aq)} + OH^-_{(aq)}$

Acids React with **Metals** and **Carbonates**

1) **Reactive metals** react with acids releasing **hydrogen gas**.
2) The metal atoms **donate electrons** to the **H^+ ions** in the acid solution. The metal atoms are **oxidised** and the H^+ ions are **reduced**.

$$\text{E.g. } Mg_{(s)} + 2H^+_{(aq)} \rightarrow Mg^{2+}_{(aq)} + H_{2(g)}$$

Oxidation Is Loss, Reduction Is Gain.

3) **Carbonates** react with acids to produce **carbon dioxide** and **water**.

$$CO_3{}^{2-}_{(aq)} + 2H^+_{(aq)} \rightarrow H_2O_{(l)} + CO_{2(g)}$$

Acids React with **Bases** and **Alkalis** Too

1) Acids produce H^+ ions when dissolved in water and bases produce OH^- ions.
2) Acids and bases **neutralise** each other to form water.
   E.g. acids and **alkalis** react like this...

$$H^+_{(aq)} + OH^-_{(aq)} \rightarrow H_2O_{(l)}$$

Remember — alkalis are bases that dissolve in water.

Most insoluble bases are **metal oxides** and they're neutralised in a similar way.

$$2H^+_{(aq)} + O^{2-}_{(s)} \rightarrow H_2O_{(l)}$$

Acids and Bases can be **Strong** or **Weak**

1) **Strong acids** dissociate (or ionise) almost completely in water — **nearly all** the H^+ ions will be released. E.g. hydrochloric acid is a strong acid:

$$HCl_{(g)} + \text{water} \rightarrow H^+_{(aq)} + Cl^-_{(aq)}$$

These are really all reversible reactions, but the equilibrium lies extremely far to the right.

2) **Strong bases** ionise almost completely in water too.
   E.g. sodium hydroxide is a strong base:

$$NaOH_{(s)} + \text{water} \rightarrow Na^+_{(aq)} + OH^-_{(aq)}$$

3) **Weak acids** (e.g. ethanoic or citric) dissociate only very **slightly** in water — so only small numbers of H^+ ions are formed. An **equilibrium** is set up which lies well over to the **left**.

$$\text{E.g. } CH_3COOH_{(aq)} \rightleftharpoons CH_3COO^-_{(aq)} + H^+_{(aq)}$$

4) **Weak bases** (e.g. ammonia) only slightly ionise in water.
   Just like with weak acids, the **equilibrium** lies well over to the **left**.

$$\text{E.g. } NH_{3(aq)} + H_2O_{(l)} \rightleftharpoons NH_4{}^+_{(aq)} + OH^-_{(aq)}$$

Acids and Bases

Acids and Bases form **Conjugate Pairs** in Water

When an acid is added to water, the equilibrium below is set up.

$$HA + H_2O \rightleftharpoons H_3O^+ + A^-$$

acid (HA), base (H_2O), acid (H_3O^+), base (A^-). Conjugate pair: H_2O and H_3O^+; conjugate pair: HA and A^-.

Don't forget — protons and H^+ ions are the same thing.

- In the **forward reaction**, HA acts as an **acid** as it **donates** a proton.
- In the **reverse reaction**, A^- acts as a **base** and **accepts** a proton from the H_3O^+ ion to form HA.

HA and A^- are called a **conjugate pair** — HA is the **conjugate acid** of A^- and A^- is the **conjugate base** of the acid, HA. **H_2O** and **H_3O^+** are a conjugate pair too.

The acid and base of a conjugate pair can be linked by an **H^+**, like this: $HA \rightleftharpoons H^+ + A^-$ or this: $H^+ + H_2O \rightleftharpoons H_3O^+$

Here's the equilibrium for aqueous HCl. Cl^- is the conjugate base of $HCl_{(aq)}$.

$$HCl_{(aq)} + H_2O_{(l)} \rightleftharpoons H_3O^+_{(aq)} + Cl^-_{(aq)}$$

acid (HCl), base (H_2O), acid (H_3O^+), base (Cl^-). Conjugate pair: H_2O and H_3O^+; conjugate pair: HCl and Cl^-.

An equilibrium with **conjugate pairs** is also set up when a **base** dissolves in water. The base B takes a proton from the water to form **BH^+** — so B is the **conjugate base** of BH^+, and BH^+ is the **conjugate acid** of B. H_2O and OH^- also form a **conjugate pair**.

$$B + H_2O \rightleftharpoons BH^+ + OH^-$$

base (B), acid (H_2O), acid (BH^+), base (OH^-). Conjugate pair: B and BH^+; conjugate pair: H_2O and OH^-.

Practice Questions

Q1 Give the Brønsted-Lowry definitions of an acid and a base.

Q2 Describe, in terms of equilibrium, how strong and weak acids differ in the way they dissolve in water.

Q3 What is the conjugate base of nitric acid, HNO_3?

Q4 Write an ionic equation to show the reaction between copper(II) oxide and hydrochloric acid.

Exam Questions

1 Magnesium completely dissolves in aqueous sulfuric acid, $H_2SO_{4\,(aq)}$.

a) Which ions are present in a solution of sulfuric acid? [1 mark]

b) Write an ionic equation for the reaction of the acid and magnesium. [1 mark]

c) What is the conjugate base of sulfuric acid? [1 mark]

d) Explain in equilibrium terms why sulfuric acid is considered a strong acid. [2 marks]

2 Hydrocyanic acid, HCN, is a weak acid.

a) Write an equation to show the equilibrium set up when it is added to water. [1 mark]

b) Use your equation to explain why HCN is a weak acid. [1 mark]

c) From your equation, identify the two conjugate pairs formed. [2 marks]

d) Which ion links conjugate pairs? [1 mark]

3 Dry ammonia gas is neutral but when it is added to water, a weakly alkaline solution forms.

a) Write an equation to show the equilibrium set up when ammonia is dissolved into water. [1 mark]

b) Is water behaving as an acid or a base in this equilibrium? Give a reason for your answer. [2 marks]

c) What species forms a conjugate pair with water in this reaction? [1 mark]

Alsatians and Bassets — keep them apart or they'll neuterise each other...

Don't confuse strong acids with concentrated acids, or weak acids with dilute acids. Strong and weak are to do with how much an acid ionises, whereas concentrated and dilute are to do with the number of moles of acid you've got per dm³. You can have a dilute strong acid, or a concentrated weak acid. It works just the same way with bases too.

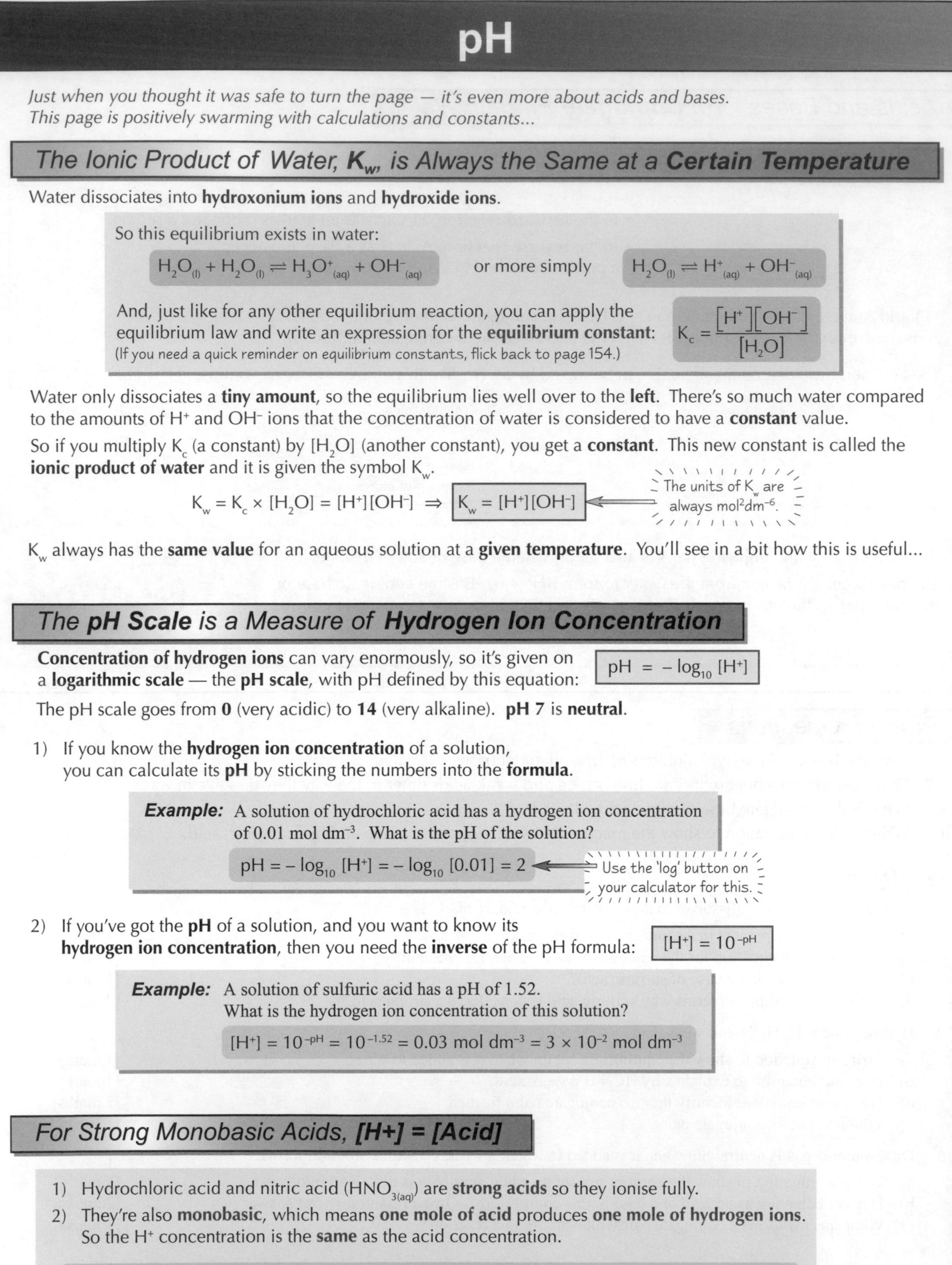

pH

Just when you thought it was safe to turn the page — it's even more about acids and bases. This page is positively swarming with calculations and constants...

The Ionic Product of Water, K_w, is Always the Same at a Certain Temperature

Water dissociates into **hydroxonium ions** and **hydroxide ions**.

So this equilibrium exists in water:

$H_2O_{(l)} + H_2O_{(l)} \rightleftharpoons H_3O^+_{(aq)} + OH^-_{(aq)}$ or more simply $H_2O_{(l)} \rightleftharpoons H^+_{(aq)} + OH^-_{(aq)}$

And, just like for any other equilibrium reaction, you can apply the equilibrium law and write an expression for the **equilibrium constant**: $K_c = \frac{[H^+][OH^-]}{[H_2O]}$

(If you need a quick reminder on equilibrium constants, flick back to page 154.)

Water only dissociates a **tiny amount**, so the equilibrium lies well over to the **left**. There's so much water compared to the amounts of H^+ and OH^- ions that the concentration of water is considered to have a **constant** value.

So if you multiply K_c (a constant) by $[H_2O]$ (another constant), you get a **constant**. This new constant is called the **ionic product of water** and it is given the symbol K_w.

$$K_w = K_c \times [H_2O] = [H^+][OH^-] \Rightarrow K_w = [H^+][OH^-]$$

The units of K_w are always mol^2dm^{-6}.

K_w always has the **same value** for an aqueous solution at a **given temperature**. You'll see in a bit how this is useful...

The pH Scale is a Measure of Hydrogen Ion Concentration

Concentration of hydrogen ions can vary enormously, so it's given on a **logarithmic scale** — the **pH scale**, with pH defined by this equation: $pH = -\log_{10}[H^+]$

The pH scale goes from **0** (very acidic) to **14** (very alkaline). **pH 7** is **neutral**.

1) If you know the **hydrogen ion concentration** of a solution, you can calculate its **pH** by sticking the numbers into the **formula**.

Example: A solution of hydrochloric acid has a hydrogen ion concentration of 0.01 mol dm^{-3}. What is the pH of the solution?

$pH = -\log_{10}[H^+] = -\log_{10}[0.01] = 2$

Use the 'log' button on your calculator for this.

2) If you've got the **pH** of a solution, and you want to know its **hydrogen ion concentration**, then you need the **inverse** of the pH formula: $[H^+] = 10^{-pH}$

Example: A solution of sulfuric acid has a pH of 1.52. What is the hydrogen ion concentration of this solution?

$[H^+] = 10^{-pH} = 10^{-1.52} = 0.03$ mol dm^{-3} $= 3 \times 10^{-2}$ mol dm^{-3}

For Strong Monobasic Acids, [H+] = [Acid]

1) Hydrochloric acid and nitric acid ($HNO_{3(aq)}$) are **strong acids** so they ionise fully.
2) They're also **monobasic**, which means **one mole of acid** produces **one mole of hydrogen ions**. So the H^+ concentration is the **same** as the acid concentration.

So for **0.1 mol dm^{-3} HCl**, $[H^+]$ is also 0.1 mol dm^{-3}. So the **pH** $= -\log_{10}[H^+] = -\log_{10} 0.1 =$ **1.0**.
Or for **0.05 mol dm^{-3} HNO_3**, $[H^+]$ is also 0.05 mol dm^{-3}, giving **pH** $= -\log_{10} 0.05 =$ **1.30**

pH

Use K_w to Find the pH of a Strong Base

1) Sodium hydroxide (NaOH) and potassium hydroxide (KOH) are **strong bases** that **fully ionise** in water.
2) They donate **one mole of OH^- ions** per mole of base.
   This means that the concentration of OH^- ions is the **same** as the **concentration of the base**.
   So for 0.02 mol dm^{-3} sodium hydroxide solution, $[OH^-]$ is also **0.02 mol dm^{-3}**.
3) But to work out the **pH** you need to know **$[H^+]$**
   — luckily this is linked to **$[OH^-]$** through the **ionic product of water, K_w**: **$K_w = [H^+][OH^-]$**
4) So if you know K_w and $[OH^-]$ for a **strong aqueous base** at a certain temperature, you can work out **$[H^+]$** and then the **pH**.

Example: The value of K_w at 298 K is 1.0×10^{-14}. Find the pH of 0.1 mol dm^{-3} NaOH at 298 K.

$$[OH^-] = 0.1 \text{ mol dm}^{-3} \Rightarrow [H^+] = \frac{K_w}{[OH^-]} = \frac{1.0\times10^{-14}}{0.1} = 1.0 \times 10^{-13} \text{ mol dm}^{-3}$$

So $pH = -\log_{10} 1.0 \times 10^{-13} =$ **13.0**

Get off the page, rate constant! You're not relevant here.

Practice Questions

Q1 Write down the formula of the ionic product of water. What are its units?

Q2 Write the formula for calculating the pH of a solution.

Q3 What is a monobasic acid? What can you assume about $[H^+]$ for a strong monobasic acid?

Q4 Explain how you'd find the pH of a strong base.

Q5 Should the rate constant be allowed to stay on this page or should he make like a tree?

Exam Questions

1 a) Explain the relationship between K_c and K_w for water. [2 marks]

b) A solution of the strong acid hydrobromic acid, HBr, has a concentration of 0.32 mol dm^{-3}. Calculate its pH. [2 marks]

c) Hydrobromic acid is a stronger acid than hydrochloric acid. Explain what that means in terms of hydrogen ions and pH. [2 marks]

2 A solution of sodium hydroxide contains 2.5 g dm^{-3}. The value of K_w at 298 K is 1.0×10^{-14}.

a) What is the molar concentration of the hydroxide ions in this solution? [2 marks]

b) Calculate the pH of this solution. [3 marks]

c) Why is this value temperature dependent? [1 mark]

3 The value of K_w, the ionic product of water, is 1.0×10^{-14} at 298 K. Calculate the pH of a 0.0370 mol dm^{-3} solution of sodium hydroxide at 298 K. [4 marks]

All I want for Christmas is a chemistry question requiring the use of logs...

You know things are getting serious when maths stuff like logs start appearing. It's fine really though, just practise a few questions and make sure you know how to use the log button on your calculator. And make sure you've learned the equation for K_w and both pH equations. And while you're up, go and make me a nice cup of coffee, lots of milk, no sugar.

More pH Calculations

More acid calculations to come, so you'll need to get that calculator warmed up... Either hold it for a couple of minutes in your armpit, or even better, warm it between your clenched buttocks. OK done that? Good stuff...

*To Find the **pH** of a **Weak Acid** you use K_a (the **Acid Dissociation Constant**)*

Weak acids **don't** ionise fully in solution, so the $[H^+]$ **isn't** the same as the acid concentration. This makes it a **bit trickier** to find their pH. You have to use yet another **equilibrium constant**, K_a.

For a weak aqueous acid, HA, you get the following equilibrium: $HA_{(aq)} \rightleftharpoons H^+_{(aq)} + A^-_{(aq)}$.

As only a **tiny amount** of HA dissociates, you can assume that $[HA_{(aq)}]_{start} = [HA_{(aq)}]_{equilibrium}$.

So if you apply the equilibrium law, you get: $K_a = \frac{[H^+][A^-]}{[HA]}$

You can also assume that **all** the **H^+ ions** come from the **acid**, so $[H^+_{(aq)}] = [A^-_{(aq)}]$.

So $K_a = \frac{[H^+]^2}{[HA]}$ ← The units of K_a are mol dm^{-3}.

Here's an example of how to use K_a to find the **pH** of a weak acid:

Example:
Calculate the hydrogen ion concentration and the pH of a 0.02 mol dm^{-3} solution of propanoic acid (CH_3CH_2COOH).
K_a for propanoic acid at this temperature is 1.30×10^{-5} mol dm^{-3}.

$K_a = \frac{[H^+]^2}{[CH_3CH_2COOH]}$ $\Rightarrow$ $[H^+]^2 = K_a[CH_3CH_2COOH] = 1.30 \times 10^{-5} \times 0.02 = 2.60 \times 10^{-7}$

$\Rightarrow$ $[H^+] = \sqrt{2.60 \times 10^{-7}}$ = **5.10×10^{-4} mol dm^{-3}**

So pH = $-\log_{10} 5.10 \times 10^{-4}$ = **3.29**

*You Might Have to Find the **Concentration** or K_a of a **Weak Acid***

You don't need to know anything new for this type of calculation. You usually just have to find **$[H^+]$** from the pH, then fiddle around with the **K_a expression** to find the missing bit of information.

Example:
The pH of an ethanoic acid (CH_3COOH) solution was 3.02 at 298 K.
Calculate the molar concentration of this solution.
The K_a of ethanoic acid is 1.75×10^{-5} mol dm^{-3} at 298 K.

$[H^+] = 10^{-pH} = 10^{-3.02} = 9.55 \times 10^{-4}$ mol dm^{-3}

$K_a = \frac{[H^+]^2}{[CH_3COOH]}$ $\Rightarrow$ $[CH_3COOH] = \frac{[H^+]^2}{K_a} = \frac{(9.55 \times 10^{-4})^2}{1.75 \times 10^{-5}}$

= **0.0521 mol dm^{-3}**

This bunny may look cute, but he can't help Horace with his revision.

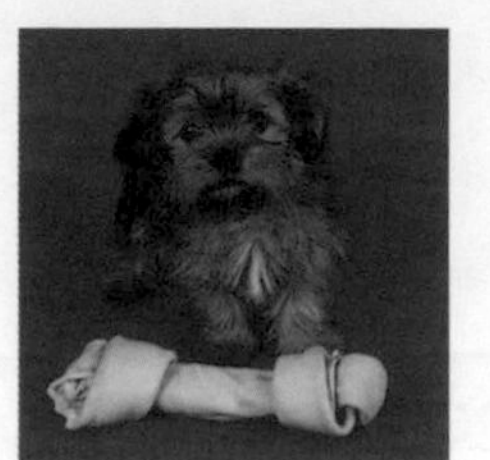

Mmmm, that does looks like a tasty bone, Rex, but I don't think it'll help with K_a calculations.

Example:
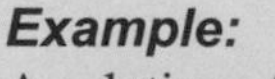
A solution of 0.162 mol dm^{-3} HCN has a pH of 5.05 at 298 K. What is the value of K_a for HCN at 298 K?

$[H^+] = 10^{-pH} = 10^{-5.05} = 8.91 \times 10^{-6}$ mol dm^{-3}

$K_a = \frac{[H^+]^2}{[HCN]} = \frac{(8.91 \times 10^{-6})^2}{0.162}$ = **4.90×10^{-10} mol dm^{-3}**

More pH Calculations

$pK_a = -log_{10} K_a$ and $K_a = 10^{-pK_a}$

pK_a is calculated from K_a in exactly the same way as pH is calculated from $[H^+]$ — and vice versa.
So if an acid has a K_a value of 1.50×10^{-7} mol dm^{-3}, its $\mathbf{pK_a = -log_{10}(1.50 \times 10^{-7}) = 6.82}$.
And if an acid has a pK_a value of 4.32, its $\mathbf{K_a = 10^{-4.32} = 4.79 \times 10^{-5}}$ **mol dm^{-3}**.

Notice how pK_a values aren't annoyingly tiny like K_a values.

Just to make things that bit more complicated, there might be a **pK_a** value in a question.
If so, you need to convert it to K_a so that you can use the **K_a expression**.

Example:
Calculate the pH of 0.050 mol dm^{-3} methanoic acid (HCOOH).
Methanoic acid has a pK_a of 3.75 at this temperature.

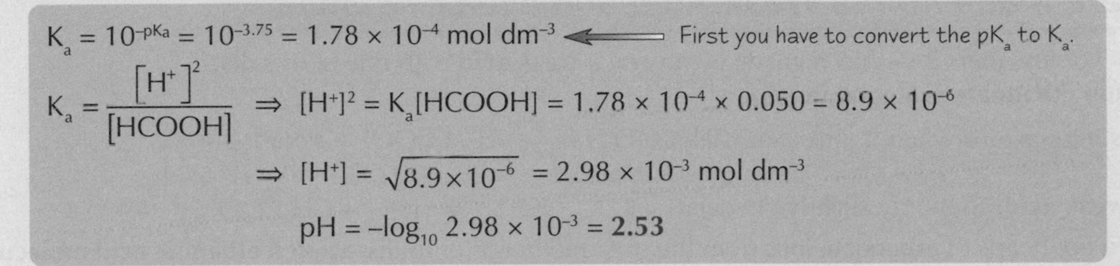

$K_a = 10^{-pK_a} = 10^{-3.75} = 1.78 \times 10^{-4}$ mol dm^{-3} ⟵ First you have to convert the pK_a to K_a.

$$K_a = \frac{[H^+]^2}{[HCOOH]} \Rightarrow [H^+]^2 = K_a[HCOOH] = 1.78 \times 10^{-4} \times 0.050 = 8.9 \times 10^{-6}$$

$$\Rightarrow [H^+] = \sqrt{8.9 \times 10^{-6}} = 2.98 \times 10^{-3} \text{ mol dm}^{-3}$$

$$pH = -log_{10} 2.98 \times 10^{-3} = \mathbf{2.53}$$

Sometimes you have to give your answer as a **pK_a** value.
In this case, you just work out the K_a value as usual and then convert it to **pK_a** — and Bob's a revision goat.

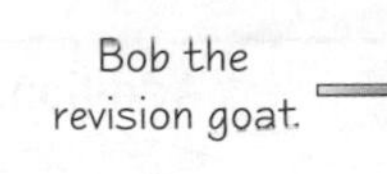

Practice Questions

Q1 What are the units of K_a?

Q2 Describe how you would calculate the pH of a weak acid from its acid dissociation constant.

Q3 How is pK_a defined?

Q4 Would you expect strong acids to have higher or lower K_a values than weak acids?

Exam Questions

1 The value of K_a for the weak acid HA, at 298 K, is 5.60×10^{-4} mol dm^{-3}.

a) Write an expression for K_a for HA. [1 mark]

b) Calculate the pH of a 0.280 mol dm^{-3} solution of HA at 298 K. [3 marks]

2 The pH of a 0.150 mol dm^{-3} solution of a weak monobasic acid, HX, is 2.65 at 298 K.

a) Calculate the value of K_a for the acid HX at 298 K. [4 marks]

b) Calculate pK_a for this acid. [2 marks]

3 Benzoic acid is a weak acid that is used as a food preservative. It has a pK_a of 4.2 at 298 K.
Find the pH of a 1.6×10^{-4} mol dm^{-3} solution of benzoic acid at 298 K. [4 marks]

Fluffy revision animals... aaawwwww...

Strong acids have high K_a values and weak acids have low K_a values. For pK_a values, it's the other way round — the stronger the acid, the lower the pK_a. If something's got p in front of it, like pH, pK_w or pK_a, it'll mean $-log_{10}$ of whatever. Oh and did you like all the cute animals on this page? Did it really make your day? Good, I'm really pleased about that.

Buffer Action

I always found buffers a bit mind-boggling. How can a solution resist becoming more acidic if you add acid to it? And why would it want to? Here's where you find out...

Buffers Resist Changes in pH

A **buffer** is a solution that **resists** changes in pH when **small** amounts of acid or alkali are added.

A buffer **doesn't** stop the pH from changing completely — it does make the changes **very slight** though.
Buffers only work for small amounts of acid or alkali — put too much in and they'll go "Waah" and not be able to cope.
You can get **acidic buffers** and **basic buffers** — but you only need to know about acidic ones.

Acidic Buffers are Made from a Weak Acid and one of its Salts

Acidic buffers have a pH of less than 7 — they're made by mixing a **weak acid** with one of its **salts**.
Ethanoic acid and **sodium ethanoate** is a good example:

The salt **fully** dissociates into its ions when it dissolves: $CH_3COO^-Na^+_{(aq)} \rightarrow CH_3COO^-_{(aq)} + Na^+_{(aq)}$.
Sodium ethanoate — Ethanoate ions

The ethanoic acid is a **weak acid**, so it only **slightly** dissociates: $CH_3COOH_{(aq)} \rightleftharpoons H^+_{(aq)} + CH_3COO^-_{(aq)}$

So in the solution you've got heaps of **ethanoate ions** from the salt, and heaps of **undissociated ethanoic acid molecules**.

Buffers work because of shifts in equilibrium (look back at p156 for more on this):

Lots of undissociated weak acid

Addition of H⁺ (acid) ←

$$CH_3COOH_{(aq)} \rightleftharpoons H^+_{(aq)} + CH_3COO^-_{(aq)}$$

Addition of OH⁻ (alkali) →

Lots of CH_3COO^-

If you add a **small** amount of **acid** the **H^+ concentration** increases. Most of the extra H^+ ions combine with CH_3COO^- ions to form CH_3COOH. This shifts the equilibrium to the **left**, reducing the H^+ concentration to close to its original value. So the **pH** doesn't change much.

The large number of CH_3COO^- ions make sure that the buffer can cope with the addition of acid.

If a **small** amount of **alkali** (e.g. NaOH) is added, the **OH^- concentration** increases. Most of the extra OH^- ions react with H^+ ions to form water — removing H^+ ions from the solution. This causes more CH_3COOH to **dissociate** to form H^+ ions — shifting the equilibrium to the **right**. The H^+ concentration increases until it's close to its original value, so the **pH** doesn't change much.

There's no problem doing this as there's absolutely loads of spare CH_3COOH molecules.

Buffer Solutions are Important in Biological Environments

1) **Cells** need a constant pH to allow the **biochemical reactions** to take place. The pH is controlled by a buffer based on the equilibrium between **dihydrogen phosphate** ions and **hydrogen phosphate** ions.

$$H_2PO_4^- \rightleftharpoons H^+ + HPO_4^{2-}$$

2) **Blood** needs to be kept at pH 7.4. It is buffered using carbonic acid. The levels of **H_2CO_3** are controlled by **respiration**.

$$H_2CO_{3(aq)} \rightleftharpoons H^+_{(aq)} + HCO_{3\,(aq)}^-$$
and
$$H_2CO_{3(aq)} \rightleftharpoons H_2O_{(l)} + CO_{2(aq)}$$

By **breathing out CO_2** the level of H_2CO_3 is reduced as it moves this **equilibrium** to the **right**.
The levels of HCO_3^- are controlled by the **kidneys** with excess being **excreted** in the urine.

3) Buffers are used in **food products** to control the pH. Changes in pH can be caused by **bacteria** and **fungi** and cause food to **deteriorate**. A common buffer is **citric acid** and **sodium citrate**. **Phosphoric acid/phosphate ions** and **benzoic acid/benzoate** ions are also used as buffers.

Buffer Action

Here's How to Calculate the pH of a Buffer Solution

Calculating the **pH** of an acidic buffer isn't too tricky. You just need to know the $\mathbf{K_a}$ of the weak acid and the **concentrations** of the weak acid and its salt. Here's how to go about it:

Example: A buffer solution contains 0.40 mol dm^{-3} methanoic acid, HCOOH, and 0.60 mol dm^{-3} sodium methanoate, $HCOO^-Na^+$.
For methanoic acid, $K_a = 1.8 \times 10^{-4}$ mol dm^{-3}. What is the pH of this buffer?

Firstly, write the expression for K_a of the weak acid:

$$HCOOH_{(aq)} \rightleftharpoons H^+_{(aq)} + HCOO^-_{(aq)} \quad \Rightarrow \quad K_a = \frac{[H^+_{(aq)}] \times [HCOO^-_{(aq)}]}{[HCOOH_{(aq)}]}$$

Remember — these all have to be equilibrium concentrations.

Then rearrange the expression and stick in the data to calculate $[H^+_{(aq)}]$:

$$[H^+_{(aq)}] = K_a \times \frac{[HCOOH_{(aq)}]}{[HCOO^-_{(aq)}]}$$

$$\Rightarrow [H^+_{(aq)}] = 1.8 \times 10^{-4} \times \frac{0.4}{0.6} = 1.20 \times 10^{-4} \text{ moldm}^{-3}$$

You have to make a **few assumptions** here:

- $HCOO^-Na^+$ is fully dissociated, so assume that the equilibrium concentration of $HCOO^-$ is the same as the initial concentration of $HCOO^-Na^+$.
- HCOOH is only slightly dissociated, so assume that its equilibrium concentration is the same as its initial concentration.

Acids and alkalis didn't mess with Jeff after he became buffer.

Finally, convert $[H^+_{(aq)}]$ to pH: $pH = -\log_{10}[H^+_{(aq)}] = -\log_{10}(1.20 \times 10^{-4}) = \mathbf{3.92}$ And that's your answer.

Practice Questions

Q1 What's a buffer solution?

Q2 Describe how a mixture of ethanoic acid and sodium ethanoate act as a buffer.

Q3 Describe how the pH of the blood is buffered.

Exam Questions

1 A buffer solution contains 0.40 mol dm^{-3} benzoic acid, C_6H_5COOH, and 0.20 mol dm^{-3} sodium benzoate, $C_6H_5COO^-Na^+$. At 25 °C, K_a for benzoic acid is 6.4×10^{-5} mol dm^{-3}.

a) Calculate the pH of the buffer solution. [3 marks]

b) Explain the effect on the buffer of adding a small quantity of dilute sulfuric acid. [3 marks]

2 A buffer was prepared by mixing solutions of butanoic acid, $CH_3(CH_2)_2COOH$, and sodium butanoate, $CH_3(CH_2)_2COO^-Na^+$, so that they had the same concentration.

a) Write a balanced chemical equation to show butanoic acid acting as a weak acid. [1 mark]

b) Given that K_a for butanoic acid is 1.5×10^{-5} mol dm^{-3}, calculate the pH of the buffer solution. [3 marks]

Old buffers are often resistant to change...

So that's how buffers work. There's a pleasing simplicity and neatness about it that I find rather elegant. Like a fine wine with a nose of berry and undertones of... OK, I'll shut up now.

pH Curves, Titrations and Indicators

If you add alkali to an acid, the pH changes in a squiggly sort of way.

Use **Titration** to Find the **Concentration** of an **Acid** or **Alkali**

Titrations let you find out **exactly** how much alkali is needed to **neutralise** a quantity of acid.

1) You measure out some **acid** of known concentration using a pipette and put it in a flask, along with some **appropriate indicator** (see below).
2) First do a rough titration — add the **alkali** to the acid using a **burette** fairly quickly to get an approximate idea of where the solution changes colour (the **end point**). Give the flask a regular **swirl**.
3) Now do an **accurate** titration. Run the alkali in to within 2 cm^3 of the end point, then add it **drop by drop**. If you don't notice exactly when the solution changes colour you've **overshot** and your result won't be accurate.
4) **Record** the amount of alkali needed to **neutralise** the acid.
   It's best to **repeat** this process a few times, making sure you get very similar answers each time (within about 0.2 cm^3 of each other).

Burette
Burettes measure different volumes and let you add the solution drop by drop.

Pipette
Pipettes measure only one volume of solution. Fill the pipette to about 3 cm above the line, then drop the level down carefully to the line.

alkali

scale

acid and indicator

You can also find out how much **acid** is needed to neutralise a quantity of **alkali**. It's exactly the same as above, but you add **acid to alkali** instead.

pH Curves Plot **pH** Against **Volume** of **Acid** or **Alkali** Added

The graphs below show pH curves for the **different combinations** of **strong and weak** monobasic acids and alkalis.

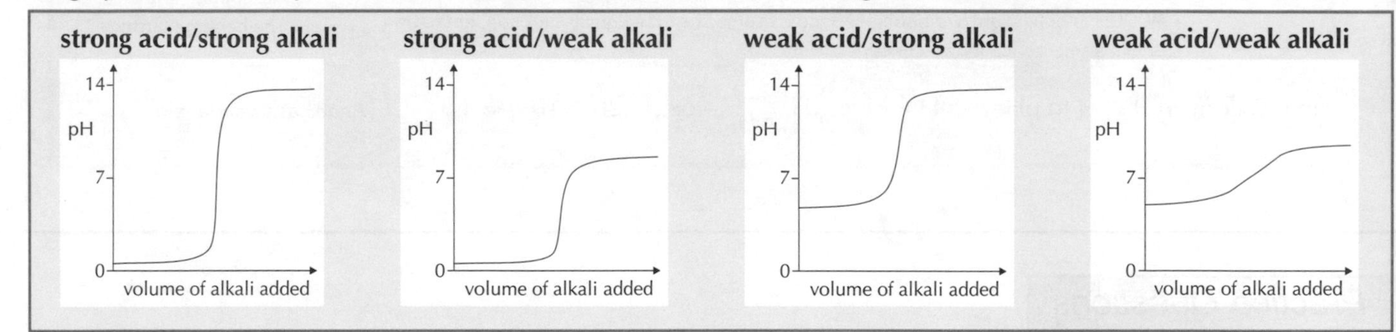

All the graphs apart from the weak acid/weak alkali graph have a bit that's almost vertical — this is the **equivalence point** or **end point**. At this point, a tiny amount of alkali causes a sudden, big change in pH — it's here that all the acid is just **neutralised**.

You don't get such a sharp change in a **weak acid/weak alkali** titration. The indicator colour changes **gradually** and it's tricky to see the exact end point. You're usually better off using a **pH meter** for this type of titration.

pH Curves can Help you Decide which **Indicator** to Use

Methyl orange and **phenolphthalein** are **indicators** that are often used for acid-base titrations. They each change colour over a **different pH range**:

Name of indicator	Colour at low pH	Approx. pH of colour change	Colour at high pH
Methyl orange	red	3.1 – 4.4	yellow
Phenolphthalein	colourless	8.3 – 10	pink

For a **strong acid/strong alkali** titration, you can use **either** of these indicators — there's a rapid pH change over the range for **both** indicators.

For a **strong acid/weak alkali** only **methyl orange** will do. The pH changes rapidly across the range for methyl orange, but not for phenolphthalein.

For a **weak acid/strong alkali**, **phenolphthalein** is the stuff to use. The pH changes rapidly over phenolphthalein's range, but not over methyl orange's.

For **weak acid/weak alkali** titrations there's no sharp pH change, so **neither** of these indicators will work.

pH Curves, Titrations and Indicators

You can Calculate **Concentrations** from Titration Data

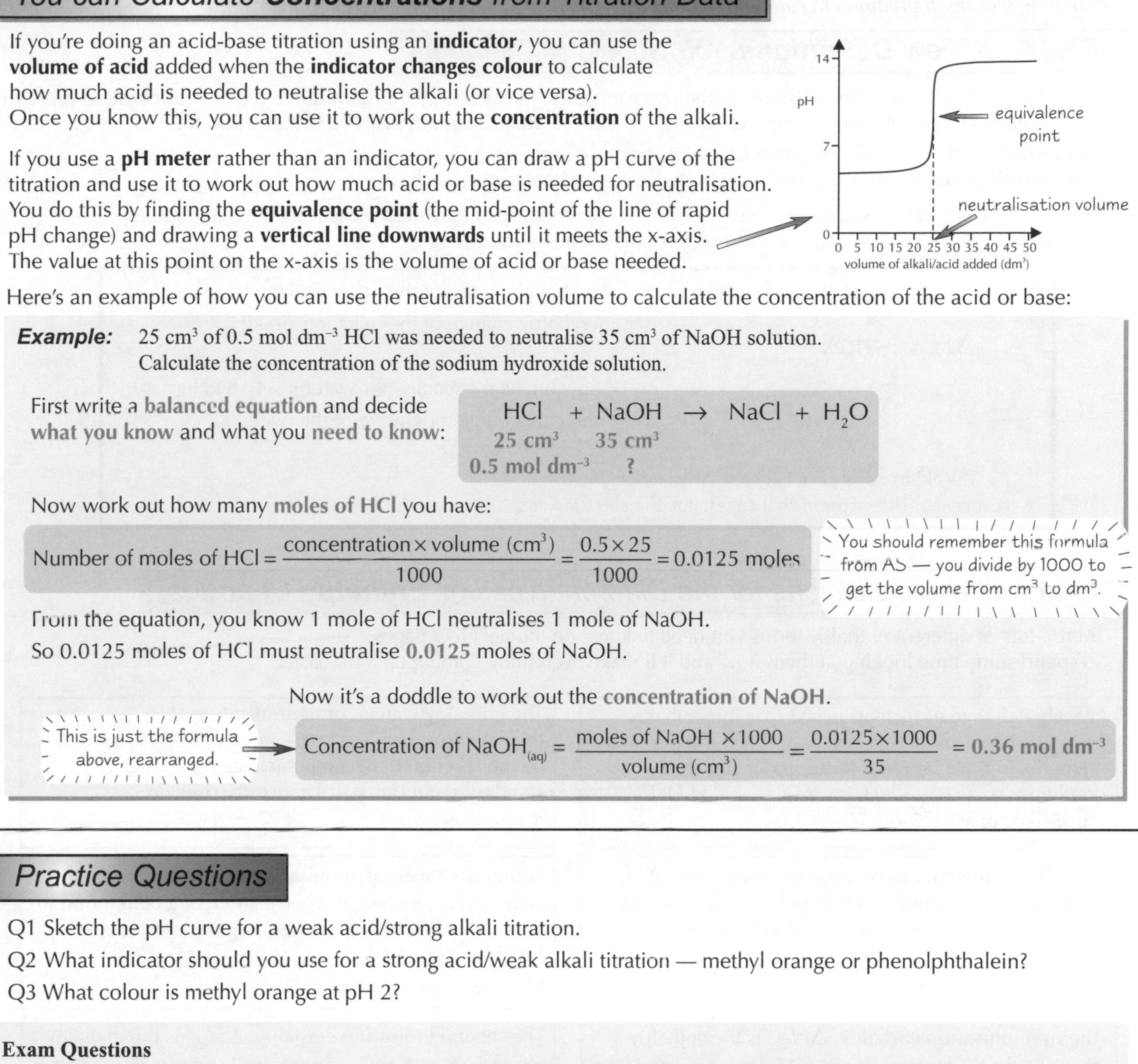

If you're doing an acid-base titration using an **indicator**, you can use the **volume of acid** added when the **indicator changes colour** to calculate how much acid is needed to neutralise the alkali (or vice versa).
Once you know this, you can use it to work out the **concentration** of the alkali.

If you use a **pH meter** rather than an indicator, you can draw a pH curve of the titration and use it to work out how much acid or base is needed for neutralisation. You do this by finding the **equivalence point** (the mid-point of the line of rapid pH change) and drawing a **vertical line downwards** until it meets the x-axis. The value at this point on the x-axis is the volume of acid or base needed.

Here's an example of how you can use the neutralisation volume to calculate the concentration of the acid or base:

Example: 25 cm³ of 0.5 mol dm⁻³ HCl was needed to neutralise 35 cm³ of NaOH solution.
Calculate the concentration of the sodium hydroxide solution.

First write a **balanced equation** and decide **what you know** and what you **need to know**:

$$HCl + NaOH \rightarrow NaCl + H_2O$$

HCl: **25 cm³**, **0.5 mol dm⁻³**; NaOH: **35 cm³**, **?**

Now work out how many **moles of HCl** you have:

$$\text{Number of moles of HCl} = \frac{\text{concentration} \times \text{volume (cm}^3)}{1000} = \frac{0.5 \times 25}{1000} = 0.0125 \text{ moles}$$

You should remember this formula from AS — you divide by 1000 to get the volume from cm³ to dm³.

From the equation, you know 1 mole of HCl neutralises 1 mole of NaOH.
So 0.0125 moles of HCl must neutralise **0.0125** moles of NaOH.

Now it's a doddle to work out the **concentration of NaOH**.

This is just the formula above, rearranged.

$$\text{Concentration of } NaOH_{(aq)} = \frac{\text{moles of NaOH} \times 1000}{\text{volume (cm}^3)} = \frac{0.0125 \times 1000}{35} = \mathbf{0.36 \text{ mol dm}^{-3}}$$

Practice Questions

Q1 Sketch the pH curve for a weak acid/strong alkali titration.

Q2 What indicator should you use for a strong acid/weak alkali titration — methyl orange or phenolphthalein?

Q3 What colour is methyl orange at pH 2?

Exam Questions

1 1.0 mol dm⁻³ NaOH is added separately to 25 cm³ samples of 1.0 mol dm⁻³ nitric acid and 1.0 mol dm⁻³ ethanoic acid. Describe two differences between the pH curves of the titrations. [2 marks]

2 A sample of 0.350 mol dm⁻³ ethanoic acid was titrated against potassium hydroxide.

a) Calculate the volume of 0.285 mol dm⁻³ potassium hydroxide required to just neutralise 25.0 cm³ of the ethanoic acid. [3 marks]

b) From the table on the right, select the best indicator for this titration, and explain your choice. [2 marks]

Name of indicator	pH range
bromophenol blue	3.0 – 4.6
methyl red	4.2 – 6.3
bromothymol blue	6.0 – 7.6
thymol blue	8.0 – 9.6

Try learning this stuff drop by drop...

Titrations involve playing with big bits of glassware that you're told not to break as they're really expensive — so you instantly become really clumsy. If you manage not to smash the burette, you'll find it easier to get accurate results if you use a dilute acid or alkali — drops of dilute acid and alkali contain fewer particles so you're less likely to overshoot.

Neutralisation and Enthalpy

I love the smell of a new section in the morning, but this one kinda merges with the last one. Never mind. Before you continue with neutralisation I'm sure you remember enthalpies from AS... well here's a quick reminder just in case...

First — A Few Definitions You Should Remember

ΔH is the symbol for **enthalpy change**. Enthalpy change is the **heat** energy transferred in a reaction at **constant pressure**.
$\Delta H^{\ominus}$ means that the enthalpy change was measured under **standard conditions (100 kPa and 298 K)**.

Exothermic reactions have a **negative ΔH** value, because heat energy is given out.
Endothermic reactions have a **positive ΔH** value, because heat energy is absorbed.

You can work out **ΔH** from a calorimeter experiment, which basically involves measuring the temperature change in some water as a result of the reaction. The formulas you use are:

$$q = mc\Delta T$$

$$\Delta H = -mc\Delta T$$

where, q = enthalpy change of the water (in kJ)
ΔH = enthalpy change of the reactants (in kJ)
m = mass of the water (in kg)
c = specific heat capacity (for water it's 4.18 kJ kg^{-1} K^{-1})
ΔT = the change in temperature of the water (K)

The change of sign is because ΔH is looking at the reactants (the water in this case), not the surroundings.

Learn These Definitions for the Different types of Enthalpy Changes

There's lots of different enthalpy terms you need to **know** on the next few pages.
So spend some time looking at them now and it'll make everything coming up a bit easier.

Enthalpy change of formation, ΔH_f, is the enthalpy change when **1 mole** of a **compound** is formed from its **elements** in their standard states under standard conditions, e.g. $2C_{(s)} + 3H_{2(g)} + \frac{1}{2}O_{2(g)} \rightarrow C_2H_5OH_{(l)}$
It's used on pages 170 and 171.

The enthalpy change of neutralisation ($\Delta H_{neutralisation}$) is the enthalpy change when **1 mole** of water is formed by the reaction between an acid and a base under **standard conditions**. It's an **exothermic** process.
It's used on page 169.

Enthalpy change of atomisation of an element, ΔH_{at}, is the enthalpy change when **1 mole** of **gaseous atoms** is formed from an element in its **standard state**, e.g. $\frac{1}{2}Cl_{2(g)} \rightarrow Cl_{(g)}$
It's used on pages 170 and 171.

Enthalpy change of atomisation of a compound, ΔH_{at}, is the enthalpy change when **1 mole** of a compound in its **standard state** is converted to **gaseous atoms**, e.g. $NaCl_{(s)} \rightarrow Na_{(g)} + Cl_{(g)}$
It's used on pages 170 and 171.

The **first ionisation enthalpy**, ΔH_{ie1}, is the enthalpy change when **1 mole** of **gaseous 1+ ions** is formed from **1 mole** of **gaseous atoms**, e.g. $Mg_{(g)} \rightarrow Mg^{+}_{(g)} + e^{-}$
It's used on pages 170 and 171.

The **second ionisation enthalpy**, ΔH_{ie2}, is the enthalpy change when **1 mole** of **gaseous 2+ ions** is formed from **1 mole** of **gaseous 1+ ions**, e.g. $Mg^{+}_{(g)} \rightarrow Mg^{2+}_{(g)} + e^{-}$
It's used on page 171.

First electron affinity, ΔH_{e1}, is the enthalpy change when **1 mole** of gaseous **1– ions** is made from **1 mole** of gaseous **atoms**, e.g. $O_{(g)} + e^{-} \rightarrow O^{-}_{(g)}$
It's used on pages 170 and 171.

Second electron affinity, ΔH_{e2}, is the enthalpy change when **1 mole** of **gaseous 2– ions** is made from **1 mole** of gaseous **1– ions**, e.g. $O^{-}_{(g)} + e^{-} \rightarrow O^{2-}_{(g)}$
It's used on page 171.

The **enthalpy change of hydration**, ΔH_{hyd}, is the enthalpy change when **1 mole** of **aqueous ions** is formed from **gaseous ions**, e.g. $Na^{+}_{(g)} \rightarrow Na^{+}_{(aq)}$
It's used on page 172.

The **enthalpy change of solution**, $\Delta H_{solution}$, is the enthalpy change when **1 mole** of **solute** is dissolved in **sufficient solvent** that no further enthalpy change occurs on further dilution, e.g. $NaCl_{(s)} \rightarrow NaCl_{(aq)}$
It's used on page 172.

Neutralisation and Enthalpy

The Enthalpy of Neutralisation can be Calculated From Experimental Data

Neutralisation always involves the reaction of **hydrogen ions** (H^+) with **hydroxide ions** (OH^-) to make **water** (H_2O). The change in energy when this happens is called the **enthalpy change of neutralisation** ($\Delta H_{neutralisation}$).

Example: 150 ml of hydrochloric acid (concentration 0.25 mol dm^{-3}) was neutralised by 150 ml of potassium hydroxide. The temperature increased by 1.71 °C. Calculate $\Delta H_{neutralisation}$.

1. The first thing to do is calculate the enthalpy change, ΔH, by plugging the numbers into $\Delta H = -mc\Delta T$.

   $\Delta H = -mc\Delta T = -0.3 \times 4.18 \times 1.71 = -2.144$ kJ

   But this is only the ΔH and **not** the $\Delta H_{neutralisation}$.

   The 0.3 comes from the mass of the solution — assume the solution has the same specific heat capacity and density (1 g/ml) as water.

2. To find the enthalpy of **neutralisation**, you need to calculate ΔH for **1 mole** of H_2O produced. To find out how many moles of water the reaction makes you can look at the **equation** for it.

   $$HCl + KOH \rightarrow KCl + H_2O$$

3. The number of moles of H_2O **made** is **equal** to the number of moles of acid used. And to work that out you can look back at the question.

   150 ml of 0.25 mol dm^{-3} hydrochloric acid has $0.25 \times 0.15 = 0.0375$ moles

   Don't forget to convert ml to dm³

   So, to make **1 mole** of H_2O:

   $\Delta H_{neutralisation} = -2.144 \div 0.0375 = -57.2$ kJ mol^{-1}

   The minus sign shows the reaction is exothermic

Weirdly, the value for **any strong acid** is about **–57 kJ mol⁻¹**. This is because all strong acids and bases completely ionise in water so essentially the reaction for each of them is the same ($H^+ + OH^- \rightarrow H_2O$).

For **weaker acids** and **alkalis** the value is **less negative** because energy is used to fully dissociate the acid or alkali meaning there's less energy released.

Practice Questions

Q1 What are the units of any enthalpy change?

Q2 What are standard conditions?

Q3 What is defined as "the enthalpy change when 1 mole of gaseous atoms is formed from a compound in its standard state"?

Q4 Why is the enthalpy change of neutralisation the same for all strong acids?

Exam Questions

1 In an experiment, 200 ml of 2.75 mol dm^{-3} H_2SO_4 was reacted with 200 ml of NaOH solution. The temperature rose by 38 °C. Assume the density of the resulting solution is 1 g ml^{-1}.

a) Write an equation for the reaction. [1 mark]

b) The specific heat capacity of water is 4.18 kJ kg^{-1} K^{-1}. Use this to calculate the enthalpy change of neutralisation for this reaction. [3 marks]

2 100 ml of 1 mol dm^{-3} HCl was added to 100 ml of 4 mol dm^{-3} KOH solution. The experiment was then repeated, replacing the acid with 100 ml of 1 mol dm^{-3} H_2SO_4 and again with 100 ml of 1 mol dm^{-3} H_3PO_4. Assume all three acids are equally strong.

a) Write equations for the three reactions. [3 marks]

b) The enthalpy change of neutralisation for any strong acid is about –57 kJ mol^{-1}. Explain which of the three experiments would give the largest temperature rise. [2 marks]

3 100 ml of a solution of hydrochloric acid with a concentration of X mol dm^{-3} was reacted with 100 ml of a solution of sodium hydroxide. The change in temperature was 3.4 K. The enthalpy change of neutralisation was –57 kJ mol^{-1}. Calculate the value of X. [4 marks]

My eyes, MY EYES — the definitions make them hurt...

The worst thing about this page is all the definitions. What's worse is it's not enough to just have a vague idea what each one means; you have to know the ins and outs — like whether it applies to gases, or to elements in their standard states. If you've forgotten, standard conditions are 298 K (otherwise known as 25 °C) and 100 kPa pressure.

Lattice Enthalpy and Born-Haber Cycles

On this page you can learn about lattice enthalpy, not lettuce enthalpy, which is the enthalpy change when 1 mole consumes salad from a veggie patch. Bu–dum cha... (that was meant to be a drum — work with me here).

Lattice Enthalpy is a Measure of Ionic Bond Strength

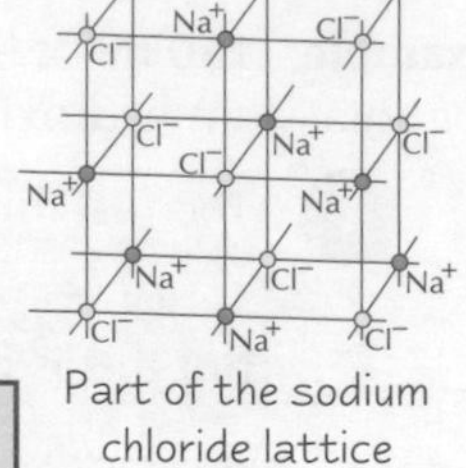

Part of the sodium chloride lattice

Remember how **ionic compounds** form regular structures called **giant ionic lattices** — and how the positive and negative ions are held together by **electrostatic attraction**. Well, when **gaseous ions** combine to make a solid lattice, energy is given out — this is called the **lattice enthalpy**. It's quite handy as it tells you how **strong** the ionic bonding is.

Here's the definition of **standard lattice enthalpy** that you need to know:

> The **standard lattice enthalpy**, $\Delta H^{\ominus}_{latt}$, is the enthalpy change when **1 mole** of a **solid ionic compound** is formed from its **gaseous ions** under standard conditions.
> It's a measure of **ionic bond strength**.

For example: $Na^+_{(g)} + Cl^-_{(g)} \rightarrow NaCl_{(s)}$ $\quad \Delta H^{\ominus} = -787$ kJmol^{-1}
$Mg^{2+}_{(g)} + O^{2-}_{(g)} \rightarrow MgO_{(s)}$ $\quad \Delta H^{\ominus} = -3791$ kJmol^{-1}

The **more negative** the lattice enthalpy, the **stronger** the bonding. So out of NaCl and MgO, **MgO** has stronger bonding.

Ionic Charge and Size Affects Lattice Enthalpy

The **higher the charge** on the ions, the **more energy** is released when an ionic lattice forms. More energy released means that the lattice enthalpy will be **more negative.** So the lattice enthalpies for compounds with **2+** or **2– ions** (e.g. Mg^{2+} or S^{2-}) are **more negative** than those with **1+** or **1– ions** (e.g. Na^+ or Cl^-).
For example, the lattice enthalpy of NaCl is only –787 kJ mol^{-1}, but the lattice enthalpy of $MgCl_2$ is –2526 kJ mol^{-1}. **MgS** has an even higher lattice enthalpy (–3299 kJ mol^{-1}) because both magnesium and sulfur ions have double charges.
Magnesium oxide has a **very exothermic** lattice enthalpy too, which means it is very resistant to heat. This makes it great as a **lining in furnaces**.

The **smaller** the **ionic radii** of the ions involved, the **more exothermic** (more negative) the **lattice enthalpy**. Smaller ions attract **more strongly** because their **charge density** is higher.

Born-Haber Cycles can be Used to Calculate Lattice Enthalpies

Hess's law says that the **total enthalpy change** of a reaction is always the **same**, no matter which route is taken.
A **lattice enthalpy** is just the **enthalpy change** when a mole of a **solid ionic compound** is formed from **gaseous ions**. You can't calculate a lattice enthalpy **directly**, so you have to use a **Born-Haber cycle** to figure out what the enthalpy change would be if you took **another**, **less direct**, **route**.

Here's how to draw a Born-Haber cycle for calculating the lattice enthalpy of **NaCl**:

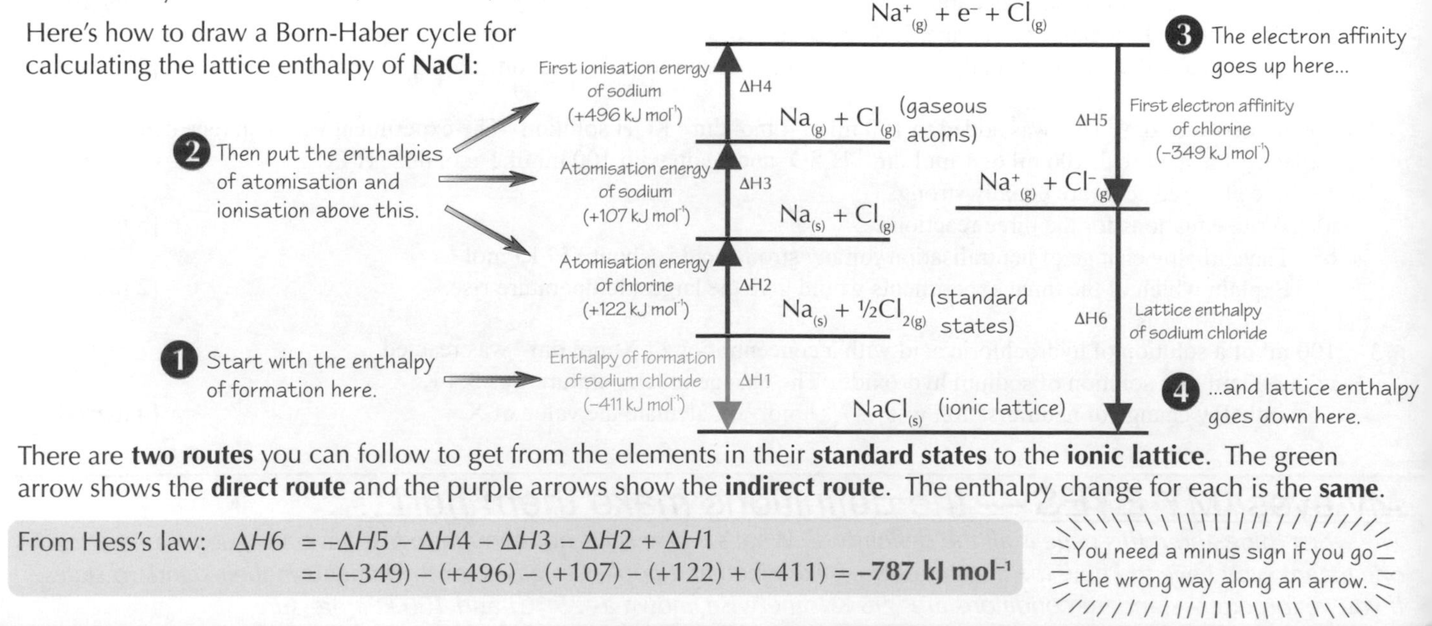

There are **two routes** you can follow to get from the elements in their **standard states** to the **ionic lattice**. The green arrow shows the **direct route** and the purple arrows show the **indirect route**. The enthalpy change for each is the **same**.

From Hess's law: $\Delta H6 = -\Delta H5 - \Delta H4 - \Delta H3 - \Delta H2 + \Delta H1$
$= -(-349) - (+496) - (+107) - (+122) + (-411) = \mathbf{-787\ kJ\ mol^{-1}}$

You need a minus sign if you go the wrong way along an arrow.

Lattice Enthalpy and Born-Haber Cycles

Calculations involving **Group 2 Elements** are a Bit **Different**

Born-Haber cycles for compounds containing **Group 2 elements** have a few **changes** from the one on the previous page. Make sure you understand what's going on so you can handle whatever compound they throw at you.

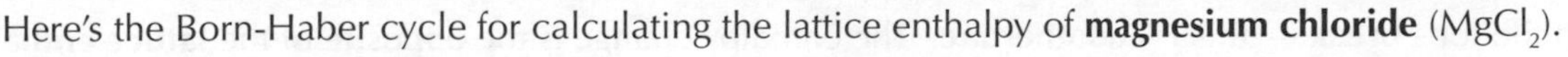

Here's the Born-Haber cycle for calculating the lattice enthalpy of **magnesium chloride** ($MgCl_2$).

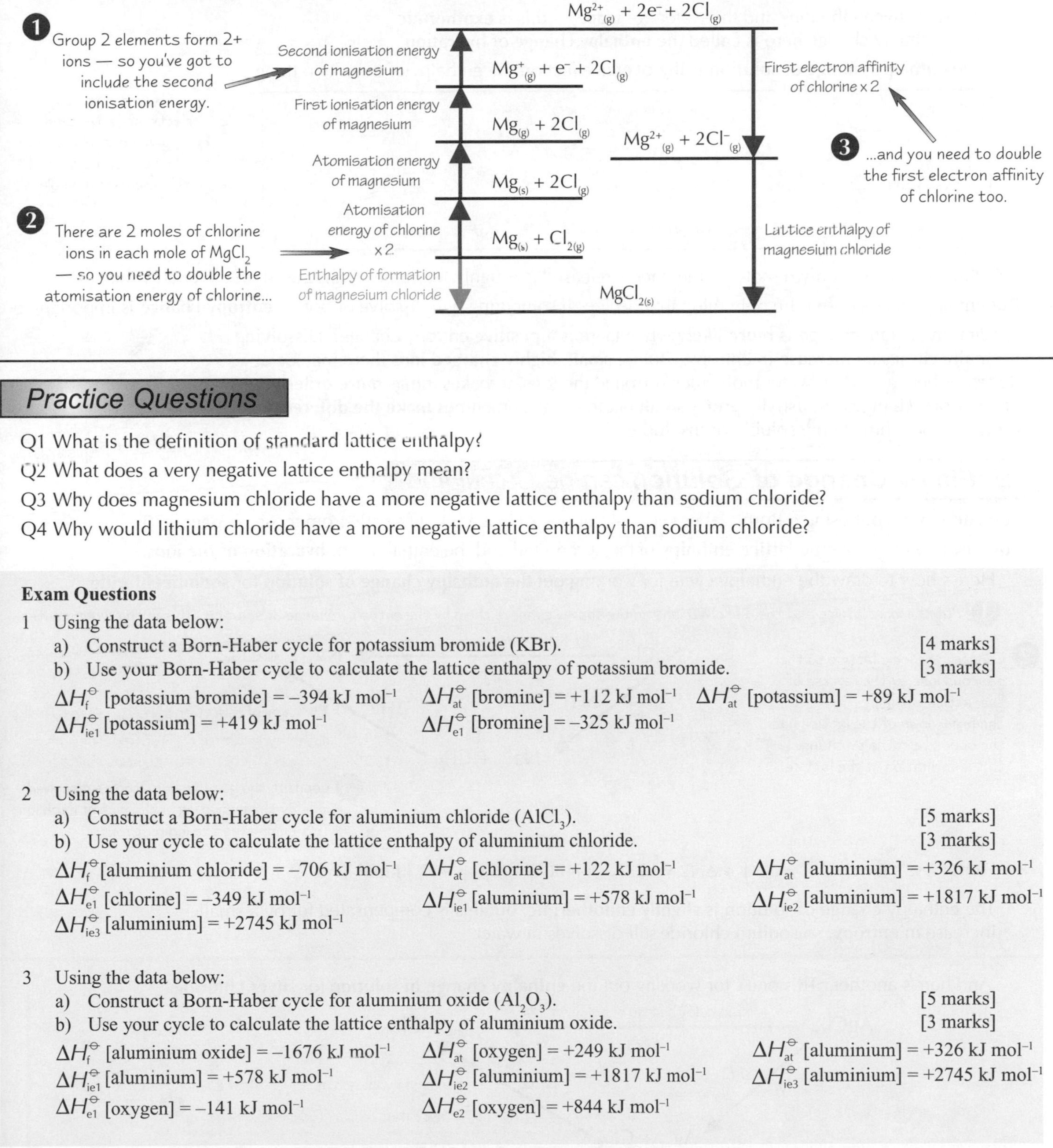

Practice Questions

Q1 What is the definition of standard lattice enthalpy?

Q2 What does a very negative lattice enthalpy mean?

Q3 Why does magnesium chloride have a more negative lattice enthalpy than sodium chloride?

Q4 Why would lithium chloride have a more negative lattice enthalpy than sodium chloride?

Exam Questions

1 Using the data below:

a) Construct a Born-Haber cycle for potassium bromide (KBr). [4 marks]

b) Use your Born-Haber cycle to calculate the lattice enthalpy of potassium bromide. [3 marks]

$\Delta H_f^\ominus$ [potassium bromide] = –394 kJ mol^{-1} $\Delta H_{at}^\ominus$ [bromine] = +112 kJ mol^{-1} $\Delta H_{at}^\ominus$ [potassium] = +89 kJ mol^{-1}

$\Delta H_{ie1}^\ominus$ [potassium] = +419 kJ mol^{-1} $\Delta H_{e1}^\ominus$ [bromine] = –325 kJ mol^{-1}

2 Using the data below:

a) Construct a Born-Haber cycle for aluminium chloride ($AlCl_3$). [5 marks]

b) Use your cycle to calculate the lattice enthalpy of aluminium chloride. [3 marks]

$\Delta H_f^\ominus$[aluminium chloride] = –706 kJ mol^{-1} $\Delta H_{at}^\ominus$ [chlorine] = +122 kJ mol^{-1} $\Delta H_{at}^\ominus$ [aluminium] = +326 kJ mol^{-1}

$\Delta H_{e1}^\ominus$ [chlorine] = –349 kJ mol^{-1} $\Delta H_{ie1}^\ominus$ [aluminium] = +578 kJ mol^{-1} $\Delta H_{ie2}^\ominus$ [aluminium] = +1817 kJ mol^{-1}

$\Delta H_{ie3}^\ominus$ [aluminium] = +2745 kJ mol^{-1}

3 Using the data below:

a) Construct a Born-Haber cycle for aluminium oxide (Al_2O_3). [5 marks]

b) Use your cycle to calculate the lattice enthalpy of aluminium oxide. [3 marks]

$\Delta H_f^\ominus$ [aluminium oxide] = –1676 kJ mol^{-1} $\Delta H_{at}^\ominus$ [oxygen] = +249 kJ mol^{-1} $\Delta H_{at}^\ominus$ [aluminium] = +326 kJ mol^{-1}

$\Delta H_{ie1}^\ominus$ [aluminium] = +578 kJ mol^{-1} $\Delta H_{ie2}^\ominus$ [aluminium] = +1817 kJ mol^{-1} $\Delta H_{ie3}^\ominus$ [aluminium] = +2745 kJ mol^{-1}

$\Delta H_{e1}^\ominus$ [oxygen] = –141 kJ mol^{-1} $\Delta H_{e2}^\ominus$ [oxygen] = +844 kJ mol^{-1}

Using Born-Haber cycles — it's just like riding a bike...

All this energy going in and out can get a bit confusing. Remember these simple rules: 1) It takes energy to break bonds, but energy is given out when bonds are made. 2) A negative ΔH means energy is given out (it's exothermic). 3) A positive ΔH means energy is taken in (it's endothermic). 4) Never return to a firework once lit.

Enthalpies of Solution

Once you know what's happening when you stir sugar into your tea, your cuppa'll be twice as enjoyable.

Dissolving Involves Enthalpy Changes

When a solid **ionic lattice** dissolves in water these **two** things happen:

1) The bonds between the ions **break** — this is **endothermic**. The enthalpy change is the **opposite** of the **lattice enthalpy**.
2) Bonds between the ions and the water are **made** — this is **exothermic**.
   The enthalpy change here is called the **enthalpy change of hydration**.
3) The **enthalpy change of solution** is the overall effect on the enthalpy of these two things.

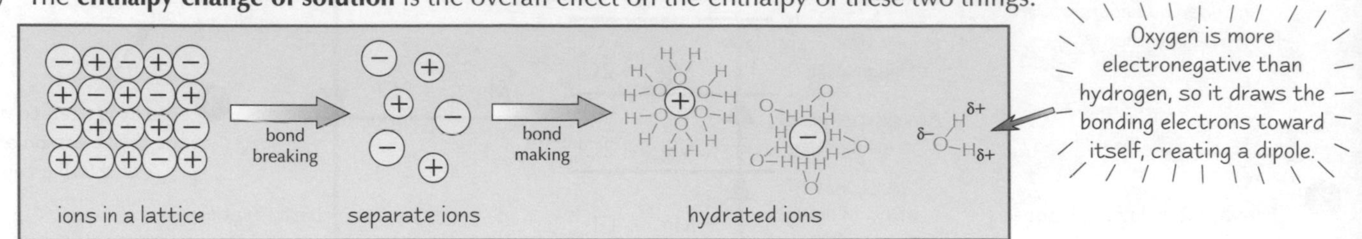

Substances generally **only** dissolve if the energy released is roughly the same, or **greater than** the energy taken in.

But enthalpy change isn't the only thing that decides if something will dissolve or not — **entropy change** is important too.

A reaction or state change is **more likely** when there is a **positive** entropy change. Dissolving normally causes an **increase** in entropy. But for **small**, **highly charged ions** there may be a **decrease** because when water molecules surround the ions, it makes things **more orderly**. The entropy changes are usually **pretty small** but they can sometimes **make the difference** between something being soluble or insoluble.

Entropy is a measure of disorder. It's covered in more detail on pages 174-175.

Enthalpy Change of Solution can be Calculated

You can work it out using a Born-Haber cycle — but one drawn a bit differently from those on page 170.

You just need to know the **lattice enthalpy** of the compound and the enthalpies of **hydration of the ions**.

Here's how to draw the enthalpy cycle for working out the **enthalpy change of solution** for **sodium chloride**.

1. Put the ionic lattice and the dissolved ions on the top — connect them by the enthalpy change of solution. This is the direct route.

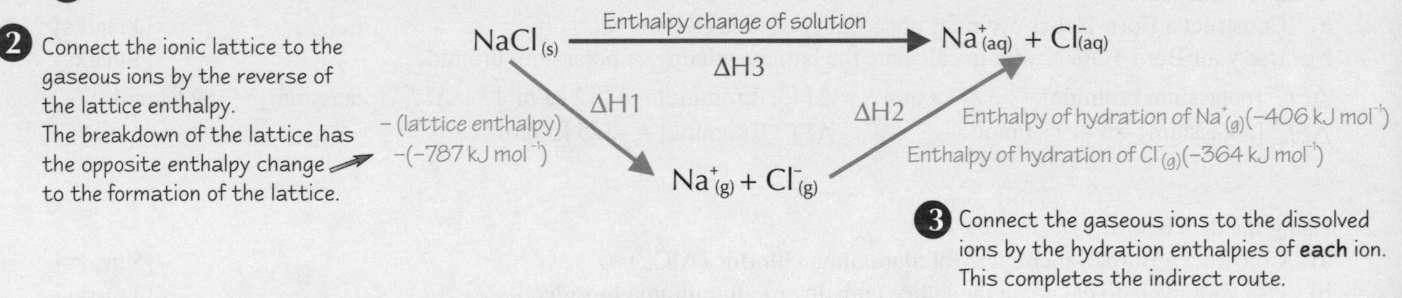

From Hess's law: $\Delta H3 = \Delta H1 + \Delta H2 = +787 + (-406 + -364) = +17$ **kJ mol⁻¹**

The enthalpy change of solution is **slightly endothermic**, but this is compensated for by a small increase in **entropy**, so sodium chloride still dissolves in water.

And here's another. This one's for working out the **enthalpy change of solution** for **silver chloride**.

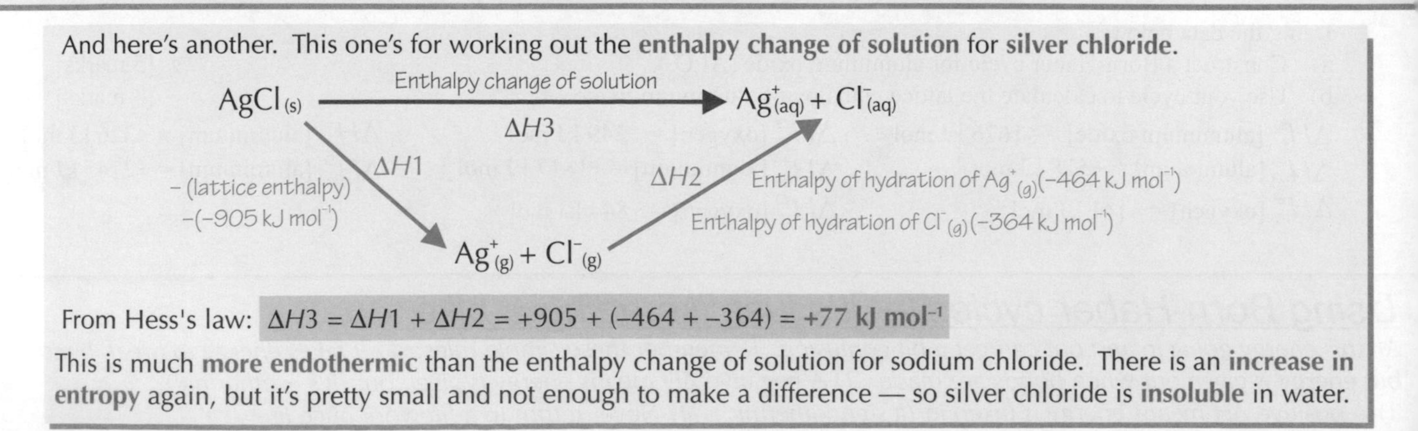

From Hess's law: $\Delta H3 = \Delta H1 + \Delta H2 = +905 + (-464 + -364) = +77$ **kJ mol⁻¹**

This is much **more endothermic** than the enthalpy change of solution for sodium chloride. There is an **increase in entropy** again, but it's pretty small and not enough to make a difference — so silver chloride is **insoluble** in water.

Enthalpies of Solution

Ionic Charge and *Ionic Radius* Affect the Enthalpy of Hydration

The **two** things that can affect the lattice enthalpy (see page 170) can also affect the enthalpy of hydration. They are the **size** and the **charge** of the ions.

Ions with a greater charge have a greater enthalpy of hydration.

Ions with a **higher charge** are better at **attracting** water molecules than those with lower charges. **More energy** is released when the bonds are **made** giving them a **more exothermic** enthalpy of hydration.

Smaller ions have a greater enthalpy of hydration.

Smaller ions have a **higher** charge density than bigger ions. They **attract** the water molecules **better** and have a **more exothermic** enthalpy of hydration.

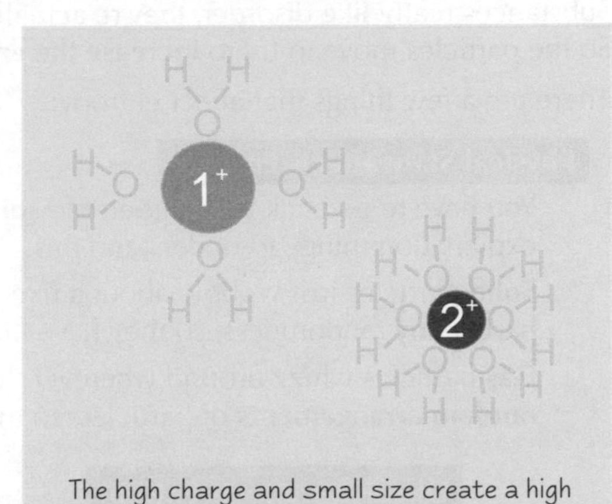

The high charge and small size create a high charge density. This creates a stronger attraction for the water molecules and gives a more exothermic enthalpy of hydration.

E.g. a magnesium ion is smaller and more charged than a sodium ion, which gives it a much bigger enthalpy of hydration.
Mg^{2+} = –1927 kJ mol^{-1}, Na^{+} = –406 kJ mol^{-1}

Practice Questions

Q1 What are the steps in producing a solution from a crystal?

Q2 Sketch a Born-Haber cycle to calculate the enthalpy change of solution of potassium bromide.

Q3 What factors affect the enthalpy of hydration of an ion?

Q4 Do soluble substances have exo- or endothermic enthalpies of solution in general?

Exam Questions

1 a) Draw an enthalpy cycle for the enthalpy change of solution of $AgF_{(s)}$. Label each enthalpy change. [4 marks]

b) Calculate the enthalpy change of solution for AgF from the following data: [2 marks]

$\Delta H^{\ominus}_{latt}[AgF_{(s)}] = -960$ kJ mol^{-1}, $\Delta H^{\ominus}_{hyd}[Ag^{+}_{(g)}] = -464$ kJ mol^{-1}, $\Delta H^{\ominus}_{hyd}[F^{-}_{(g)}] = -506$ kJ mol^{-1}.

2 a) Draw an enthalpy cycle for the enthalpy change of solution of $SrF_{2(s)}$. Label each enthalpy change. [4 marks]

b) Calculate the enthalpy change of solution for SrF_2 from the following data: [2 marks]

$\Delta H^{\ominus}_{latt}[SrF_{2(s)}] = -2492$ kJ mol^{-1}, $\Delta H^{\ominus}_{hyd}[Sr^{2+}_{(g)}] = -1480$ kJ mol^{-1}, $\Delta H^{\ominus}_{hyd}[F^{-}_{(g)}] = -506$ kJ mol^{-1}.

3 Show that the enthalpy change of solution for $MgCl_{2(s)}$ is –122 kJ mol^{-1}, given that: [3 marks]

$\Delta H^{\ominus}_{latt}[MgCl_{2(s)}] = -2526$ kJ mol^{-1}, $\Delta H^{\ominus}_{hyd}[Mg^{2+}_{(g)}] = -1920$ kJ mol^{-1}, $\Delta H^{\ominus}_{hyd}[Cl^{-}_{(g)}] = -364$ kJ mol^{-1}.

Enthalpy change of solution of the Wicked Witch of the West = 8745 kJ mol^{-1}...

Compared to the ones on page 170, these enthalpy cycles are an absolute breeze. You've got to make sure the definitions are firmly fixed in your mind though. Don't forget that a positive enthalpy change doesn't mean the stuff definitely won't dissolve — there might be an entropy change that'll make up for it. The delights of entropy are on the next two pages.

Free-Energy Change and Entropy Change

Free energy — I could do with a bit of that. My gas bill is astronomical.

Entropy Tells you How Much **Disorder** there is

Entropy is a measure of the **number of ways** that **particles** can be **arranged** and the **number of ways** that the **energy** can be shared out between the particles.

Substances really **like** disorder, they're actually more **energetically stable** when there's more disorder. So the particles move to try to **increase the entropy**.

There are a few things that affect entropy:

Physical State affects Entropy

You have to go back to the good old **solid-liquid-gas** particle explanation thingy to understand this.

Solid particles just wobble about a fixed point — there's **hardly any** randomness, so they have the **lowest entropy**.

Gas particles whizz around wherever they like. They've got the most **random arrangements** of particles, so they have the **highest entropy**.

Solid → Liquid → Gas

ORDERED — SOME DISORDER — RANDOM

Increasing disorder

Increasing entropy

Dissolving affects Entropy

Dissolving a solid also increases its entropy — dissolved particles can **move freely** as they're no longer held in one place.

More Particles means More Entropy

It makes sense — the more particles you've got, the **more ways** they and their energy can be **arranged** — so in a reaction like $N_2O_{4(g)} \rightarrow 2NO_{2(g)}$, entropy increases because the **number of moles** increases.

Reactions **Won't** Happen Unless the **Total Entropy Change is Positive**

During a reaction, there's an entropy change between the **reactants and products** — the entropy change of the **system**. The entropy of the **surroundings** changes too (because **energy** is transferred to or from the system).
The **TOTAL entropy change** is the sum of the entropy changes of the **system** and the **surroundings**.

The units of entropy are $J\ K^{-1} mol^{-1}$

$$\Delta S_{total} = \Delta S_{system} + \Delta S_{surroundings}$$

This equation isn't much use unless you know ΔS_{system} and $\Delta S_{surroundings}$. Luckily, there are formulas for them too:

This is just the difference between the entropies of the reactants and products. → $\Delta S_{system} = S_{products} - S_{reactants}$ and $\Delta S_{surroundings} = -\frac{\Delta H}{T}$ ← ΔH = enthalpy change (in $J\ mol^{-1}$), T = temperature (in K)

Example: Calculate the total entropy change for the reaction of ammonia and hydrogen chloride under standard conditions.

$$NH_{3(g)} + HCl_{(g)} \rightarrow NH_4Cl_{(s)} \qquad \Delta H^{\ominus} = -315\ kJ\ mol^{-1}\ (at\ 298\ K)$$

$$S^{\ominus}[NH_{3(g)}] = 192.3\ J\ K^{-1} mol^{-1},\ S^{\ominus}[HCl_{(g)}] = 186.8\ J\ K^{-1} mol^{-1},\ S^{\ominus}[NH_4Cl_{(s)}] = 94.6\ J\ K^{-1} mol^{-1}$$

First find the entropy change of the **system**:

$$\Delta S^{\ominus}_{system} = S^{\ominus}_{products} - S^{\ominus}_{reactants} = 94.6 - (192.3 + 186.8) = \mathbf{-284.5\ J\ K^{-1} mol^{-1}}$$

← This shows a negative change in entropy. It's not surprising as 2 moles of gas have combined to form 1 mole of solid.

Now find the entropy change of the **surroundings**:

$$\Delta H^{\ominus} = -315\ kJ\ mol^{-1} = -315 \times 10^3\ J\ mol^{-1}$$

← Put $\Delta H^{\ominus}$ in the right units.

$$\Delta S^{\ominus}_{surroundings} = -\frac{\Delta H^{\ominus}}{T} = \frac{-(-315 \times 10^3)}{298} = \mathbf{+1057\ J\ K^{-1} mol^{-1}}$$

Finally you can find the **total** entropy

$$\Delta S^{\ominus}_{total} = \Delta S^{\ominus}_{system} + \Delta S^{\ominus}_{surroundings} = -284.5 + (+1057) = \mathbf{+772.5\ J\ K^{-1} mol^{-1}}$$

The total entropy has **increased**. The entropy increase in the surroundings was big enough to make up for the entropy decrease in the system.

Free-Energy Change and Entropy Change

Entropy Increase May Explain Spontaneous Endothermic Reactions

A spontaneous (or feasible) change is one that'll **just happen** by itself — you don't need to give it energy.

But the weird thing is, some **endothermic** reactions are **spontaneous**. You'd normally have to supply **energy** to make an endothermic reaction happen, but if the **entropy** increases enough, the reaction will happen by itself.

1) Water evaporates at room temperature. This change needs **energy** to break the bonds between the molecules — but because it's **changing state** (from a liquid to a gas), the entropy increases.
2) The reaction of sodium hydrogencarbonate with hydrochloric acid is a **spontaneous endothermic reaction**. Again there's an **increase in entropy**.

$$NaHCO_{3(s)} + H^+_{(aq)} \rightarrow Na^+_{(aq)} + CO_{2(g)} + H_2O_{(l)}$$

1 mole solid, 1 mole aqueous ions, 1 mole aqueous ions, 1 mole gas, 1 mole liquid

The product has more particles — and gases and liquids have more entropy than solids too.

For Spontaneous Reactions ΔG must be Negative or Zero

The tendency of a process to take place is dependent on three things — the **entropy**, **ΔS**, the **enthalpy**, **ΔH**, and the **temperature**, ***T***. When you put all these things **together** you get the **free energy change**, **ΔG**, and it tells you if a reaction is **feasible** or not.

Of course, there's a formula for it:

$$\Delta G = \Delta H - T\Delta S_{system}$$

ΔH = enthalpy change (in J mol^{-1})
T = temperature (in K)

Even if ΔG shows that a reaction is theoretically feasible, it might have a really high activation energy and be so slow that you wouldn't notice it happening at all.

Example: Calculate the free energy change for the following reaction at 298 K.

$$MgCO_{3(g)} \rightarrow MgO_{(s)} + CO_{2(g)} \qquad \Delta H^{\ominus} = +117\,000 \text{ J mol}^{-1}, \ \Delta S^{\ominus}_{system} = +175 \text{ J mol}^{-1}$$

$$\Delta G = \Delta H - T\Delta S_{system} = +117\,000 - (298 \times (+175)) = \mathbf{+64\,850 \text{ J mol}^{-1}}$$

ΔG is positive — so the reaction isn't feasible at this temperature.

Practice Questions

Q1 What does the term 'entropy' mean?

Q2 In each of the following pairs choose the one with the greater entropy value.
a) 1 mole of $NaCl_{(aq)}$ and 1 mole of $NaCl_{(s)}$
b) 1 mole of $Br_{2(l)}$ and 1 mole of $Br_{2(g)}$
c) 1 mole of $Br_{2(g)}$ and 2 moles of $Br_{2(g)}$

Q3 Write down the formulas for:
a) total entropy change,
b) entropy change of the surroundings,
c) free energy change.

Exam Questions

1 a) Based on just the equation, predict whether the reaction below is likely to be spontaneous. Give a reason for your answer.

$$Mg_{(s)} + ½O_{2(g)} \rightarrow MgO_{(s)}$$ [2 marks]

b) Use the data on the right to calculate the entropy change for the system above. [3 marks]

c) Does the result of the calculation indicate that the reaction will be spontaneous? Give a reason for your answer. [2 marks]

Substance	Entropy — standard conditions (J K^{-1} mol^{-1})
$Mg_{(s)}$	32.7
$½O_{2(g)}$	102.5
$MgO_{(s)}$	26.9

2 $S^{\ominus}[H_2O_{(l)}] = 70$ J K^{-1} mol^{-1}, $S^{\ominus}[H_2O_{(s)}] = 48$ J K^{-1} mol^{-1}, $\Delta H^{\ominus} = -6$ kJ mol^{-1}

For the reaction $H_2O_{(l)} \rightarrow H_2O_{(s)}$:

a) Calculate the total entropy change at i) 250 K ii) 300 K [5 marks]

b) Will this reaction be spontaneous at 250 K or 300 K? Explain your answer. [2 marks]

Being neat and tidy is against the laws of nature...

There's a scary amount of scary looking formulas on these pages. They aren't too hard to use, but watch out for your units. Make sure the temperature's in kelvin — if you're given one in °C, you need to add 273 to change it to kelvin. And check that all your enthalpy and entropy values involve joules, not kilojoules (so J mol^{-1}, not kJ mol^{-1}, etc.).

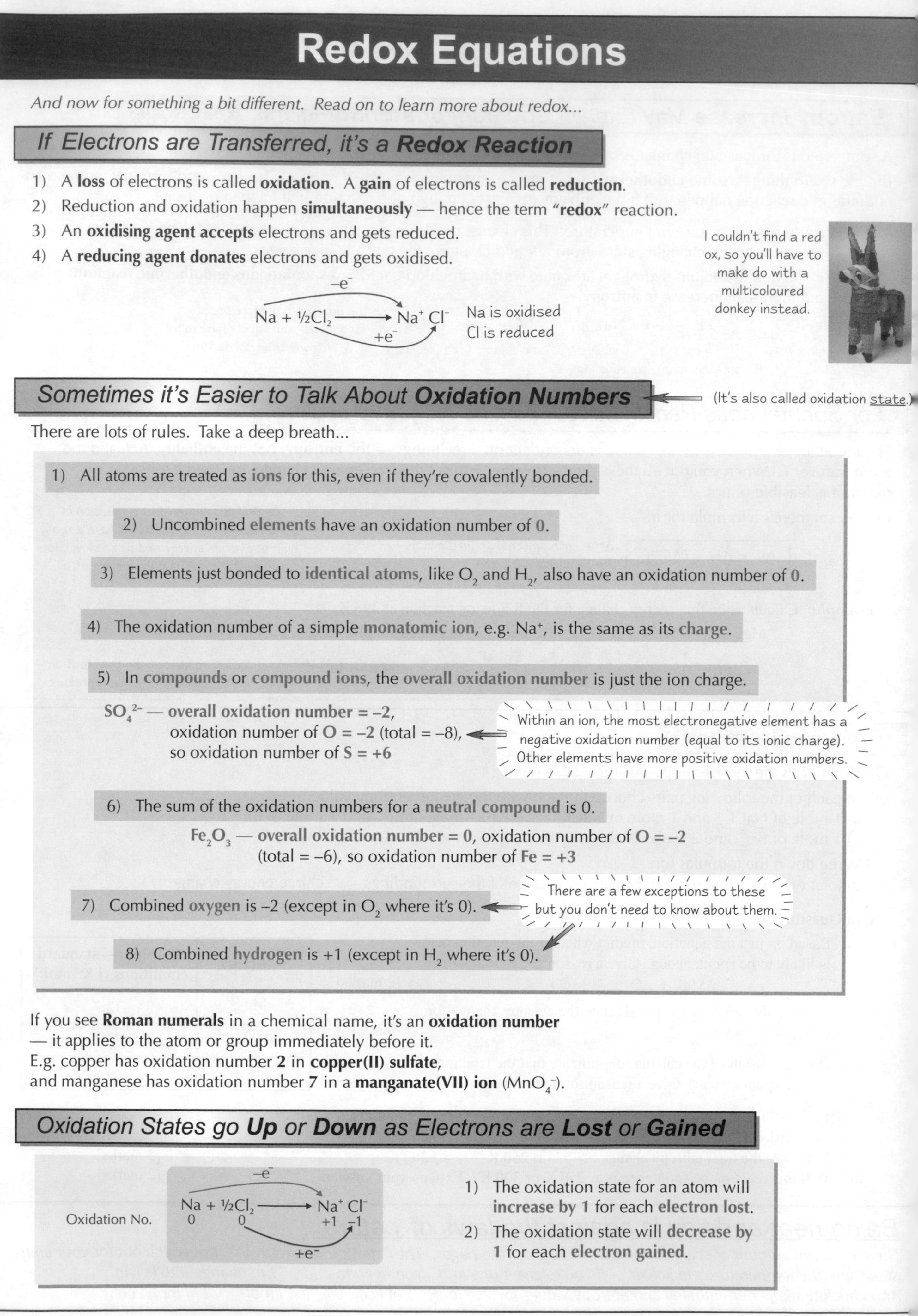

Redox Equations

And now for something a bit different. Read on to learn more about redox...

If Electrons are Transferred, it's a **Redox Reaction**

1) A **loss** of electrons is called **oxidation**. A **gain** of electrons is called **reduction**.
2) Reduction and oxidation happen **simultaneously** — hence the term "**redox**" reaction.
3) An **oxidising agent accepts** electrons and gets reduced.
4) A **reducing agent donates** electrons and gets oxidised.

$$Na + \frac{1}{2}Cl_2 \longrightarrow Na^+ Cl^-$$

$-e^-$ (Na → Na^+), $+e^-$ (Cl → Cl^-)

Na is oxidised
Cl is reduced

I couldn't find a red ox, so you'll have to make do with a multicoloured donkey instead.

Sometimes it's Easier to Talk About **Oxidation Numbers**

(It's also called oxidation state.)

There are lots of rules. Take a deep breath...

1) All atoms are treated as **ions** for this, even if they're covalently bonded.

2) Uncombined **elements** have an oxidation number of **0**.

3) Elements just bonded to **identical atoms**, like O_2 and H_2, also have an oxidation number of **0**.

4) The oxidation number of a simple **monatomic ion**, e.g. Na^+, is the same as its **charge**.

5) In **compounds** or **compound ions**, the **overall oxidation number** is just the ion charge.

SO_4^{2-} — **overall oxidation number = –2**,
oxidation number of **O = –2** (total = –8),
so oxidation number of **S = +6**

Within an ion, the most electronegative element has a negative oxidation number (equal to its ionic charge). Other elements have more positive oxidation numbers.

6) The sum of the oxidation numbers for a **neutral compound** is 0.

Fe_2O_3 — **overall oxidation number = 0**, oxidation number of **O = –2** (total = –6), so oxidation number of **Fe = +3**

7) Combined **oxygen** is –2 (except in O_2 where it's 0).

8) Combined **hydrogen** is +1 (except in H_2 where it's 0).

There are a few exceptions to these but you don't need to know about them.

If you see **Roman numerals** in a chemical name, it's an **oxidation number** — it applies to the atom or group immediately before it.
E.g. copper has oxidation number **2** in **copper(II) sulfate**,
and manganese has oxidation number **7** in a **manganate(VII) ion** (MnO_4^-).

Oxidation States go **Up** or **Down** as Electrons are **Lost** or **Gained**

1) The oxidation state for an atom will **increase by 1** for each **electron lost**.
2) The oxidation state will **decrease by 1** for each **electron gained**.

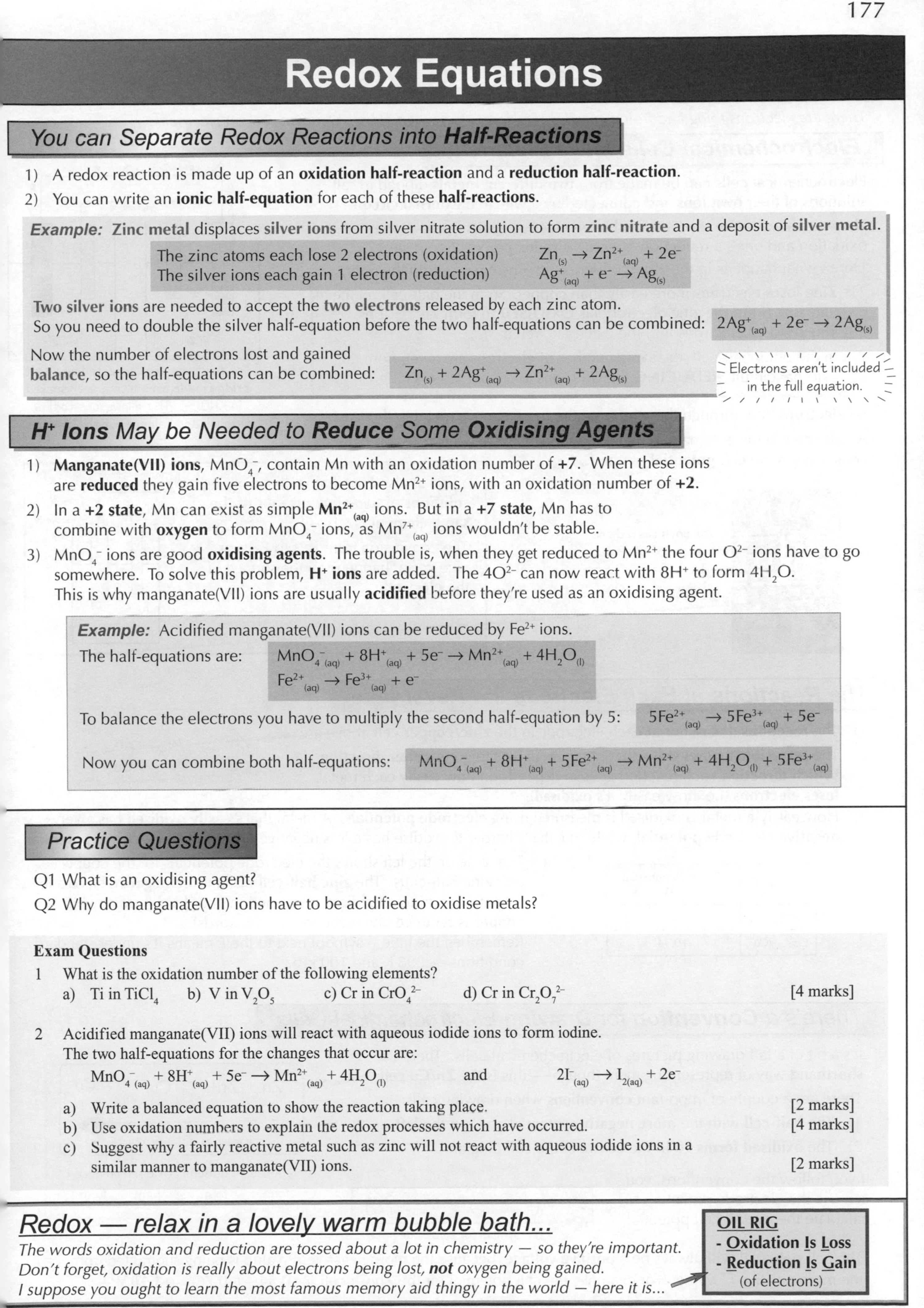

Redox Equations

You can Separate Redox Reactions into *Half-Reactions*

1) A redox reaction is made up of an **oxidation half-reaction** and a **reduction half-reaction.**
2) You can write an **ionic half-equation** for each of these **half-reactions**.

Example: **Zinc metal** displaces **silver ions** from silver nitrate solution to form **zinc nitrate** and a deposit of **silver metal.**

The zinc atoms each lose 2 electrons (oxidation) $Zn_{(s)} \rightarrow Zn^{2+}_{(aq)} + 2e^-$

The silver ions each gain 1 electron (reduction) $Ag^+_{(aq)} + e^- \rightarrow Ag_{(s)}$

Two silver ions are needed to accept the **two electrons** released by each zinc atom.
So you need to double the silver half-equation before the two half-equations can be combined: $2Ag^+_{(aq)} + 2e^- \rightarrow 2Ag_{(s)}$

Now the number of electrons lost and gained **balance**, so the half-equations can be combined: $Zn_{(s)} + 2Ag^+_{(aq)} \rightarrow Zn^{2+}_{(aq)} + 2Ag_{(s)}$

Electrons aren't included in the full equation.

H⁺ Ions May be Needed to *Reduce* Some *Oxidising Agents*

1) **Manganate(VII) ions**, MnO_4^-, contain Mn with an oxidation number of **+7**. When these ions are **reduced** they gain five electrons to become Mn^{2+} ions, with an oxidation number of **+2.**
2) In a **+2 state**, Mn can exist as simple $\mathbf{Mn^{2+}_{(aq)}}$ ions. But in a **+7 state**, Mn has to combine with **oxygen** to form MnO_4^- ions, as $Mn^{7+}_{(aq)}$ ions wouldn't be stable.
3) MnO_4^- ions are good **oxidising agents**. The trouble is, when they get reduced to Mn^{2+} the four O^{2-} ions have to go somewhere. To solve this problem, $\mathbf{H^+}$ **ions** are added. The $4O^{2-}$ can now react with $8H^+$ to form $4H_2O$. This is why manganate(VII) ions are usually **acidified** before they're used as an oxidising agent.

Example: Acidified manganate(VII) ions can be reduced by Fe^{2+} ions.

The half-equations are:

$$MnO^-_{4\,(aq)} + 8H^+_{(aq)} + 5e^- \rightarrow Mn^{2+}_{(aq)} + 4H_2O_{(l)}$$

$$Fe^{2+}_{(aq)} \rightarrow Fe^{3+}_{(aq)} + e^-$$

To balance the electrons you have to multiply the second half-equation by 5: $5Fe^{2+}_{(aq)} \rightarrow 5Fe^{3+}_{(aq)} + 5e^-$

Now you can combine both half-equations: $MnO^-_{4\,(aq)} + 8H^+_{(aq)} + 5Fe^{2+}_{(aq)} \rightarrow Mn^{2+}_{(aq)} + 4H_2O_{(l)} + 5Fe^{3+}_{(aq)}$

Practice Questions

Q1 What is an oxidising agent?

Q2 Why do manganate(VII) ions have to be acidified to oxidise metals?

Exam Questions

1 What is the oxidation number of the following elements?

a) Ti in $TiCl_4$ b) V in V_2O_5 c) Cr in CrO_4^{2-} d) Cr in $Cr_2O_7^{2-}$ [4 marks]

2 Acidified manganate(VII) ions will react with aqueous iodide ions to form iodine.
The two half-equations for the changes that occur are:

$MnO^-_{4\,(aq)} + 8H^+_{(aq)} + 5e^- \rightarrow Mn^{2+}_{(aq)} + 4H_2O_{(l)}$ and $2I^-_{(aq)} \rightarrow I_{2(aq)} + 2e^-$

a) Write a balanced equation to show the reaction taking place. [2 marks]

b) Use oxidation numbers to explain the redox processes which have occurred. [4 marks]

c) Suggest why a fairly reactive metal such as zinc will not react with aqueous iodide ions in a similar manner to manganate(VII) ions. [2 marks]

Redox — relax in a lovely warm bubble bath...

The words oxidation and reduction are tossed about a lot in chemistry — so they're important. Don't forget, oxidation is really about electrons being lost, ***not*** *oxygen being gained. I suppose you ought to learn the most famous memory aid thingy in the world — here it is...*

OIL RIG
- Oxidation Is Loss
- Reduction Is Gain
(of electrons)

Electrode Potentials

There are electrons toing and froing in redox reactions. And when electrons move, you get electricity.

Electrochemical Cells Make *Electricity*

Electrochemical cells can be made from **two different metals** dipped in salt solutions of their **own ions** and connected by a wire (the **external circuit**).

There are always **two** reactions within an electrochemical cell — one's an oxidation and one's a reduction — so it's a **redox process** (see page 176).

Here's what happens in the **zinc/copper** electrochemical cell on the right:

1) Zinc **loses electrons** more easily than copper. So in the half-cell on the left, zinc (from the zinc electrode) is **OXIDISED** to form $Zn^{2+}_{(aq)}$ ions. This releases electrons into the external circuit.
2) In the other half-cell, the **same number of electrons** are taken from the external circuit, **REDUCING** the Cu^{2+} ions to copper atoms.

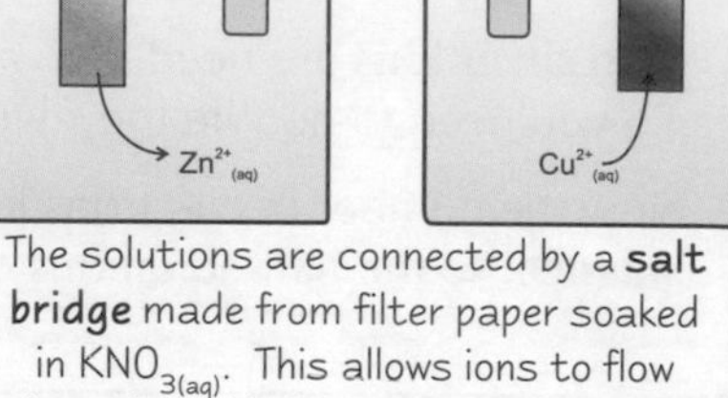

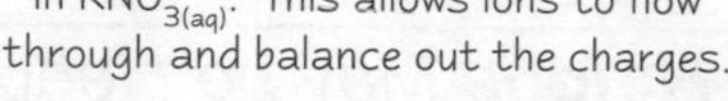

The solutions are connected by a **salt bridge** made from filter paper soaked in $KNO_{3(aq)}$. This allows ions to flow through and balance out the charges.

So **electrons** flow through the wire from the most reactive metal to the least.

A voltmeter in the external circuit shows the **voltage** between the two half-cells. This is the **cell potential** or **e.m.f.**, E_{cell}.

The boys tested the strength of the bridge, whilst the girls just stood and watched.

You can also have half-cells involving **solutions of two aqueous ions of the same element**, such as $Fe^{2+}_{(aq)}/Fe^{3+}_{(aq)}$. The conversion from Fe^{2+} to Fe^{3+}, or vice versa, happens on the surface of the **electrode**.

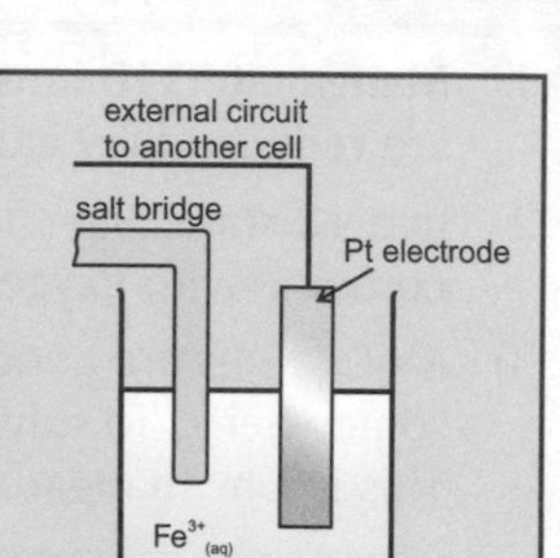

The *Reactions* at *Each Electrode* are *Reversible*

1) The **reactions** that occur at each electrode in the **zinc/copper cell** above are:

$$Zn^{2+}_{(aq)} + 2e^- \rightleftharpoons Zn_{(s)}$$
$$Cu^{2+}_{(aq)} + 2e^- \rightleftharpoons Cu_{(s)}$$

2) The **reversible arrows** show that both reactions can go in **either direction**. **Which direction** each reaction goes in depends on **how easily** each metal **loses electrons** (i.e. how easily it's **oxidised**).
3) How easily a metal is oxidised is measured using **electrode potentials**. A metal that's **easily oxidised** has a very **negative electrode potential**, while one that's harder to oxidise has a less negative or **a positive electrode potential**.

Half-cell	Electrode potential E° (V)
$Zn^{2+}_{(aq)}/Zn_{(s)}$	–0.76
$Cu^{2+}_{(aq)}/Cu_{(s)}$	+0.34

4) The table on the left shows the electrode potentials for the copper and zinc half-cells. The **zinc half-cell** has a **more negative** electrode potential, so **zinc is oxidised** (the reaction goes **backwards**), while **copper is reduced** (the reaction goes **forwards**).
Remember, the little $\circ$ symbol next to the E means it's under standard conditions — 298 K and 100 kPa.

There's a *Convention* for *Drawing Electrochemical Cells*

It's a bit of a faff drawing pictures of electrochemical cells. There's a **shorthand** way of representing them though — this is the **Zn/Cu cell**:

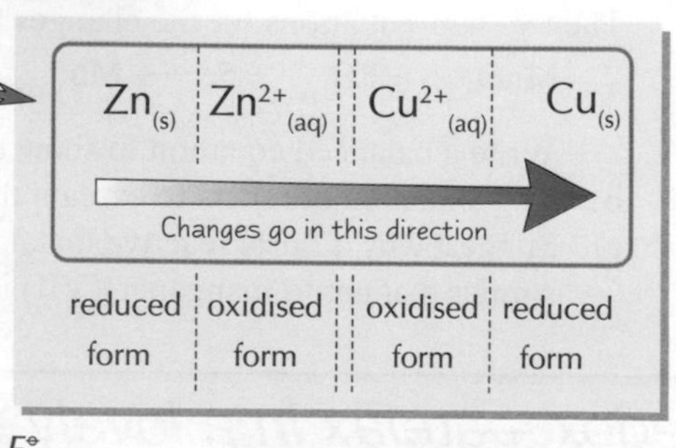

There are a couple of important **conventions** when drawing cells:

1) The **half-cell** with the **more negative** potential goes on the **left**.
2) The **oxidised forms** go in the **centre** of the cell diagram.

If you follow the conventions, you can use the electrode potentials to calculate the overall cell potential.

$$E^\circ_{cell} = (E^\circ_{\text{right hand side}} - E^\circ_{\text{left hand side}})$$

The symbol for electrode potential is E°.

The cell potential will always be a **positive voltage**, because the more negative E° value is being subtracted from the more positive E° value. For example, the cell potential for the Zn/Cu cell = +0.34 – (–0.76) = **+1.10 V**

Electrode Potentials

Electrode Potentials are Measured Against *Standard Hydrogen Electrodes*

You measure the electrode potential of a half-cell against a **standard hydrogen electrode**.

> The **standard electrode potential** $E^\ominus$ of a half-cell is the **voltage measured** under **standard conditions** when the **half-cell** is connected to a **standard hydrogen electrode**.

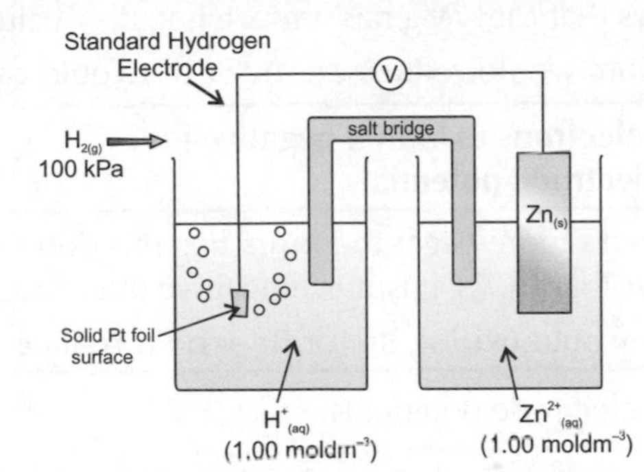

Standard conditions are:
1) Any solution must have a concentration of 1.00 mol dm^{-3}
2) The temperature must be 298 K (25 °C)
3) The pressure must be 100 kPa

1) The **standard hydrogen electrode** is always shown on the **left** — it doesn't matter whether or not the other half-cell has a more positive value. The standard hydrogen electrode half-cell has a value of **0.00 V**.
2) The whole cell potential = $E^\ominus_{\text{right-hand side}} - E^\ominus_{\text{left-hand side}}$.
   $E^\ominus_{\text{left-hand side}}$ = 0.00 V, so the **voltage reading** will be equal to $E^\ominus_{\text{right-hand side}}$.
   This reading could be **positive** or **negative**, depending which way the **electrons flow**.
3) In an electrochemical cell, the half-cell with the **most negative** standard electrode potential is the one in which **oxidation** happens.

Practice Questions

Q1 $Fe^{3+} + e^- \rightleftharpoons Fe^{2+}$, $E^\ominus = +0.77$ V $Mn^{3+} + e^- \rightleftharpoons Mn^{2+}$, $E^\ominus = +1.48$ V
Calculate the standard cell potential for the above system.

Q2 What's the definition of standard electrode potential?

Q3 List the three standard conditions used when measuring standard electrode potentials.

Exam Questions

1 A cell is made up of a lead and an iron plate, dipped in solutions of lead(II) nitrate and iron(II) nitrate respectively and connected by a salt bridge. The electrode potentials for the two electrodes are:

$Fe^{2+}_{(aq)} + 2e^- \rightleftharpoons Fe_{(s)}$ $E^\ominus = -0.44$ V $Pb^{2+}_{(aq)} + 2e^- \rightleftharpoons Pb_{(s)}$ $E^\ominus = -0.13$ V

a) Which metal becomes oxidised in the cell? Explain your answer. [2 marks]
b) Which half-cell releases electrons into the circuit? Explain your answer. [2 marks]
c) Find the standard cell potential of this cell. [1 mark]

2 An electrochemical cell containing a zinc half-cell and a silver half-cell was set up using a potassium nitrate salt bridge. The cell potential at 25 °C was measured to be 1.40 V.

$Zn^{2+}_{(aq)} + 2e^- \rightleftharpoons Zn_{(s)}$ $E^\ominus = -0.76$ V $Ag^+_{(aq)} + e^- \rightleftharpoons Ag_{(s)}$ $E^\ominus = +0.80$ V

a) Use the standard electrode potentials given to calculate the standard cell potential for a zinc-silver cell. [1 mark]
b) Suggest two possible reasons why the actual cell potential was different from the value calculated in part (a). [2 marks]
c) Write an equation for the overall cell reaction. [1 mark]
d) Which half-cell released the electrons into the circuit? Why is this? [1 mark]

Half-equations are not straightforward reactions — they're reversible...

You've just got to think long and hard about this stuff. The metal on the left-hand electrode disappears off into the solution, leaving its electrons behind. This makes the left-hand electrode the negative one. So the right-hand electrode's got to be the positive one. It makes sense if you think about it. This electrode gives up electrons to turn the positive ions into atoms.

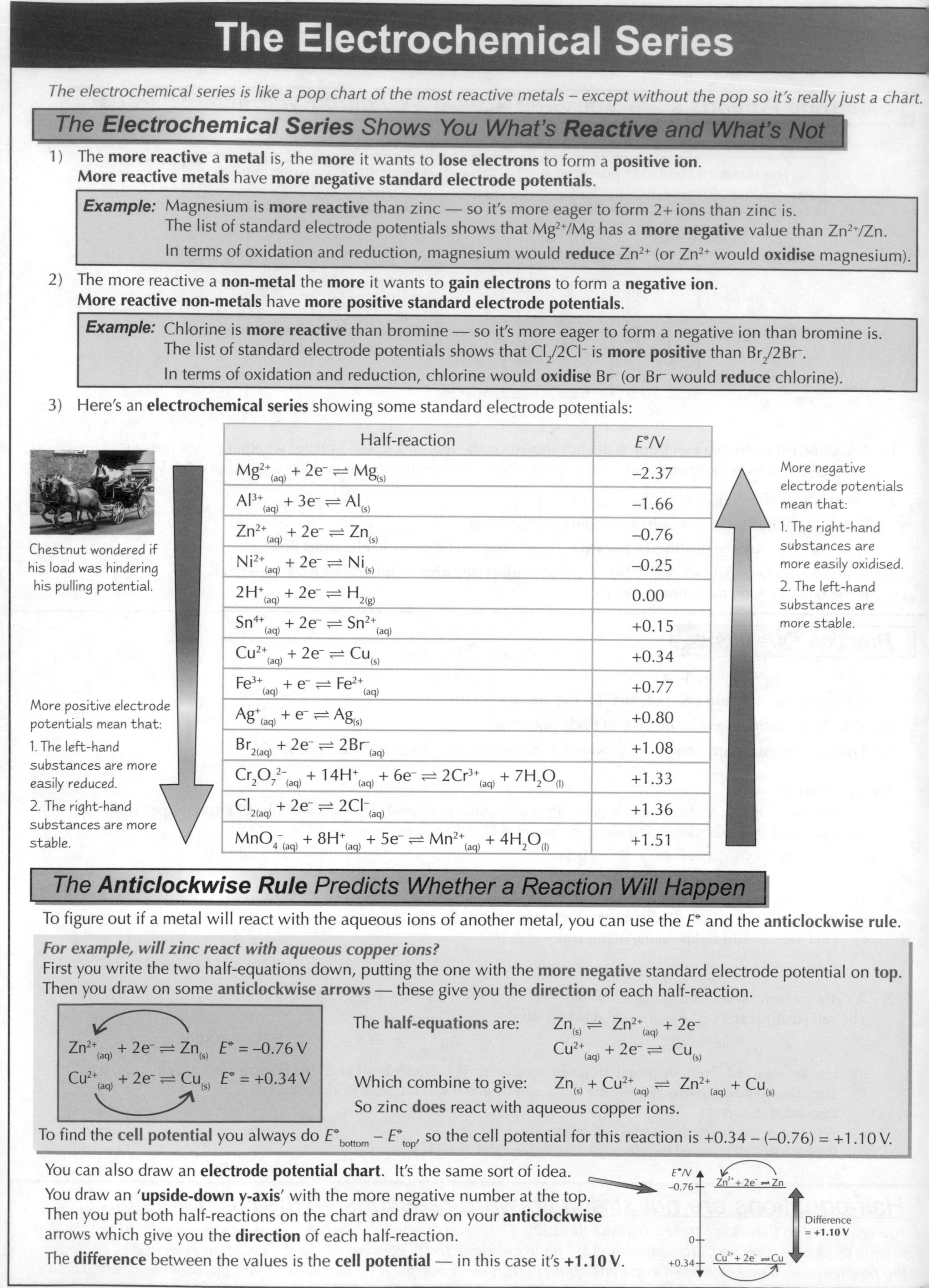

The Electrochemical Series

The electrochemical series is like a pop chart of the most reactive metals – except without the pop so it's really just a chart.

The Electrochemical Series Shows You What's Reactive and What's Not

1) The **more reactive** a **metal** is, the **more** it wants to **lose electrons** to form a **positive ion**.
More reactive metals have **more negative standard electrode potentials**.

> ***Example:*** Magnesium is **more reactive** than zinc — so it's more eager to form 2+ ions than zinc is.
> The list of standard electrode potentials shows that Mg^{2+}/Mg has a **more negative** value than Zn^{2+}/Zn.
> In terms of oxidation and reduction, magnesium would **reduce** Zn^{2+} (or Zn^{2+} would **oxidise** magnesium).

2) The more reactive a **non-metal** the **more** it wants to **gain electrons** to form a **negative ion**.
More reactive non-metals have **more positive standard electrode potentials**.

> ***Example:*** Chlorine is **more reactive** than bromine — so it's more eager to form a negative ion than bromine is.
> The list of standard electrode potentials shows that $Cl_2/2Cl^-$ is **more positive** than $Br_2/2Br^-$.
> In terms of oxidation and reduction, chlorine would **oxidise** Br^- (or Br^- would **reduce** chlorine).

3) Here's an **electrochemical series** showing some standard electrode potentials:

Chestnut wondered if his load was hindering his pulling potential.

Half-reaction	E°/V
$Mg^{2+}_{(aq)} + 2e^- \rightleftharpoons Mg_{(s)}$	–2.37
$Al^{3+}_{(aq)} + 3e^- \rightleftharpoons Al_{(s)}$	–1.66
$Zn^{2+}_{(aq)} + 2e^- \rightleftharpoons Zn_{(s)}$	–0.76
$Ni^{2+}_{(aq)} + 2e^- \rightleftharpoons Ni_{(s)}$	–0.25
$2H^+_{(aq)} + 2e^- \rightleftharpoons H_{2(g)}$	0.00
$Sn^{4+}_{(aq)} + 2e^- \rightleftharpoons Sn^{2+}_{(aq)}$	+0.15
$Cu^{2+}_{(aq)} + 2e^- \rightleftharpoons Cu_{(s)}$	+0.34
$Fe^{3+}_{(aq)} + e^- \rightleftharpoons Fe^{2+}_{(aq)}$	+0.77
$Ag^+_{(aq)} + e^- \rightleftharpoons Ag_{(s)}$	+0.80
$Br_{2(aq)} + 2e^- \rightleftharpoons 2Br^-_{(aq)}$	+1.08
$Cr_2O_7^{2-}{}_{(aq)} + 14H^+_{(aq)} + 6e^- \rightleftharpoons 2Cr^{3+}_{(aq)} + 7H_2O_{(l)}$	+1.33
$Cl_{2(aq)} + 2e^- \rightleftharpoons 2Cl^-_{(aq)}$	+1.36
$MnO_4^-{}_{(aq)} + 8H^+_{(aq)} + 5e^- \rightleftharpoons Mn^{2+}_{(aq)} + 4H_2O_{(l)}$	+1.51

More positive electrode potentials mean that:
1. The left-hand substances are more easily reduced.
2. The right-hand substances are more stable.

More negative electrode potentials mean that:
1. The right-hand substances are more easily oxidised.
2. The left-hand substances are more stable.

The Anticlockwise Rule Predicts Whether a Reaction Will Happen

To figure out if a metal will react with the aqueous ions of another metal, you can use the E° and the **anticlockwise rule**.

For example, will zinc react with aqueous copper ions?
First you write the two half-equations down, putting the one with the **more negative** standard electrode potential on **top**.
Then you draw on some **anticlockwise arrows** — these give you the **direction** of each half-reaction.

$Zn^{2+}_{(aq)} + 2e^- \rightleftharpoons Zn_{(s)}$ $E^\circ = -0.76$ V
$Cu^{2+}_{(aq)} + 2e^- \rightleftharpoons Cu_{(s)}$ $E^\circ = +0.34$ V

The **half-equations** are: $Zn_{(s)} \rightleftharpoons Zn^{2+}_{(aq)} + 2e^-$
$Cu^{2+}_{(aq)} + 2e^- \rightleftharpoons Cu_{(s)}$

Which combine to give: $Zn_{(s)} + Cu^{2+}_{(aq)} \rightleftharpoons Zn^{2+}_{(aq)} + Cu_{(s)}$
So zinc **does** react with aqueous copper ions.

To find the **cell potential** you always do $E^\circ_{bottom} - E^\circ_{top}$, so the cell potential for this reaction is +0.34 – (–0.76) = +1.10 V.

You can also draw an **electrode potential chart**. It's the same sort of idea.
You draw an '**upside-down y-axis**' with the more negative number at the top.
Then you put both half-reactions on the chart and draw on your **anticlockwise** arrows which give you the **direction** of each half-reaction.
The **difference** between the values is the **cell potential** — in this case it's **+1.10 V**.

The Electrochemical Series

Sometimes the Prediction is Wrong

A **prediction** using E° and the anticlockwise rule only states if a reaction is **possible** under **standard conditions**. The prediction might be **wrong if...**

...the conditions are not standard

1) Changing the **concentration** (or temperature) of the solution can cause the electrode potential to **change.**
2) For example the zinc/copper cell has these half equations in equilibrium...

$$Zn_{(s)} \rightleftharpoons Zn^{2+}_{(aq)} + 2e^-$$
$$Cu^{2+}_{(aq)} + 2e^- \rightleftharpoons Cu_{(s)}$$

3) ...if you **increase** the concentration of Zn^{2+}, the **equilibrium** will shift to the **left**, **reducing** the ease of **electron loss**. The whole cell potential will be lower.
4) ...if you **increase** the concentration of Cu^{2+}, the **equilibrium** will shift to the **right**, **increasing** the ease of **electron gain**. The whole cell potential will be higher.

Gary was hopeful, but Sue's high activation energy meant it was never going to happen

...the reaction kinetics are not favourable

1) The **rate of a reaction** may be so **slow** that the reaction might **not appear** to happen.
2) If a reaction has a **high activation energy**, this may stop it happening.

Practice Questions

Q1 Cu is less reactive than Pb.
Which half-reaction has a more negative standard electrode potential, $Pb^{2+} + 2e^- \rightleftharpoons Pb$ or $Cu^{2+} + 2e^- \rightleftharpoons Cu$?

Q2 Use electrode potentials to show that magnesium will reduce Zn^{2+}.

Q3 What is the anticlockwise rule used for? Outline how you use it.

Q4 Use the table on the opposite page to predict whether or not Zn^{2+} ions can oxidise Fe^{2+} ions to Fe^{3+} ions.

Exam Question

1 Use E° values quoted on the opposite page to determine the outcome of mixing the following solutions.
If there is a reaction, determine the E° value and write the equation. If there isn't a reaction, state this and explain why.

a) Zinc metal and Ni^{2+} ions [2 marks]
b) Acidified MnO_4^- ions and Sn^{2+} ions [2 marks]
c) $Br_{2(aq)}$ and acidified $Cr_2O_7^{2-}$ ions [2 marks]
d) Silver ions and Fe^{2+} ions [2 marks]

2 Potassium manganate(VII), $KMnO_4$, and potassium dichromate $K_2Cr_2O_7$, are both used as oxidising agents.

a) From their electrode potentials, which would you predict is the stronger oxidising agent? Explain why. [2 marks]
b) Write equations to show each oxidising agent reacting with a solution of Fe^{2+} ions. [2 marks]
c) Calculate the cell potential for each reaction. [2 marks]

3 A cell is set up with copper and nickel electrodes in 1 mol dm^{-3} solutions of their ions, Cu^{2+} and Ni^{2+}, connected by a salt bridge.

a) Write equations for the reactions that occur in each half-cell. [2 marks]
b) Find the voltage of the cell. [1 mark]
c) What is the overall equation for this reaction? [1 mark]
d) How would the voltage of the cell change if:
   i) A more dilute copper solution was used? [1 mark]
   ii) A more concentrated nickel solution was used? [1 mark]

The forward reaction that happens is the one with the most positive E° value...

To see if a reaction will happen, you basically find the two half-equations in the electrochemical series and check whether you can draw anticlockwise arrows on them to get from your reactants to your products. If you can — great. The reaction will have a positive electrode potential, so it should happen. If you can't — well, it ain't gonna work.

Storage and Fuel Cells

Yet more electrochemical reactions on these pages but you're nearly at the end of the section so keep going...

Energy Storage Cells are Like Electrochemical Cells

Energy storage cells (fancy name for a battery) have been around for ages and modern ones **work** just like an **electrochemical cell**. For example the nickel-iron cell was developed way back at the start of the 1900s and is often used as a back-up power supply because it can be repeatedly charged and is very robust. You can work out the **voltage** produced by these **cells** by using the **electrode potentials** of the substances used in the cell.

There are **lots** of different cells and you **won't** be asked to remember the E° for the reactions, but you might be **asked** to work out the **cell potential** or **cell voltage** for a given cell...so here's an example I prepared earlier.

Example

The nickel-iron cell has a nickel oxide hydroxide (NiOOH) cathode and an iron (Fe) anode with potassium hydroxide as the electrolyte. Using the half equations given:

a) write out the full equation for the reaction.
b) calculate the cell voltage produced by the nickel-iron cell.

$Fe + 2OH^- \rightarrow Fe(OH)_2 + 2e^-$	$E^\circ = -0.44$ V
$NiOOH + H_2O + e^- \rightarrow Ni(OH)_2 + OH^-$	$E^\circ = +0.76$ V

The **overall** reaction is...

$2NiOOH + 2H_2O + Fe \rightarrow 2Ni(OH)_2 + Fe(OH)_2$

For the first part you have to **combine** the two half equations together. The e^- and the OH^- are not shown because they cancel each other out.

To calculate the **cell voltage** you use the **same formula** for working out the **cell potential** (page 180).

So the **cell voltage** $= E^\circ_{bottom} - E^\circ_{top}$
$= +0.76 - (-0.44)$
$= 1.2$ V

Fuel Cells Generate Electricity from Reacting a Fuel with an Oxidant

A **fuel cell** produces electricity by reacting a **fuel**, usually hydrogen, with an **oxidant**, which is most likely to be oxygen.

1) At the **anode** the platinum catalyst **splits** the H_2 into protons and electrons

2) The **polymer electrolyte membrane** (PEM) **only** allows the H^+ across and this **forces** the e^- to travel **around** the circuit to get to the cathode.

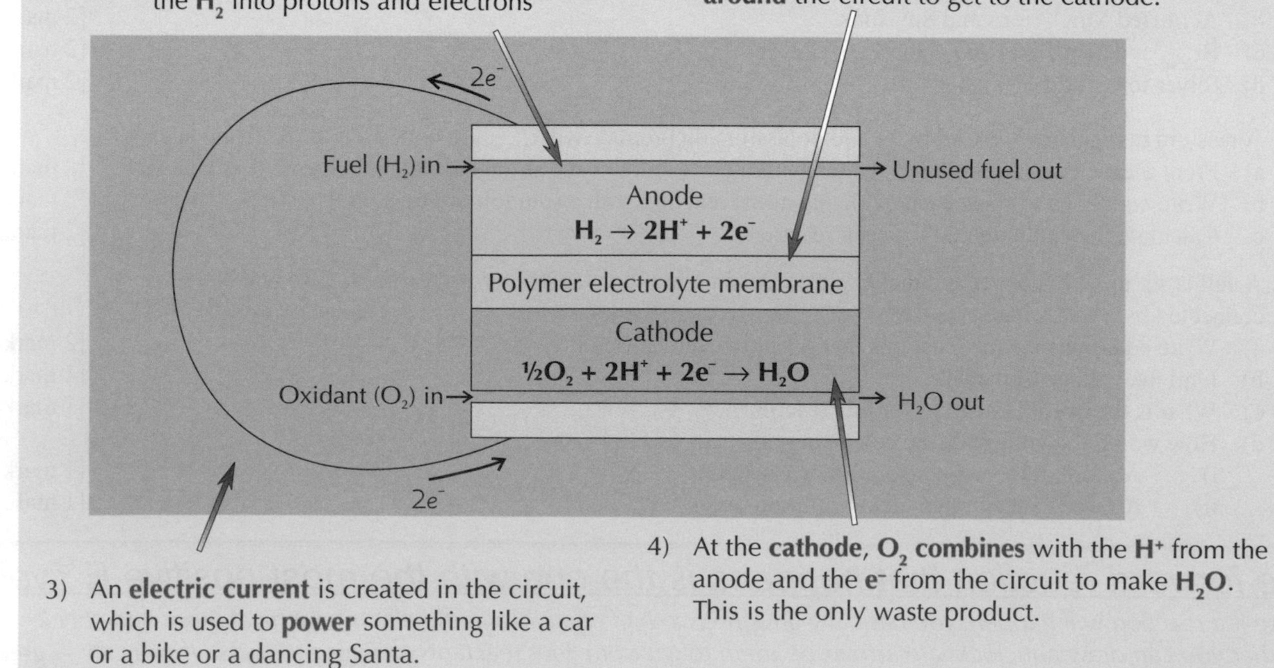

3) An **electric current** is created in the circuit, which is used to **power** something like a car or a bike or a dancing Santa.

4) At the **cathode**, O_2 **combines** with the H^+ from the anode and the e^- from the circuit to make H_2O. This is the only waste product.

Storage and Fuel Cells

Fuel Cell Vehicles use Fuel Cells For Power

Fuel cell vehicles are, unsurprisingly, electric vehicles powered by fuel cells as opposed to petrol or diesel. Using a **hydrogen** fuel cell does give FCVs some **important advantages** over regular cars.

1) They produce a lot **less pollution** because the only waste product is water.
2) The fuel cell is twice as **efficient** at converting fuel to power vs a petrol engine.

Fuel cells don't just use hydrogen — they can **also** be powered by **hydrogen-rich** fuels.

1) Hydrogen-rich fuels include **methanol**, **natural gas**, or **petrol**.
2) These are **converted** into **hydrogen gas** by a **reformer** before being used in the fuel cell.
3) Hydrogen rich fuels only release **small amounts** of **pollutants** and CO_2 when used in a fuel cell compared to burning them in a conventional engine.

Hydrogen Fuel Cells still have some Problems

Hydrogen might sound like the perfect fuel but there are some other things to think about...

... the fuel cells are not easy to make

1) The **platinum** catalysts and membrane are **expensive**.
2) The **production** of a fuel cell involves the use of **toxic chemicals**, which you need to dispose of afterwards.
3) Fuel cells only have a **limited life span** so need to be replaced, which means new ones have to be made, and old ones disposed of. **Disposing** of a fuel cell is an **expensive** process because of the chemicals they contain and the need to recycle some of the materials.

... storing and transporting hydrogen can be a pain

1) If you store it as a **gas** it is very **explosive**.
2) If you try to store it as a **liquid** you need really **expensive fridges** because it has such a low boiling point.
3) You can also store it **adsorbed** to the **surface** of a solid like charcoal or **absorbed** into a material like palladium but...
...these can be **very expensive** and often have a **limited life span**.

Adsorption is when something forms a layer on a surface
Absorption is when something is taken up by another substance

Martin had seen the chocolate sundae and was ready to adsorb it to his face.

... manufacturing hydrogen takes energy

1) **Most** hydrogen is currently produced from **reacting natural gas** with steam, which **produces carbon dioxide** as a waste product.
2) Not only is one of the **reactants** a **fossil fuel**, but fossil fuels are also used to **heat** the process.
3) Hydrogen can be produced by the **electrolysis** of water but the **large** amounts of **electricity** needed are produced by conventional **power stations** using fossil fuels.
4) **Hydrogen** is described as a **energy carrier** and not an energy source because it requires energy to make it.

Storage and Fuel Cells

A Hydrogen Economy Uses Hydrogen Fuel For Its Energy

Lots of people think that hydrogen will be really important in the future. They think there will be a **hydrogen economy** instead of an **oil economy** where all the **energy needs** of cars, buildings and electronics will be **powered** by **hydrogen fuel cells**.

Sounds great but there are a few things to overcome first...

1) People **accepting** hydrogen as a fuel.

Many people have concerns about the **safety** and **reliability** of hydrogen as a fuel. Most people are happy filling their cars with petrol but might think doing the same with hydrogen is more dangerous.

2) The **cost** of the new system.

If hydrogen is going to be the future it needs to be as cheap or **cheaper** than **existing** energy systems to convince people to change. The **infrastructure** for hydrogen fuel supplies would also be **very expensive** to set up.

3) Clean, renewable energy systems needed to **produce** the hydrogen are also **expensive**.

Hydrogen is an **energy carrier**, which means it needs an energy source to make it. So if you want to make **clean hydrogen** you need to make it using a **clean energy source**, e.g. solar or wind energy. Unfortunately these methods are **expensive** and currently don't supply large amounts of energy.

Sam pondered bottling horse wind to make hydrogen. It was cheap but often not clean.

Practice Questions

Q1 What are the half-equations for the reactions at each electrode in a hydrogen-oxygen fuel cell?

Q2 Give two ways that hydrogen fuel could be stored.

Q3 What is meant by a hydrogen-rich fuel?

Q4 Give two advantages of FCVs over conventional cars.

Exam Questions

1 a) Sketch a diagram showing the structure and operation of a hydrogen-oxygen fuel cell. Include the relevant half-equations. [5 marks]

b) Label the site of oxidation and the site of reduction on the diagram. [2 marks]

2 a) Fuel cell vehicles use hydrogen as a fuel. Give one advantage and one disadvantage of fuel cells over conventional petrol engines. [2 marks]

b) It is possible that in the future we will move towards a 'hydrogen economy'. Explain what this means. [2 marks]

c) Outline three issues that will need to be addressed before a hydrogen economy could be implemented. [3 marks]

3 The half equations for a lead-acid battery are as follows:

$Pb + HSO_4^- \rightleftharpoons PbSO_4 + H^+ + 2e^-$ $E^\circ = +0.35$

$PbO_2 + 3H^+ + HSO_4^- + 2e^- \rightleftharpoons PbSO_4 + 2H_2O$ $E^\circ = +1.68$

a) Write out the overall equation for the reaction in its simplest form. [2 marks]

b) Calculate the voltage produced by the cell. [1 mark]

In the olden days they used donkey-powered mills. They were called...wait for it...

...mule cells. Buddum tish. Oh dear, I'm really struggling today. Anyway, the hydrogen future is not upon us yet. It may have some advantages over the oil we use today but there are a lot of other things to overcome first. So maybe it will happen and maybe it won't. All we really know is it's the end of the section and that's got to be a good thing. Hurrah!!

Properties of Transition Elements

The transition elements are the metallic ones that sit slap bang in the middle of the periodic table. Thanks to their weird electronic structure, they make pretty-coloured solutions, and get involved in all sorts of fancy reactions.

Transition Elements are Found in the d-Block

The **d-block** is the block of elements in the middle of the periodic table.
Most of the elements in the d-block are **transition elements**.
You only need to know about the transition elements in the first row of the d-block — the ones from **titanium to copper**.

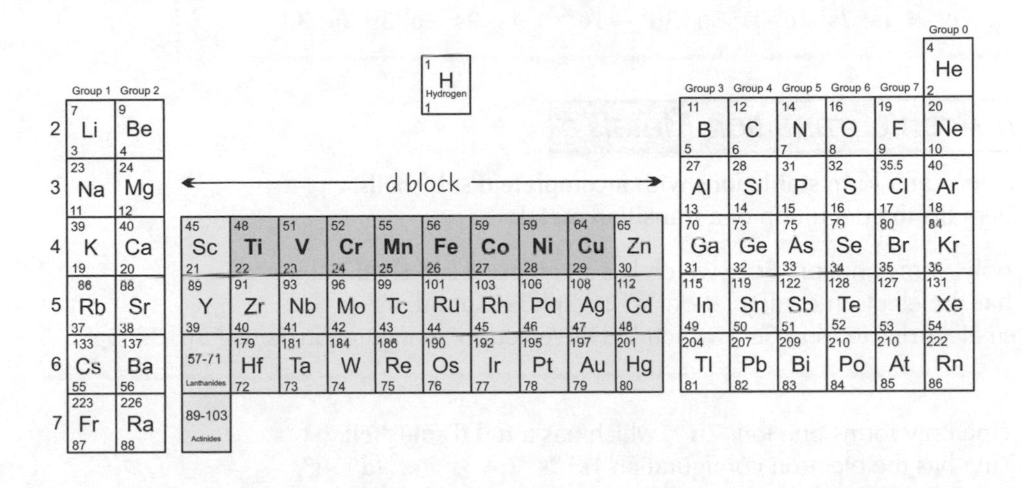

You Need to Know the Electron Configurations of the Transition Elements

1) Make sure you can write down the **electron configurations** of **all** the **Period 4 d-block elements** in sub-shell notation. Have a look at your AS notes if you've forgotten the details of how to do this. Here are a couple of examples:

   V = $1s^2\ 2s^2\ 2p^6\ 3s^2\ 3p^6\ 3d^3\ 4s^2$ **Co = $1s^2\ 2s^2\ 2p^6\ 3s^2\ 3p^6\ 3d^7\ 4s^2$**

   The 4s electrons fill up before the 3d electrons.
   But chromium and copper are a trifle odd — see below.

2) Here's the definition of a transition element:

   A **transition element** is one that can form **at least one stable ion** with an **incomplete d sub-shell.**

3) A d sub-shell can hold **10** electrons. So transition metals must form **at least one ion** that has **between 1 and 9 electrons** in the d sub-shell. All the Period 4 d-block elements are transition metals apart from **scandium** and **zinc**. This diagram shows the 3d and 4s sub-shells of these elements:

		3d					4s
Ti	[Ar]	↑	↑				↑↓
V	[Ar]	↑	↑	↑			↑↓
Cr	[Ar]	↑	↑	↑	↑	↑	↑
Mn	[Ar]	↑	↑	↑	↑	↑	↑↓
Fe	[Ar]	↑↓	↑	↑	↑	↑	↑↓
Co	[Ar]	↑↓	↑↓	↑	↑	↑	↑↓
Ni	[Ar]	↑↓	↑↓	↑↓	↑	↑	↑↓
Cu	[Ar]	↑↓	↑↓	↑↓	↑↓	↑↓	↑

The 3d orbitals are occupied singly at first. They only double up when they have to.

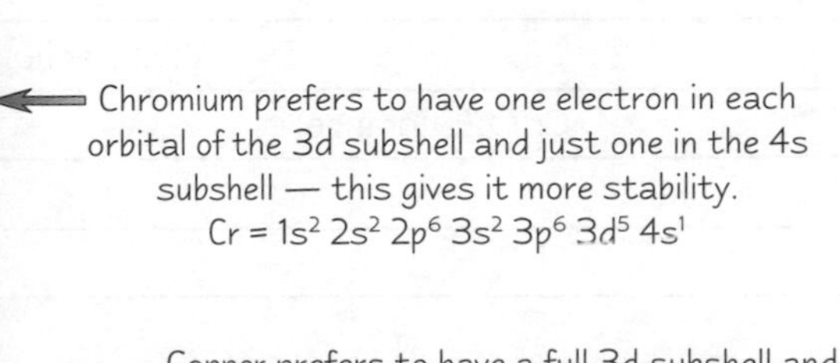

Chromium prefers to have one electron in each orbital of the 3d subshell and just one in the 4s subshell — this gives it more stability.
Cr = $1s^2\ 2s^2\ 2p^6\ 3s^2\ 3p^6\ 3d^5\ 4s^1$

Copper prefers to have a full 3d subshell and just one electron in the 4s subshell — it's more stable that way. Cu = $1s^2\ 2s^2\ 2p^6\ 3s^2\ 3p^6\ 3d^{10}\ 4s^1$
Copper forms a stable Cu^{2+} ion by losing 2 electrons. The Cu^{2+} ion has an incomplete d sub-shell.

4) It's because of their **incomplete d sub-shells** that the transition elements have some **special chemical properties**. There's more about this on the next page.

Properties of Transition Elements

When Ions are Formed, the s Electrons are Removed First

When transition elements form **positive** ions, the **s electrons** are removed **first**, **then** the d electrons.

1) Iron can form Fe^{2+} ions and Fe^{3+} ions.
2) When it forms 2+ ions, it loses **both its 4s electrons.**
   $Fe = 1s^2\ 2s^2\ 2p^6\ 3s^2\ 3p^6\ 3d^6\ 4s^2 \rightarrow Fe^{2+} = 1s^2\ 2s^2\ 2p^6\ 3s^2\ 3p^6\ 3d^6$
3) Only once the 4s electrons are removed can a **3d electron** be removed.
   E.g. **$Fe^{2+} = 1s^2\ 2s^2\ 2p^6\ 3s^2\ 3p^6\ 3d^6 \rightarrow Fe^{3+} = 1s^2\ 2s^2\ 2p^6\ 3s^2\ 3p^6\ 3d^5$**

Sc and Zn Aren't Transition Metals

Scandium and zinc can't form **stable ions** with **incomplete d sub-shells.**
So neither of them fits the definition of a **transition metal.**

Scandium only forms one ion, **Sc^{3+}**, which has an **empty d sub-shell.**
Scandium has the electron configuration **$1s^2\ 2s^2\ 2p^6\ 3s^2\ 3p^6\ 3d^1\ 4s^2$**.
It loses three electrons to form Sc^{3+}, which has the electron configuration **$1s^2\ 2s^2\ 2p^6\ 3s^2\ 3p^6$**.

Zinc only forms one ion, **Zn^{2+}**, which has a **full d sub-shell.**
Zinc has the electron configuration **$1s^2\ 2s^2\ 2p^6\ 3s^2\ 3p^6\ 3d^{10}\ 4s^2$**.
When it forms Zn^{2+} it loses 2 electrons, both from the 4s sub-shell — so it keeps its full 3d sub-shell.

Transition Elements have Special Chemical Properties

1) Transition elements can form **complex ions** — see page 188.
   For example, iron forms a **complex ion with water** — $[Fe(H_2O)_6]^{2+}$.
2) They can exist in **variable oxidation states.**
   For example, iron can exist in the **+2** oxidation state as **Fe^{2+}** ions and in the **+3** oxidation state as **Fe^{3+}** ions.
3) They form **coloured ions.** E.g. Fe^{2+} ions are **pale green** and Fe^{3+} ions are **yellow.**
4) Transition metals and their compounds make **good catalysts** because they can **change oxidation states** by gaining or losing electrons within their **d orbitals.** This means they can **transfer electrons** to **speed up** reactions.

- **Iron** is the catalyst used in the **Haber process** to produce ammonia.
- **Vanadium(V) oxide, V_2O_5,** is the catalyst used in the **contact process** to make sulfuric acid.
- **Nickel** is the catalyst used to **harden margarine.**

Some common **coloured** ions and **oxidation states** are shown below. The colours refer to the **aqueous ions.**

oxidation state +7	+6	+5	+4	+3	+2
		VO_2^+ (yellow)	VO^{2+} (blue)	V^{3+} (green)	V^{2+} (violet)
	$Cr_2O_7^{2-}$ (orange)			Cr^{3+} (green)	
MnO_4^- (purple)					Mn^{2+} (pale pink)
				Fe^{3+} (yellow)	Fe^{2+} (pale green)
					Co^{2+} (pink)
					Ni^{2+} (green)
					Cu^{2+} (pale blue)
				Ti^{3+} (purple)	Ti^{2+} (violet)

These elements show **variable** oxidation states because the **energy levels** of the 4s and the 3d sub-shells are **very close** to one another. So different numbers of electrons can be gained or lost using fairly **similar** amounts of energy.

Properties of Transition Elements

Transition Metals Hydroxides are Brightly Coloured Precipitates

When you mix a solution of **transition metal ions** with **sodium hydroxide solution** you get a **coloured precipitate**.

You need to know the equations for the following reactions, and the colours of the hydroxide precipitates:

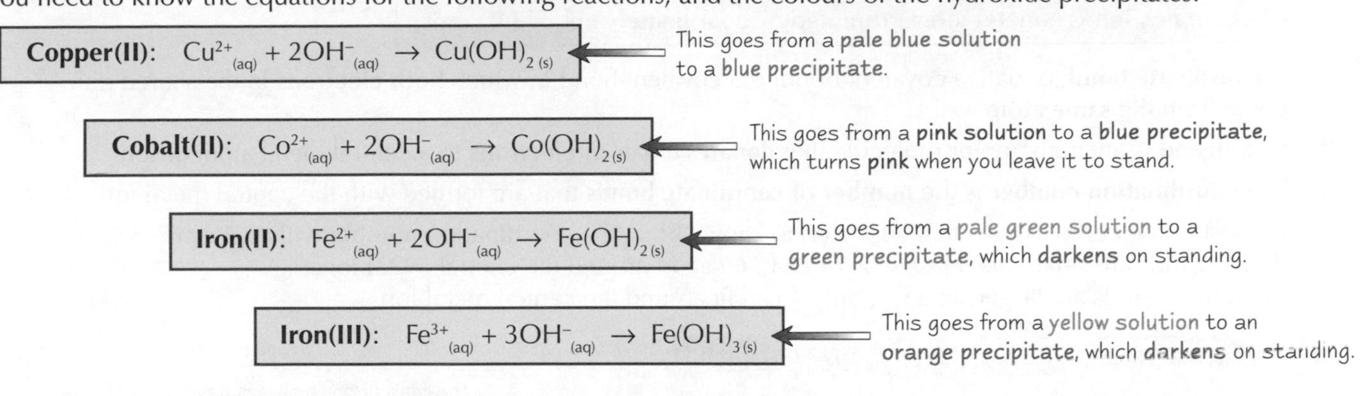

If you know which **transition metal** ion produces which **colour** precipitate, you can use these reactions to **identify** a transition metal ion solution.

Practice Questions

Q1 What's the definition of a transition metal?

Q2 Give the electron configuration of: (a) a vanadium atom, (b) a V^{2+} ion.

Q3 State four chemical properties which are characteristic of transition elements.

Q4 Write an equation for the reaction of iron(II) ions with hydroxide ions.
Describe the colour change that occurs during this reaction.

Exam Questions

1 When solid copper(I) sulfate is added to water, a blue solution forms with a red-brown precipitate of copper metal.

a) Give the electron configuration of copper(I) ions. [1 mark]

b) Does the formation of copper(I) ions show copper acting as a transition metal? Explain your answer. [2 marks]

c) Identify the blue solution. [1 mark]

2 Manganese and iron sit next to each other in the periodic table. Both are transition metals.
Their most stable ions are Mn^{2+} and Fe^{3+} respectively.

a) Write a balanced equation for the reaction of Fe^{3+} ions with hydroxide ions in solution. [2 marks]

b) What is the oxidation state of manganese in the following compounds:

i) $KMnO_4$ [1 mark]

ii) MnO_2 [1 mark]

iii) Mn_2O_7 [1 mark]

c) Sodium hydroxide solution is added to a test tube containing $FeCl_3$ solution.
Describe the change that you would expect to observe in the test tube as the sodium hydroxide is added. [2 marks]

3 Aluminium and iron are the two most common metals in the Earth's crust.
Both can form an ion with a 3+ charge.

a) i) Give the electron configuration of the Fe^{3+} ion. [1 mark]

ii) Give the electron configuration of the Al^{3+} ion. [1 mark]

b) With reference to oxidation states and colours of compounds,
explain why iron is a typical transition metal and aluminium is not. [4 marks]

4s electrons — like rats leaving a sinking ship...

Have a quick read of the electronic configuration stuff in your AS notes if it's been pushed to a little corner of your mind labelled, "Well, I won't be needing that again in a hurry". It should come flooding back pretty quickly. And don't forget to learn all the metal ion/hydroxide reactions at the top of this page — plus the colour changes that go with them...

Complex Ions

Transition metals are always forming complex ions. These aren't as complicated as they sound, though. Honest.

Complex Ions are ***Metal Ions*** Surrounded by ***Ligands***

A **complex ion** is a **metal ion** surrounded by **coordinately bonded ligands**.

1) A **coordinate bond** (or dative covalent bond) is a covalent bond in which **both electrons** in the shared pair come from the **same atom**.
2) So a **ligand** is an atom, ion or molecule that **donates a pair of electrons** to a central metal atom or ion.
3) The **coordination number** is the **number** of **coordinate bonds** that are formed with the central metal ion.
4) In most of the complex ions that you need to know about, the coordination number will be **4** or **6**.
If the ligands are **small**, like H_2O, CN^- or NH_3, **6** can fit around the central metal ion.
But if the ligands are **larger**, like Cl^-, **only 4** can fit around the central metal ion.

6 COORDINATE BONDS MEAN AN <u>OCTAHEDRAL</u> SHAPE

Here are a few examples:

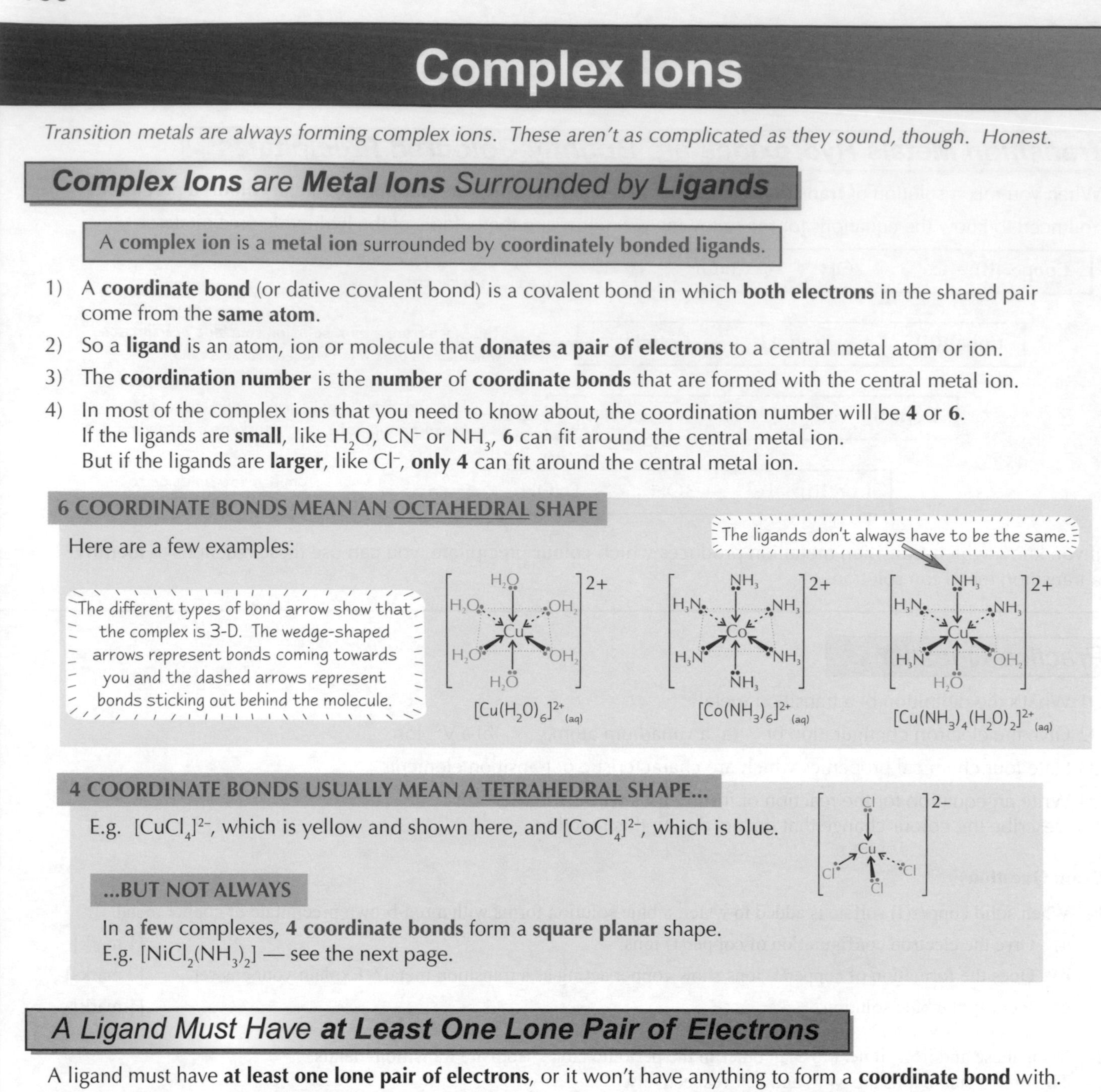

4 COORDINATE BONDS USUALLY MEAN A <u>TETRAHEDRAL</u> SHAPE...

E.g. $[CuCl_4]^{2-}$, which is yellow and shown here, and $[CoCl_4]^{2-}$, which is blue.

...BUT NOT ALWAYS

In a **few complexes**, **4 coordinate bonds** form a **square planar** shape.
E.g. $[NiCl_2(NH_3)_2]$ — see the next page.

A Ligand Must Have ***at Least One Lone Pair of Electrons***

A ligand must have **at least one lone pair of electrons**, or it won't have anything to form a **coordinate bond** with.

- Ligands that have **one lone pair** available for bonding are called **monodentate** — e.g. $H_2\ddot{O}$, $\ddot{N}H_3$, $\ddot{Cl}^-$, $\ddot{C}N^-$.
- Ligands with **two lone pairs** are called **bidentate** — e.g. ethane-1,2-diamine: $\ddot{N}H_2CH_2CH_2\ddot{N}H_2$.
Bidentate ligands can each form **two coordinate bonds** with a metal ion.
- Ligands with **more than two lone pairs** are called **multidentate**.

You might see ethane-1,2-diamine abbreviated to "en".

Complex Ions Can Show ***Optical Isomerism***

Optical isomerism is a type of **stereoisomerism** (see the next page). With complex ions, it happens when an ion can exist in **two non-superimposable mirror images**.

This happens when **three bidentate ligands**, such as ethane-1,2-diamine, $H_2NCH_2CH_2NH_2$, use the lone pairs on **both** nitrogen atoms to coordinately bond with **nickel**.

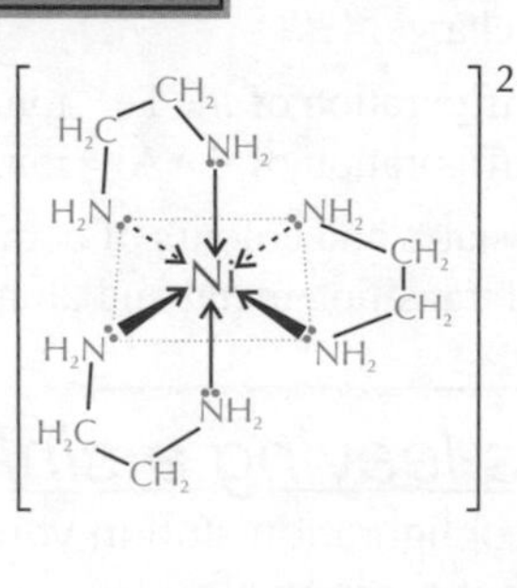

Complex Ions

Complex Ions Can Show Cis-Trans Isomerism

Cis-trans isomerism is a special case of **E/Z isomerism** (see page 130). When there are only **two different groups** involved, you can use the ***cis-trans* naming system**. **Square planar** complex ions that have **two pairs** of ligands show **cis-trans isomerism**. $[NiCl_2(NH_3)_2]$ is an example of this:

E-Z is another type of **stereoisomerism**. In stereoisomerism, the atoms are joined together in the same order, but they have different orientations in space.

Cis isomers have the same groups on the same sides.

cis-$[NiCl_2(NH_3)_2]$

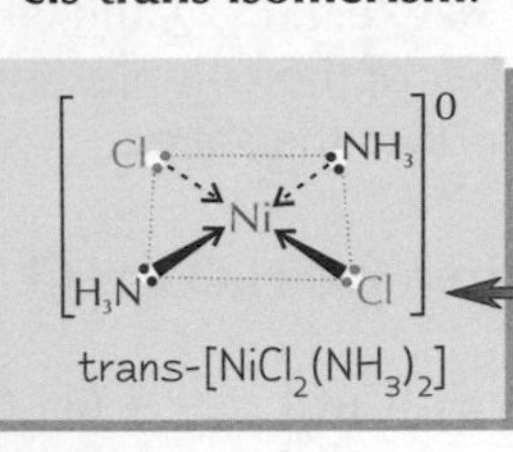

trans-$[NiCl_2(NH_3)_2]$

Trans isomers have the same groups diagonally across from each other.

Cis-platin Can Bind to DNA in Cancer Cells

Cis-platin is a complex of platinum(II) with two chloride ions and two ammonia molecules in a square planar shape. It is used as an anti-cancer drug.

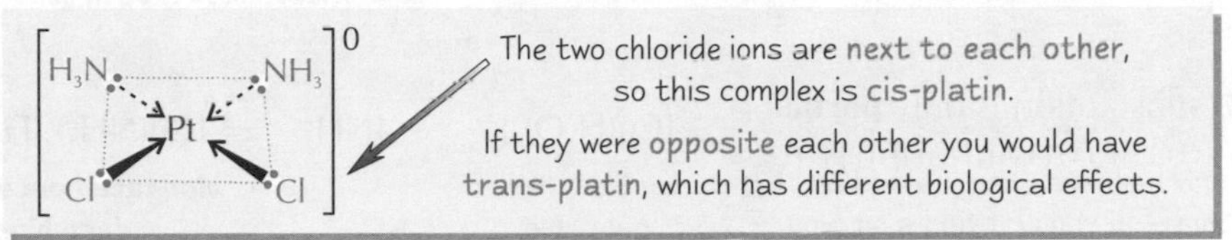

This is how it works:

1) The two **chlorine ligands** are very easy to displace. So the cis-platin loses them, and bonds to two **nitrogen atoms** on the **DNA molecule** inside the **cancerous cell** instead.
2) This **block** on its DNA prevents the cancerous cell from **reproducing** by division. The cell will **die**, since it is unable to repair the damage.

Practice Questions

Q1 Explain what the term 'coordination number' means in relation to a complex ion.

Q2 Draw the shape of the complex ion $[Co(NH_3)_6]^{3+}$. Name the shape.

Q3 What is meant by the term 'bidentate ligand'? Give an example of one.

Q4 Draw the cis and trans isomers of the complex $[NiBr_2(NH_3)_2]$.

Exam Questions

1 When potassium cyanide is added to iron(II) chloride solution, the complex ion $[Fe(CN)_6]^{4-}$ is produced.

a) What is meant by the term 'complex ion'? [1 mark]

b) Explain how the cyanide ions bond with the central iron ion. [1 mark]

c) Draw a diagram to show the structure of the complex ion. [1 mark]

2 Iron(III) can form the complex ion $[Fe(C_2O_4)_3]^{3-}$ with three ethanedioate ions. The ethanedioate ion is a bidentate ligand. Its structure is shown on the right.

a) Explain the term 'bidentate ligand'. [2 marks]

b) What is the coordination number of the $[Fe(C_2O_4)_3]^{3-}$ complex? [1 mark]

c) Use your answer from part (b) to suggest what shape the $[Fe(C_2O_4)_3]^{3-}$ complex is. [1 mark]

3 Cis-platin is a platinum-based complex ion that is used as an anti-cancer drug.

a) Draw the structure of cis-platin. [3 marks]

b) Explain how cis-platin works as a cancer treatment. [4 marks]

Put your hands up — we've got you surrounded...

You'll never get transition metal ions floating around by themselves in a solution — they'll always be surrounded by other molecules. It's kind of like what'd happen if you put a dish of sweets in a room of eight (or eighteen) year-olds. When you're drawing complex ions, you should always include some wedge-shaped bonds to show that it's 3-D.

Substitution Reactions

There are more equations on this page than the number of elephants you can fit in a Mini.

***Ligands** can **Exchange** Places with One Another*

One ligand can be **swapped** for another ligand — this is **ligand substitution**. It pretty much always causes a **colour change**.

1) If the ligands are of **similar size**, e.g. H_2O and NH_3, then the **coordination number** of the complex ion **doesn't change**, and neither does the **shape**.

$$[Co(H_2O)_6]^{2+}_{(aq)} + 6NH_{3(aq)} \rightleftharpoons [Co(NH_3)_6]^{2+}_{(aq)} + 6H_2O_{(l)}$$

octahedral, pink → octahedral, pale brown

2) If the ligands are **different sizes**, e.g. H_2O and Cl^-, there's a **change of coordination number** and a **change of shape**.

$$[Cu(H_2O)_6]^{2+}_{(aq)} + 4Cl^-_{(aq)} \rightleftharpoons [CuCl_4]^{2-}_{(aq)} + 6H_2O_{(l)}$$

octahedral, pale blue → tetrahedral, yellow

$$[Co(H_2O)_6]^{2+}_{(aq)} + 4Cl^-_{(aq)} \rightleftharpoons [CoCl_4]^{2-}_{(aq)} + 6H_2O_{(l)}$$

octahedral, pink → tetrahedral, blue

The forward reaction is endothermic, so the equilibrium can be shifted to the right-hand side by heating. The equilibrium will also shift to the right if you add more concentrated hydrochloric acid. Adding water to this equilibrium shifts it back to the left.

3) Sometimes the substitution is only **partial**.

$$[Cu(H_2O)_6]^{2+}_{(aq)} + 4NH_{3(aq)} \rightleftharpoons [Cu(NH_3)_4(H_2O)_2]^{2+}_{(aq)} + 4H_2O_{(l)}$$

octahedral, pale blue → **elongated octahedral, deep blue**

This reaction only happens when you add an excess of ammonia — if you just add a bit, you get a blue precipitate of $[Cu(H_2O)_4(OH)_2]$ instead.

*Fe^{2+} in **Haemoglobin** Allows **Oxygen** to be Carried in the Blood*

1) **Haemoglobin** contains Fe^{2+} ions. The Fe^{2+} ions form **6 coordinate bonds**. Four of the **lone pairs** come from nitrogen atoms within a circular part of a molecule called **'haem'**. A fifth lone pair comes from a nitrogen atom on a protein (**globin**). The last position is the important one — this has a **water ligand** attached to the **iron**.

Haem is a multidentate ligand.

2) In the lungs the oxygen concentration is high, so the water ligand is **substituted** for an **oxygen molecule** (**O_2**), forming **oxyhaemoglobin**. This is carried around the body and when it gets to a place where oxygen is needed the oxygen molecule is exchanged for a water molecule again.

3) If **carbon monoxide** (**CO**) is inhaled, the **haemoglobin** swaps its **water** ligand for a **carbon monoxide** ligand, forming **carboxyhaemoglobin**. This is bad news because carbon monoxide is a **strong** ligand and **doesn't** readily exchange with oxygen or water ligands, meaning the haemoglobin **can't transport oxygen** any more.

Stability Constants** are Special **Equilibrium Constants

The **stability constant of a complex ion** is just what it sounds like — it tells you how stable a complex ion is in solution. Here's the official definition:

The **stability constant**, **K_{stab}**, of a **complex ion** is the **equilibrium constant** for the **formation** of the complex ion from its **constituent ions** in solution.

If you want a reminder of how equilibrium constants work, look back at page 154.

Example: Write an expression for the stability constant for the formation of the complex ion $[Fe(CN)_6]^{4-}$

Here is the equation for the formation of this ion in solution: $Fe^{2+}_{(aq)} + 6CN^-_{(aq)} \rightleftharpoons [Fe(CN)_6]^{4-}_{(aq)}$

So the stability constant for this reaction is: $K_{stab} = \dfrac{[(Fe(CN)_6)^{4-}]}{[Fe^{2+}][CN^-]^6}$

The square brackets in the K_{stab} expression mean 'the concentrations of the ions'. Don't mix them up with the square brackets you'd use in the formula of a complex ion — they just keep everything together.

Substitution Reactions

K_{stab} Can Tell You If a **Ligand Substitution Reaction** Will Happen

1) All ligand substitution reactions are **reversible** — so you can write a ligand substitution reaction as an **equilibrium**.
2) The **equilibrium constant** will also be the **stability constant**, or K_{stab}, for this ligand substitution reaction.
3) If the complex that you start with **only** has **water** ligands, **don't** include $[H_2O]$ in the stability constant expression. Since all the ions are in solution, there's so much water around that the few extra molecules produced don't alter its concentration much — it's practically constant.
4) The size of the stability constant tells you how stable the new complex ion is, and how likely it is to form.

Example: Write an expression for the stability constant, K_{stab}, of the following ligand substitution reaction:

$$[Cu(H_2O)_6]^{2+}_{(aq)} + 4Cl^-_{(aq)} \rightleftharpoons [CuCl_4]^{2-}_{(aq)} + 6H_2O_{(l)}$$

$$K_{stab} = \frac{[(CuCl_4)^{2-}]}{[(Cu(H_2O)_6)^{2+}][Cl^-]^4}$$

Remember that you don't need to include $[H_2O]$ in the expression.

The **stability constant** for the reaction in the example is 4.2×10^5 dm^{12} mol^{-4} at 291 K.
This is a **large** stability constant — which tells you that the $[CuCl_4]^{2-}$ complex ion is **very stable**.
So if you add **chloride ions** to a solution containing $[Cu(H_2O)_6]^{2+}$ ions, it's very likely that a **ligand substitution** reaction will happen, and you'll end up with $[CuCl_4]^{2-}$ ions.

Practice Questions

Q1 Give an example of a ligand substitution reaction that involves a change of coordination number.

Q2 What is the coordination number of the Fe^{2+} ion in the haemoglobin complex?

Q3 What does the size of the stability constant for a particular ligand substitution reaction tell you?

Exam Questions

1 a) Write an expression for the stability constant for the formation of the complex ion $[Co(NH_3)_6]^{2+}$. [1 mark]

b) Write an expression for the stability constant of the following ligand substitution reaction:

$[Cu(H_2O)_6]^{2+}_{(aq)} + 6CN^-_{(aq)} \rightleftharpoons [Cu(CN)_6]^{4-}_{(aq)} + 6H_2O_{(l)}$ [1 mark]

2 A sample of copper(II) sulfate powder is dissolved in pure water, giving a pale blue solution.

a) Give the formula of the complex ion that is present in the pale blue solution. [1 mark]

b) When an excess of ammonia is added to the solution, its colour changes to deep blue.

i) Write a balanced equation for the ligand substitution reaction that has taken place. [2 marks]

ii) Write an expression for the stability constant of this ligand substitution reaction. [2 marks]

3 Haemoglobin is a complex ion that is found in the blood. It consists of an Fe^{2+} ion bonded to four nitrogen atoms from a haem ring, one nitrogen atom from a protein called globin, and one water molecule.

a) When blood passes through the lungs, a ligand substitution reaction occurs.

i) Which of the ligands in the haemoglobin complex is replaced, and by what? [2 marks]

ii) Why is this ligand substitution reaction important? [1 mark]

b) Explain how inhaling carbon monoxide can damage the human body. [3 marks]

My friend suffers from Nativity Play Phobia — he's got a stable complex...

Four things to do with this page — One: learn what a ligand substitution reaction is, and why haemoglobin's an important example of one. Two: learn the definition of the stability constant of a complex ion. Three: make sure you can write an expression for K_{stab} for the formation of an ion, and a ligand substitution reaction. Four: fold it into a lovely origami crane.

Redox Reactions and Transition Elements

Transition elements love to swap electrons around, so they're always getting involved in redox reactions. That makes them handy for doing redox titrations — which are like acid-base titrations, but different (you don't need indicators for a start).

Transition Elements are Used as Oxidising and Reducing Agents

1) Transition elements can exist in many different **oxidation states** (see page 186).
2) They can **change** oxidation state by **gaining** or **losing electrons** in **redox reactions** (see page 176).
3) The ability to gain or lose electrons easily makes transition metal ions good **oxidising** or **reducing agents.**

Here are a couple of examples:

Acidified **potassium manganate(VII)** solution, $KMnO_{4\,(aq)}$, is used as an **oxidising agent**.
It contains **manganate(VII) ions** (MnO_4^-), in which manganese has an oxidation state of +7.
They can be reduced to Mn^{2+} ions during a **redox reaction.**

Example: The oxidation of Fe^{2+} to Fe^{3+} by manganate(VII) ions in solution.

Half equations:

$MnO_4^- + 8H^+ + 5e^- \rightarrow Mn^{2+} + 4H_2O$ — Manganese is **reduced**

$5Fe^{2+} \rightarrow 5Fe^{3+} + 5e^-$ — Iron is **oxidised**

$$MnO_4^- + 8H^+ + 5Fe^{2+} \rightarrow Mn^{2+} + 4H_2O + 5Fe^{3+}$$

$MnO_{4\,(aq)}^-$ is **purple**. $[Mn(H_2O)_6]^{2+}_{(aq)}$ is **colourless**. During this reaction, you'll see a colour change from **purple** to **colourless**.

Acidified **potassium dichromate** solution, $K_2Cr_2O_{7\,(aq)}$, is another **oxidising agent**.
It contains **dichromate(VI) ions** ($Cr_2O_7^{2-}$), in which chromium has an oxidation state of **+6**.
They can be reduced to Cr^{3+} ions during a **redox reaction.**

Example: The oxidation of Zn to Zn^{2+} by dichromate(VI) ions in solution.

Half equations:

$Cr_2O_7^{2-} + 14H^+ + 6e^- \rightarrow 2Cr^{3+} + 7H_2O$ — Chromium is **reduced**

$3Zn \rightarrow 3Zn^{2+} + 6e^-$ — Zinc is **oxidised**

$$Cr_2O_7^{2-} + 14H^+ + 3Zn \rightarrow 2Cr^{3+} + 7H_2O + 3Zn^{2+}$$

$Cr_2O_{7\,(aq)}^{2-}$ is **orange**. $[Cr(H_2O)_6]^{3+}_{(aq)}$ is violet, but usually looks **green**. During this reaction, you'll see a colour change from **orange** to **green**.

Titrations Using Transition Element Ions are Redox Titrations

Redox titrations are used to find out how much **oxidising agent** is needed to **exactly** react with a quantity of **reducing agent.** You need to know the **concentration** of either the oxidising agent or the reducing agent. Then you can use the titration results to work out the concentration of the other.

You need to know about redox titrations in which **manganate(VII) ions** (MnO_4^-) are the oxidising agents.

1) First you measure out a quantity of **reducing agent**, e.g. aqueous Fe^{2+} ions, using a pipette, and put it in a conical flask.
2) You then add some **dilute sulfuric acid** to the flask — this is an excess, so you don't have to be too exact (about 20 cm³ should do it). The acid is added to make sure there are plenty of H^+ ions to allow the oxidising agent to be reduced.
3) Now you add the aqueous MnO_4^- (the **oxidising agent**) to the reducing agent using a **burette**, **swirling** the conical flask as you do so.
4) You stop when the mixture in the flask **just** becomes tainted with the colour of the MnO_4^- (the **end point**) and record the volume of the oxidising agent added. This is the **rough titration**.
5) Now you do some **accurate titrations**. You need to do a few until you get **two or more** readings that are **within 0.20 cm³** of each other.

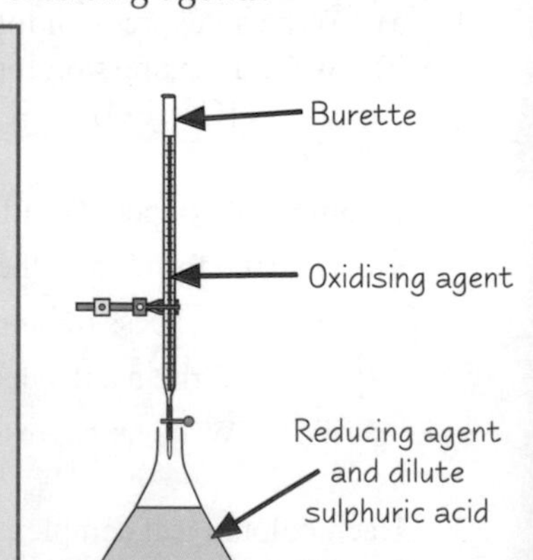

You can also do titrations the **other way round** — adding the reducing agent to the oxidising agent.

The Sharp Colour Change Tells You when the Reaction's Just Been Completed

1) **Manganate(VII) ions** (MnO_4^-) in **aqueous potassium manganate(VII)** ($KMnO_4$) are **purple**. When they're added to the reducing agent, they start reacting. This reaction will continue until **all** of the reducing agent is used up.
2) The **very next drop** into the flask will give the mixture the **purple colour of the oxidising agent**. The trick is to spot **exactly** when this happens. (You could use a coloured reducing agent and a colourless oxidising agent instead — then you'd be watching for the moment that the colour in the flask disappears.)

Redox Reactions and Transition Elements

You Can Calculate the Concentration of a Reagent from the Titration Results

Example: 27.5 cm^3 of 0.020 $moldm^{-3}$ aqueous potassium manganate(VII) reacted with 25.0 cm^3 of acidified iron(II) sulfate solution. Calculate the concentration of Fe^{2+} ions in the solution.

$$MnO_{4\,(aq)}^- + 8H^+_{(aq)} + 5Fe^{2+}_{(aq)} \rightarrow Mn^{2+}_{(aq)} + 4H_2O_{(l)} + 5Fe^{3+}_{(aq)}$$

1) Work out the number of **moles of MnO_4^- ions** added to the flask.

$$\text{Number of moles } MnO_4^- \text{ added} = \frac{\text{concentration} \times \text{volume}}{1000} = \frac{0.020 \times 27.5}{1000} = 5.50 \times 10^{-4} \text{ moles}$$

2) Look at the balanced equation to find how many moles of **Fe^{2+}** react with **every mole** of MnO_4^-. Then you can work out the **number of moles of Fe^{2+}** in the flask.

5 moles of Fe^{2+} react with 1 mole of MnO_4^-. So moles of Fe^{2+} = $5.50 \times 10^{-4} \times 5 = 2.75 \times 10^{-3}$ moles.

3) Work out the **number of moles of Fe^{2+}** that would be in 1000 cm^3 (1 dm) of solution — this is the **concentration**.

25.0 cm^3 of solution contained 2.75×10^{-3} moles of Fe^{2+}.

1000 cm^3 of solution would contain $\frac{(2.75 \times 10^{-3}) \times 1000}{25.0} = 0.11$ moles of Fe^{2+}.

So the concentration of Fe^{2+} is **0.11 $moldm^{-3}$**.

Manganate 007, licensed to oxidise.

Practice Questions

Q1 Write a half equation to show manganate(VII) ions acting as an oxidising agent.

Q2 What is the change in the oxidation state of manganese during this reaction?

Q3 If you carry out a redox titration by slowly adding aqueous MnO_4^- ions to aqueous Fe^{2+} ions, how can you tell that you've reached the end point?

Q4 Why is dilute acid added to the reaction mixture in redox titrations involving MnO_4^- ions?

Exam Questions

1 Steel wool contains a high percentage of iron, and a small amount of carbon.
A 1.3 g piece of steel wool was dissolved in 50 cm^3 of aqueous sulfuric acid.
The resulting solution was titrated with 0.4 mol dm^{-3} of potassium manganate(VII) solution.
11.5 cm^3 of the potassium manganate(VII) solution was needed to oxidise all of the iron(II) ions to iron(III).

a) Write a balanced equation for the reaction between the manganate(VII) ions and the iron(II) ions. [3 marks]

b) Calculate the number of moles of iron(II) ions present in the original solution. [3 marks]

c) Calculate the percentage of iron present in the steel wool.
Give your answer to one decimal place. [3 marks]

2 A 10 cm^3 sample of 0.5 mol dm^{-3} $SnCl_2$ solution was titrated with acidified potassium manganate(VII) solution.
Exactly 20 cm^3 of 0.1 mol dm^{-3} potassium manganate(VII) solution was needed to fully oxidise the tin(II) chloride.

a) What type of reaction is this? [1 mark]

b) How many moles of tin(II) chloride were present in the 10 cm^3 sample? [2 marks]

c) How many moles of potassium manganate(VII) were needed to fully oxidise the tin(II) chloride? [2 marks]

The half equation for acidified MnO_4^- acting as an oxidising agent is: $MnO_4^- + 8H^+ + 5e^- \rightarrow Mn^{2+} + 4H_2O$

d) Find the oxidation state of the oxidised tin ions present in the solution at the end of the titration. [4 marks]

And how many moles does it take to change a light bulb...

...two, one to change the bulb, and another to ask "Why do we need light bulbs? We're moles — most of the time that we're underground, we keep our eyes shut. We've mostly been using our senses of touch and smell to find our way around anyway. And we're not on mains, so the electricity must be costing a packet. We haven't thought this through properly..."

Iodine-Sodium Thiosulfate Titrations

This is another example of a redox titration — it's a handy little reaction that you can use to find the concentration of an oxidising agent. And since it's a titration, that also means a few more calculations to get to grips with...

Iodine-Sodium Thiosulfate Titrations are Dead Handy

Iodine-sodium thiosulfate titrations are a way of finding the concentration of an **oxidising agent**.
The **more concentrated** an oxidising agent is, the **more ions will be oxidised** by a certain volume of it.
So here's how you can find out the concentration of a solution of the oxidising agent **potassium iodate(V)**:

STAGE 1: Use a sample of oxidising agent to oxidise as much iodide as possible.

1) Measure out a certain volume of **potassium iodate(V)** solution (**KIO_3**) (the oxidising agent) — say **25 cm³**.
2) Add this to an excess of acidified **potassium iodide** solution (**KI**).
   The iodate(V) ions in the potassium iodate(V) solution **oxidise** some of the **iodide ions** to **iodine**.

$$IO_3^-{}_{(aq)} + 5I^-{}_{(aq)} + 6H^+{}_{(aq)} \rightarrow 3I_{2\,(aq)} + 3H_2O$$

STAGE 2: Find out how many moles of iodine have been produced.

You do this by **titrating** the resulting solution with **sodium thiosulfate** (**$Na_2S_2O_3$**).
(You need to know the concentration of the sodium thiosulfate solution.)
The iodine in the solution reacts with **thiosulfate ions** like this:

$$I_2 + 2S_2O_3^{2-} \rightarrow 2I^- + S_4O_6^{2-}$$

Titration of Iodine with Sodium Thiosulfate

1) Take the flask containing the solution that was produced in Stage 1.
2) From a burette, add sodium thiosulfate solution to the flask drop by drop.
3) It's hard to see the end point, so when the iodine colour fades to a pale yellow, add 2 cm³ of starch solution (to detect the presence of iodine). The solution in the conical flask will go dark blue, showing there's still some iodine there.
4) Add sodium thiosulfate <u>one drop at a time</u> until the blue colour disappears.
5) When this happens, it means all the iodine has <u>just</u> been reacted.
6) Now you can <u>calculate</u> the number of moles of iodine in the solution.

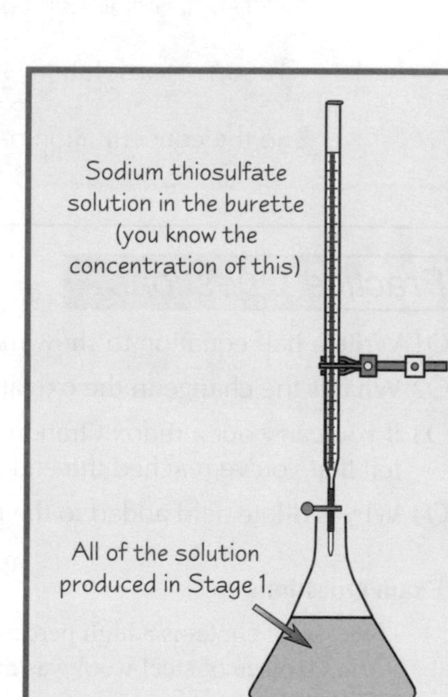

Here's how you'd do the titration calculation to find the **number of moles of iodine** produced in Stage 1.

Example

The iodine in the solution produced in Stage 1 reacted fully with 11.1 cm³ of 0.12 mol dm⁻³ thiosulfate solution.

$$I_2 + 2S_2O_3^{2-} \rightarrow 2I^- + S_4O_6^{2-}$$

11.1 cm³
0.12 mol dm⁻³

Number of moles of thiosulfate $= \frac{\text{concentration} \times \text{volume (cm}^3\text{)}}{1000} = \frac{0.12 \times 11.1}{1000} = \mathbf{1.332 \times 10^{-3}}$ **moles**

1 mole of iodine reacts with **2 moles** of thiosulfate.

So number of **moles of iodine** in the solution $= 1.332 \times 10^{-3} \div 2 = \mathbf{6.66 \times 10^{-4}}$ **moles**

Iodine-Sodium Thiosulfate Titrations

STAGE 3: Calculate the concentration of the oxidising agent.

1) Now you look back at your original equation: $IO_{3\,(aq)}^- + 5I^-_{(aq)} + 6H^+_{(aq)} \rightarrow 3I_{2\,(aq)} + 3H_2O$
2) 25 cm³ of potassium iodate(V) solution produced **6.66 × 10⁻⁴ moles of iodine**.
   The equation shows that **one mole** of iodate(V) ions will produce **three moles** of iodine.
3) That means there must have been **6.66 × 10⁻⁴ ÷ 3 = 2.22 × 10⁻⁴ moles of iodate(V) ions** in the original solution.
   So now it's straightforward to find the **concentration** of the potassium iodate(V) solution, which is what you're after:

$$\text{number of moles} = \frac{\text{concentration} \times \text{volume (cm}^3)}{1000} \quad \Rightarrow \quad 2.22\times10^{-4} = \frac{\text{concentration}\times 25}{1000}$$

⇒ concentration of potassium iodate(V) solution = 0.0089 mol dm⁻³

Practice Questions

Q1 How can an iodine-sodium thiosulfate titration help you to work out the concentration of an oxidising agent?

Q2 How many moles of thiosulfate ions react with one mole of iodine molecules?

Q3 What is added during an iodine-sodium thiosulfate titration to make the end point easier to see?

Q4 Describe the colour change at the end point of the iodine-sodium thiosulfate titration.

Exam Questions

1 10 cm³ of potassium iodate(V) solution was reacted with excess acidified potassium iodide solution.
All of the resulting solution was titrated with 0.15 mol dm⁻³ sodium thiosulfate solution.
It fully reacted with 24.0 cm³ of the sodium thiosulfate solution.

a) Write an equation showing how iodine is formed in the reaction between iodate(V) ions and iodide ions in acidic solution. [2 marks]

b) How many moles of thiosulfate ions were there in 24.0 cm³ of the sodium thiosulfate solution? [1 mark]

c) In the titration, iodine reacted with sodium thiosulfate according to this equation:

$$I_{2(aq)} + 2Na_2S_2O_{3(aq)} \rightarrow 2NaI_{(aq)} + Na_2S_4O_{6(aq)}$$

Calculate the number of moles of iodine that reacted with the sodium thiosulfate solution. [1 mark]

d) How many moles of iodate(V) ions produce 1 mole of iodine from potassium iodide? [1 mark]

e) What was the concentration of the potassium iodate(V) solution? [2 marks]

2 An 18 cm³ sample of potassium manganate(VII) solution was reacted with an excess of acidified potassium iodide solution. The resulting solution was titrated with 0.3 mol dm⁻³ sodium thiosulfate solution.
12.5 cm³ of sodium thiosulfate solution were needed to fully react with the iodine.

When they were mixed, the manganate(VII) ions reacted with the iodide ions according to this equation:

$$2MnO^-_{4\,(aq)} + 10I^-_{(aq)} + 16H^+ \rightarrow 5I_{2(aq)} + 8H_2O_{(aq)} + 2Mn^{2+}_{(aq)}$$

During the titration, the iodine reacted with sodium thiosulfate according to this equation:

$$I_{2(aq)} + 2Na_2S_2O_{3(aq)} \rightarrow 2NaI_{(aq)} + Na_2S_4O_{6(aq)}$$

Calculate the concentration of the potassium manganate(VII) solution. [5 marks]

Two vowels went out for dinner — they had an iodate...

This might seem like quite a faff — you do a redox reaction to release iodine, titrate the iodine solution, do a sum to find the iodine concentration, write an equation, then do another sum to work out the concentration of something else. The thing is though, it does work, and you do have to know how. If you're rusty on the calculations, look back to page 167.

AS Answers

Unit 1: Module 1 — Atoms and Reactions

Page 5 — The Atom

1)a) Similarity — They've all got the same number of protons/ electrons. **[1 mark]**
Difference — They all have different numbers of neutrons. **[1 mark]**

b) 1 proton **[1 mark]**, 1 neutron (2 – 1) **[1 mark]**, 1 electron **[1 mark]**.

c) ^3H. **[1 mark]**
Since tritium has 2 neutrons in the nucleus and also 1 proton, it has a mass number of 3. You could also write ^3_1H but you don't really need the atomic number.

2)a) (i) Same number of electrons. **[1 mark]**
$^{32}_{16}S^{2-}$ has 16 + 2 = 18 electrons. $^{40}_{18}Ar$ has 18 electrons too. **[1 mark]**
(ii) Same number of protons. **[1 mark]**
Each has 16 protons (the atomic number of S must always be the same) **[1 mark]**.
(iii) Same number of neutrons. **[1 mark]**
$^{40}_{18}Ar$ has 40 – 18 = 22 neutrons. $^{42}_{20}Ca$ has 42 – 20 = 22 neutrons. **[1 mark]**

b) **A** and **C**. **[1 mark]** They have the same number of protons but different numbers of neutrons. **[1 mark]**
It doesn't matter that they have a different number of electrons because they are still the same element.

Page 7 — Atomic Models

1)a) Bohr knew that if an electron was freely orbiting the nucleus it would spiral into it, causing the atom to collapse **[1 mark]**. His model only allowed electrons to be in fixed shells and not in between them **[1 mark]**.

b) When an electron moves from one shell to another electromagnetic radiation is emitted or absorbed **[1 mark]**.

c) Atoms react in order to gain full shells of electrons **[1 mark]**. Noble gases have full shells and so do not react **[1 mark]**. (Alternatively: a full shell of electrons makes an atom stable **[1 mark]**; noble gases have full shells and do not react because they are stable **[1 mark]**.)

Page 9 — Relative Mass

1)a) First multiply each relative abundance by the relative mass —
120.8 × 63 = 7610.4, 54.0 × 65 = 3510.0
Next add up the products: 7610.4 + 3510.0 = 11 120.4 **[1 mark]**
Now divide by the total abundance (120.8 + 54.0 = 174.8)

$A_r(Cu) = \frac{11\,120.4}{174.8} \approx$ **63.6** **[1 mark]**

You can check your answer by seeing if A_r(Cu) is in between 63 and 65 (the lowest and highest relative isotopic masses).

b) A sample of copper is a mixture of 2 isotopes in different abundances **[1 mark]**. The weighted average mass of these isotopes isn't a whole number **[1 mark]**.

2) You use pretty much the same method here as for question 1)a).
93.11 × 39 = 3631.29, 0.12 × 40 = 4.8, 6.77 × 41 = 277.57
3631.29 + 4.8 + 277.57 = 3913.66 **[1 mark]**
This time you divide by 100 because they're percentages.

$A_r(K) = \frac{3913.66}{100} \approx$ **39.14** **[1 mark]**

Again check your answer's between the lowest and highest relative isotopic masses, 39 and 41. A_r(K) is closer to 39 because most of the sample (93.11 %) is made up of this isotope.

Page 11 — The Mole

1) M of CH_3COOH = (2 × 12) + (4 × 1) + (2 × 16) = 60 g mol^{-1} **[1 mark]**
so mass of 0.36 moles = 60 × 0.36 = **21.6 g [1 mark]**

2) No. of moles = $\frac{0.25 \times 60}{1000}$ = 0.015 moles H_2SO_4 **[1 mark]**
M of H_2SO_4 = (2 × 1) + (1 × 32) + (4 × 16) = 98 g mol^{-1}
Mass of 0.015 H_2SO_4 = 98 x 0.015 = **1.47 g [1 mark]**

3) M of C_3H_8 = (3 × 12) + (8 × 1) = 44 g mol^{-1}
No. of moles of C_3H_8 = $\frac{88}{44}$ = 2 moles **[1 mark]**
At r.t.p. 1 mole of gas occupies 24 dm^3
so 2 moles of gas occupies 2 x 24 = **48 dm^3 [1 mark]**

Page 13 — Empirical and Molecular Formulas

1) Assume you've got 100 g of the compound so you can turn the % straight into mass.
No. of moles of C = $\frac{92.3}{12}$ = 7.69 moles
No. of moles of H = $\frac{7.7}{1}$ = 7.7 moles **[1 mark]**
Divide both by the smallest number, in this case 7.69.
So ratio C : H = 1 : 1
So, the empirical formula = CH **[1 mark]**
The empirical mass = 12 + 1 = 13
No. of empirical units in molecule = $\frac{78}{13}$ = 6
So the molecular formula = C_6H_6 **[1 mark]**

2) The magnesium is burning, so it's reacting with oxygen and the product is magnesium oxide.
First work out the number of moles of each element.
No. of moles Mg = $\frac{1.2}{24}$ = 0.05 moles
Mass of O is everything that isn't Mg: 2 – 1.2 = 0.8 g
No. of moles O = $\frac{0.8}{16}$ = 0.05 moles **[1 mark]**
Ratio Mg : O = 0.05 : 0.05
Divide both by the smallest number, in this case 0.05.
So ratio Mg : O = 1 : 1
So the empirical formula is MgO **[1 mark]**

3) First calculate the no. of moles of each product and then the mass of C and H:
No. of moles of CO_2 = $\frac{33}{44}$ = 0.75 moles
Mass of C = 0.75 × 12 = 9 g
No. of moles of H_2O = $\frac{10.8}{18}$ = 0.6 moles
0.6 moles H_2O = 1.2 moles H
Mass of H = 1.2 × 1 = 1.2 g **[1 mark]**
Organic acids contain C, H and O, so the rest of the mass must be O.
Mass of O = 19.8 – (9 + 1.2) = 9.6 g
No. of moles of O = $\frac{9.6}{16}$ = 0.6 moles **[1 mark]**
Mole ratio = C : H : O = 0.75 : 1.2 : 0.6
Divide by smallest 1.25 : 2 : 1
This isn't a whole number ratio, so you have to multiply them all up until it is. Multiply them all by 4.
So, mole ratio = C : H : O = 5 : 8 : 4
Empirical formula = $\mathbf{C_5H_8O_4}$ **[1 mark]**
Empirical mass = (12 × 5) + (1 × 8) + (16 × 4) = 132 g
This is the same as what we're told the molecular mass is, so the molecular formula is also $\mathbf{C_5H_8O_4}$. **[1 mark]**

AS Answers

Page 15 — Equations and Calculations

1) M of $C_2H_5Cl = (2 \times 12) + (5 \times 1) + (1 \times 35.5) = 64.5\ g\ mol^{-1}$ **[1 mark]**

Number of moles of $C_2H_5Cl = \frac{258}{64.5} = 4$ moles **[1 mark]**

From the equation, 1 mole C_2H_5Cl is made from 1 mole C_2H_4 so, 4 moles C_2H_5Cl is made from 4 moles C_2H_4. **[1 mark]**

M of $C_2H_4 = (2 \times 12) + (4 \times 1) = 28\ g\ mol^{-1}$
so, the mass of 4 moles $C_2H_4 = 4 \times 28 = 112$ g **[1 mark]**

2)a) M of $CaCO_3 = 40 + 12 + (3 \times 16) = 100\ g\ mol^{-1}$

Number of moles of $CaCO_3 = \frac{15}{100} = 0.15$ moles

From the equation, 1 mole $CaCO_3$ produces 1 mole CaO so, 0.15 moles of $CaCO_3$ produces 0.15 moles of CaO. **[1 mark]**

M of $CaO = 40 + 16 = 56\ g\ mol^{-1}$ **[1 mark]**
so, mass of 0.15 moles of CaO = $56 \times 0.15 = 8.4$ g **[1 mark]**

b) From the equation, 1 mole $CaCO_3$ produces 1 mole CO_2 so, 0.15 moles of $CaCO_3$ produces 0.15 moles of CO_2. **[1 mark]**

1 mole gas occupies 24 dm^3, **[1 mark]**
so, 0.15 moles occupies = $24 \times 0.15 = 3.6\ dm^3$ **[1 mark]**

3) On the LHS, you need 2 each of K and I, so use 2KI
The final equation is: $2KI + Pb(NO_3)_2 \rightarrow PbI_2 + 2KNO_3$ **[1 mark]**

In this equation, the NO_3 group remains unchanged, so it makes balancing much easier if you treat it as one indivisible lump.

Page 17 — Acids, Bases and Salts

1)a) $CaCO_{3(s)} + 2HClO_{4(aq)} \rightarrow Ca(ClO_4)_{2(aq)} + H_2O_{(l)} + CO_{2(g)}$
[1 mark for the state symbols, 1 mark for all the correct formulas and 1 mark for the correct balance.]

b) i) $2Li_{(s)} + 2H^+_{(aq)} \rightarrow 2Li^+_{(aq)} + H_{2(g)}$
[1 mark for the correct formulas, 1 mark for the correct balance.]

The SO_4^{2-} ions are left out of the ionic equation — they're spectator ions that don't get involved in the reaction.

ii) $2KOH_{(aq)} + H_2SO_{4(aq)} \rightarrow K_2SO_{4(aq)} + 2H_2O_{(l)}$
[1 mark for the correct formulas, 1 mark for the correct balance.]

iii) $2NH_{3(aq)} + H_2SO_{4(aq)} \rightarrow (NH_4)_2SO_{4(aq)}$
[1 mark for the correct formulas, 1 mark for the correct balance.]

2)a) M of $CaSO_4 = 40 + 32 + (4 \times 16) = 136\ g\ mol^{-1}$ **[1 mark]**
no. moles = 1.133 g/136 = 0.00833 moles **[1 mark]**

b) mass of water = difference in mass between hydrated and anhydrous salt = 1.883 – 1.133 = 0.750 g **[1 mark]**

c) no. moles of water = mass/molar mass = 0.750/18 = 0.04167 **[1 mark]**

X = ratio of no. moles water to no. moles salt = 0.04167/0.00833 = 5.002 **[1 mark]**
Rounded to nearest integer X = 5 **[1 mark]**

Page 19 — Titrations

1) First write down what you know:

$CH_3COOH + NaOH \rightarrow CH_3COONa + H_2O$
25.4 cm^3 14.6 cm^3
? 0.5 M

Number of moles of NaOH = $\frac{0.5 \times 14.6}{1000} = 0.0073$ moles **[1 mark]**

From the equation, you know 1 mole of NaOH neutralises 1 mole of CH_3COOH, so if you've used 0.0073 moles NaOH you must have neutralised 0.0073 moles CH_3COOH. **[1 mark]**

Concentration of $CH_3COOH = \frac{0.0073 \times 1000}{25.4} =$ **0.287 M** **[1 mark]**

2) First write down what you know again:

$CaCO_3 + H_2SO_4 \rightarrow CaSO_4 + H_2O + CO_2$
0.75 g 0.25 M

M of $CaCO_3 = 40 + 12 + (3 \times 16) = 100\ g\ mol^{-1}$ **[1 mark]**

Number of moles of $CaCO_3 = \frac{0.75}{100} = 7.5 \times 10^{-3}$ moles **[1 mark]**

From the equation, 1 mole $CaCO_3$ reacts with 1 mole H_2SO_4 so, 7.5×10^{-3} moles $CaCO_3$ reacts with 7.5×10^{-3} moles H_2SO_4. **[1 mark]**

The volume needed is = $\frac{(7.5 \times 10^{-3}) \times 1000}{0.25} = 30\ cm^3$ **[1 mark]**

If the question mentions concentration or molarities, you can bet your last clean pair of underwear that you'll need to use the formula

$$\text{number of moles} = \frac{\text{concentration} \times \text{volume}}{1000}$$

Just make sure the volume's in cm^3 though.

Page 21 — Oxidation and Reduction

1)a) $H_2SO_4\ (aq) + 8HI\ (aq) \rightarrow H_2S\ (g) + 4I_2\ (s) + 4H_2O\ (l)$ **[1 mark]**

b) Ox. No. of S in $H_2SO_4 = +6$ **[1 mark]**
Ox. No. of S in $H_2S = -2$ **[1 mark]**

c) Iodide **[1 mark]** — it donates electrons / its oxidation number increases **[1 mark]**

Unit 1: Module 2 — Electrons, Bonding and Structure

Page 23 — Electronic Structure

1)a) K atom: $1s^2\ 2s^2\ 2p^6\ 3s^2\ 3p^6\ 4s^1$ **[1 mark]**
K^+ ion: $1s^2\ 2s^2\ 2p^6\ 3s^2\ 3p^6$ **[1 mark]**

b) $1s^2\ 2s^2\ 2p^4$ **[1 mark]**

c) The outer shell electrons in potassium and oxygen can get close to the outer shells of other atoms, so they can be transferred or shared **[1 mark]**. The inner shell electrons are tightly held and shielded from the electrons in other atoms/molecules **[1 mark]**.

2)a) $1s^2\ 2s^2\ 2p^6\ 3s^2\ 3p^6\ 3d^5\ 4s^2$. **[1 mark]**

b) Germanium ($1s^2\ 2s^2\ 2p^6\ 3s^2\ 3p^6\ 3d^{10}\ 4s^2\ 4p^2$). **[1 mark]**
(The 4p sub-shell is partly filled, so it must be a p block element.)

AS Answers

c) Ar (atom) **[1 mark]**, K^+ (positive ion) **[1 mark]**, Cl^- (negative ion) **[1 mark]**. *You also could have suggested Ca^{2+}, S^{2-} or P^{3-}.*

d) $1s^2\ 2s^2\ 2p^6$ **[1 mark]**

Page 25 — Ionisation Energies

1)a) Group 3 **[1 mark]**
There are three electrons removed before the first big jump in energy.

b) The electrons are being removed from an increasing positive charge **[1 mark]** so more energy is needed to remove an electron / the force of attraction that has to be broken is greater **[1 mark]**.

c) When an electron is removed from a different shell there is a big increase in the energy required (since that shell is closer to the nucleus) **[1 mark]**.

d) There are 3 shells (because there are 2 big jumps in energy) **[1 mark]**.

Page 27 — Ionic Bonding

1)a)

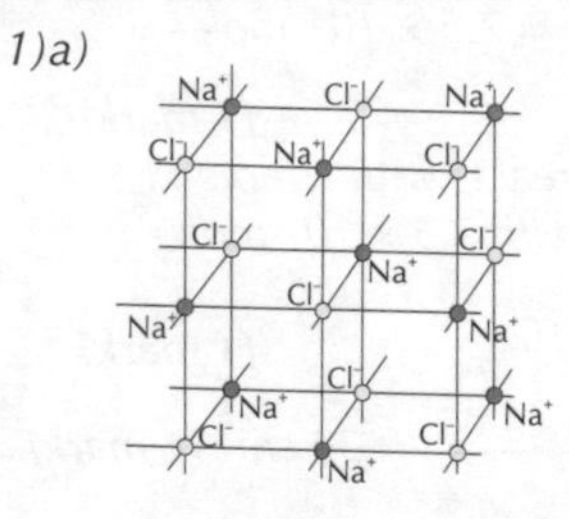

Your diagram should show the following —
- cubic structure with ions at corners **[1 mark]**
- sodium ions and chloride ions labelled **[1 mark]**
- alternating sodium ions and chloride ions **[1 mark]**

b) giant ionic/crystal (lattice) **[1 mark]**

c) You'd expect it to have a high melting point **[1 mark]**. Because there are strong bonds between the ions **[1 mark]** due to the electrostatic forces **[1 mark]**. A lot of energy is required to overcome these bonds **[1 mark]**.

2)a) Electrons move from one atom to another **[1 mark]**.
Any correct examples of ions, one positive, one negative. E.g. Na^+, Cl^-. **[2 x 1 mark]**

b) In a solid, ions are held in place by strong ionic bonds **[1 mark]**.
When the solid is heated to melting point, the ions gain enough energy **[1 mark]** to overcome the forces of attraction enough to become mobile **[1 mark]** and so carry charge (and hence electricity) through the substance **[1 mark]**.

Page 29 — Covalent Bonding

1)a) Covalent **[1 mark]**

b)

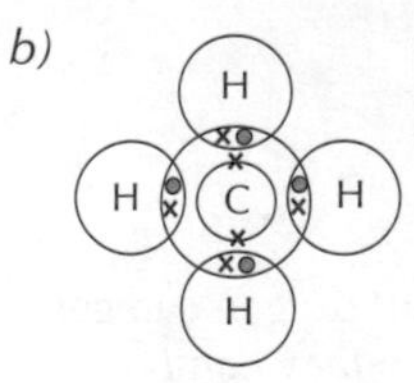

Your diagram should show the following —
- a completely correct electron arrangement **[1 mark]**
- all 4 overlaps correct (one dot + one cross in each) **[1 mark]**

2 a) Dative covalent/coordinate bond **[1 mark]**

b) One atom **[1 mark]** donates a pair of/both the electrons to the bond **[1 mark]**.

Page 31 — Giant Molecular Structures & Metallic Bonding

1)a) Giant covalent lattice **[1 mark]**

b)

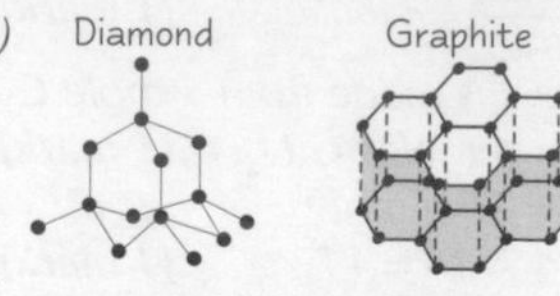

[1 mark for each correctly drawn diagram]
Diamond's a bit awkward to draw without it looking like a bunch of ballet dancing spiders — just make sure each central carbon is connected to four others.

c) Diamond has electrons in localised covalent bonds **[1 mark]**, so is a poor electrical conductor **[1 mark]**. Graphite has delocalised electrons between the sheets **[1 mark]** which can flow, so is a good electrical conductor **[1 mark]**.

2)

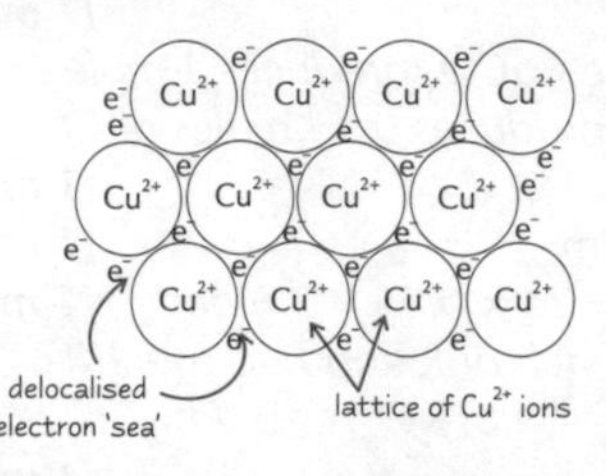

[2 marks for reasonable diagram showing closely packed Cu^{2+} ions and a sea of delocalised electrons]
Metallic bonding results from the attraction between positive metal ions **[1 mark]** and a sea of delocalised electrons **[1 mark]**.

Page 33 — Shapes of Molecules

1) a) NCl_3 **[1 mark]** BCl_3 **[1 mark]**

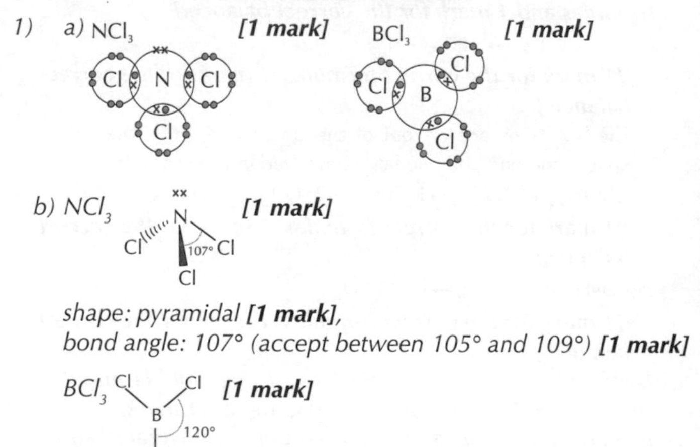

b) NCl_3 **[1 mark]**

shape: pyramidal **[1 mark]**,
bond angle: 107° (accept between 105° and 109°) **[1 mark]**

BCl_3 **[1 mark]**

(It must be a reasonable "Y" shaped molecule.)
shape: trigonal planar **[1 mark]**,
bond angle: 120° exactly **[1 mark]**

c) BCl_3 has three electron pairs only around B. **[1 mark]**
NCl_3 has four electron pairs around N **[1 mark]**, including one lone pair. **[1 mark]**

AS Answers

Page 35 — Electronegativity and Intermolecular Forces

1)a) *The power of an atom to withdraw electron density* ***[1 mark]*** *from a covalent bond* ***[1 mark]*** *OR the ability of an atom to attract the bonding electrons* ***[1 mark]*** *in a covalent bond* ***[1 mark]****.*

b) *(i)* Br — Br *(ii)* $O^{\delta-}$, $\delta+$ H, H $\delta+$

(iii) $N^{\delta-}$, H $\delta+$, H $\delta+$, H $\delta+$

[1 mark for correct shape, 1 mark for bond polarities on H_2O, 1 mark for bond polarities on NH_3.]

2)a) *Van der Waals OR instantaneous/temporary dipole-induced dipole OR dispersion forces.*
Permanent dipole-dipole interactions/forces.
Hydrogen bonding.
(Permanent dipole-induced dipole interactions.)
[1 mark each for any three]

b) *(i) More energy* ***[1 mark]*** *is needed to break the hydrogen bonds between water molecules* ***[1 mark]****.*

(ii)

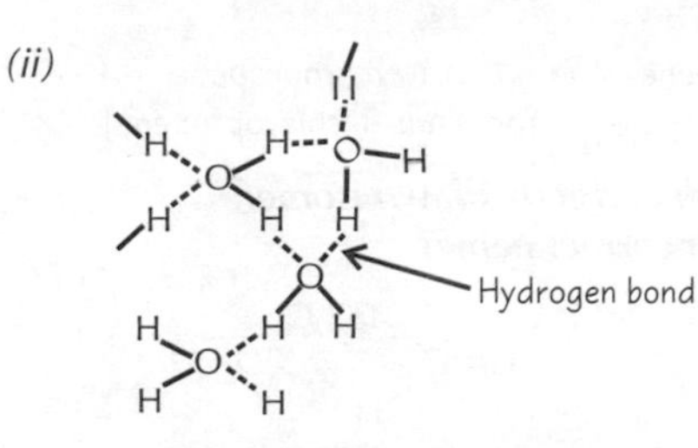

Your diagram should show the following —
- *Labelled hydrogen bonds between the water molecules* ***[1 mark]****.*
- *At least two hydrogen bonds between an oxygen atom and a hydrogen atom on adjacent molecules* ***[1 mark]****.*

Page 37 — Van der Waals Forces

1) a) *A — Ionic* *B — (Simple) molecular*
C — Metallic *D — Giant molecular (macromolecular)*
[1 mark for each]

b) *(i) Diamond — D* *(ii) Aluminium — C*
(iii) Sodium chloride — A *(iv) Iodine — B*
[2 marks for all four correct. 1 mark for two correct.]

2) ***Magnesium*** *has a metallic crystal lattice (it has metallic bonding)* ***[1 mark]****. It has a sea of electrons/delocalised electrons/freely moving electrons* ***[1 mark]****, which allow it to conduct electricity in the solid or liquid state* ***[1 mark]****.*
Sodium chloride *has a (giant) ionic lattice* ***[1 mark]****.*
It doesn't conduct electricity when it's solid ***[1 mark]*** *because its ions don't move freely, but vibrate about a fixed point* ***[1 mark]****. Sodium chloride conducts electricity when liquid/molten* ***[1 mark]*** *or in aqueous solution* ***[1 mark]*** *because it has freely moving ions (not electrons)* ***[1 mark]****.*
Graphite *is giant covalent/macromolecular* ***[1 mark]****. It has delocalised/freely moving electrons* ***[1 mark]*** *within the layers. It conducts electricity along the layers in the solid state* ***[1 mark]****.*

Unit 1: Module 3 — The Periodic Table

Page 39 — The Periodic Table

1)a) *Sodium* $1s^2\ 2s^2\ 2p^6\ 3s^1$ ***[1 mark]***
b) *s block* ***[1 mark]***
c) *Bromine* $1s^2\ 2s^2\ 2p^6\ 3s^2\ 3p^6\ 3d^{10}\ 4s^2\ 4p^5$ ***[1 mark]***
d) *p block* ***[1 mark]***

Page 41 — Periodic Trends

1) *Increasing number of protons means a stronger pull from the positively charged nucleus* ***[1 mark]*** *making it harder to remove an electron from the outer shell* ***[1 mark]****.*
There are no extra inner electrons to add to the shielding effect ***[1 mark]****.*

2) *Mg has more delocalised electrons per atom* ***[1 mark]*** *and the ion has a greater charge density* ***[1 mark]****. This gives Mg a stronger metal-metal bond, resulting in a higher boiling point* ***[1 mark]****.*

3)a) *Si has a giant covalent lattice structure* ***[1 mark]*** *consisting of very strong covalent bonds* ***[1 mark]****.*
b) *Sulfur (S_8) has a larger molecule than phosphorus (P_4)* ***[1 mark]****, which results in larger van der Waals forces of attraction between molecules* ***[1 mark]****.*

4) *The atomic radius decreases across the period from left to right* ***[1 mark]****. The number of protons increases, so nuclear charge increases* ***[1 mark]****, meaning electrons are pulled closer to the nucleus* ***[1 mark]****. The electrons are all added to the same outer shell, so there's little effect on shielding* ***[1 mark]****.*

Page 43 — Group 2 — The Alkaline Earth Metals

1) *First ionisation energy of Ca is smaller* ***[1 mark]*** *because Ca has (one) more electron shell(s)* ***[1 mark]****.*
This reduces the attraction between the nucleus and the outer electrons OR increases shielding effect ***[1 mark]****.*
The outer shell of Ca is also further from the nucleus ***[1 mark]****.*

2)a) $2Ca_{(s)} + O_{2(g)} \rightarrow 2CaO_{(s)}$ ***[1 mark]***
b) *From 0 to +2* ***[1 mark]***
c) *White* ***[1 mark]*** *solid* ***[1 mark]***
d) *Ionic* ***[1 mark]***
...because as everybody who's anybody knows, Group 2 compounds (including oxides) are generally white ionic solids.

3)a) *Y* ***[1 mark]***
b) *Y has the largest radius* ***[1 mark]*** *so it will be furthest down the group / have the smallest ionisation energy* ***[1 mark]****.*

Page 45 — Group 7 — The Halogens

1)a) $I_2 + 2At^- \rightarrow 2I^- + At_2$ ***[1 mark]***
b) *The (sodium) astatide* ***[1 mark]***

2) *Aqueous solutions of both halides are tested* ***[1 mark]****.*
First, some dilute nitric acid is added ***[1 mark]****.*
Sodium chloride *— silver nitrate gives white precipitate which dissolves in dilute ammonia solution* ***[1 mark]****.*
$Ag^+ + Cl^- \rightarrow AgCl$ ***[1 mark]***
Sodium bromide *— silver nitrate gives cream precipitate which is only soluble in concentrated ammonia solution* ***[1 mark]****.*
$Ag^+ + Br^- \rightarrow AgBr$ ***[1 mark]***

3)a) *The melting and boiling points of the halogens increase down the group* ***[1 mark]****. Iodine is a solid at r.t.p.* ***[1 mark]****, so you would expect that astatine is also a solid at r.t.p.* ***[1 mark]****.*

AS Answers

b) *AgI is insoluble in concentrated ammonia solution **[1 mark]**. The solubility of halides in ammonia solution decreases down the group **[1 mark]**, so you would expect AgAt **NOT** to dissolve either **[1 mark]**.*
Question 2 is the kind of question that could completely throw you if you're not really clued up on the facts. If you really know p45, then in part b) you'll go, "Ah - ha!!! Solubility of silver halides in ammonia decreases down the group..." If not, you basically won't have a clue. The moral is... it really is just about learning all the facts. Boring, but true.

Page 47 — Disproportionation and Water Treatment

1) a) *$2OH^- + Cl_2 \rightarrow OCl^- + Cl^- + H_2O$ **[1 mark]***
*Disproportionation is simultaneous oxidation and reduction of an element in a reaction **[1 mark]**. Cl_2 has been reduced to Cl^- **[1 mark]** and oxidised to OCl^- **[1 mark]**.*
b) *E.g. water treatment, bleaching paper, bleaching textiles, cleaning **[1 mark for each of two sensible applications]**.*

2)a) *$2I^- + ClO^- + H_2O \rightarrow I_2 + Cl^- + 2OH^-$*
[2 marks — 1 for correct formulas, 1 for balancing the equation]
b) *Iodine: −1 to 0 — oxidation **[1 mark]***
*Chlorine: +1 to −1 — reduction **[1 mark]***
c) *Violet/pink **[1 mark]***
The colour formed would be due to the iodine.

Unit 2: Module 1 — Basic Concepts and Hydrocarbons

Page 49 — Basic Stuff

1)a)

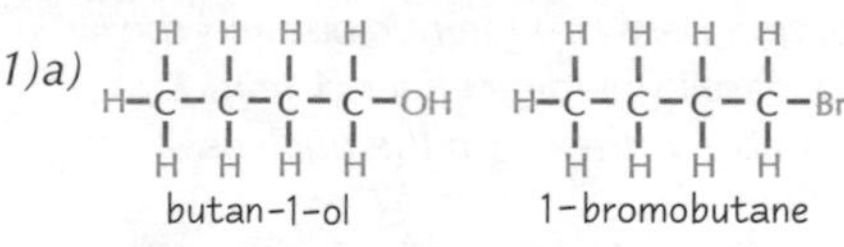

[1 mark for each correct structure]

b) *–OH (hydroxyl) **[1 mark]**.*
*It could be attached to the first or second carbon OR butan-2-ol also exists OR because the position of the –OH group affects its chemistry **[1 mark]**.*

2) *A = pentane **[1 mark]***
*B = methylbutane **[1 mark for methyl, 1 mark for butane]***
There's only actually one type of methylbutane. You can't have 1-methylbutane — it'd be exactly the same as pentane.
C = 2,2-dimethylpropane
[1 mark for 2,2-, 1 mark methyl, 1 mark for propane]

Page 51 — Isomerism

1)a) (i)

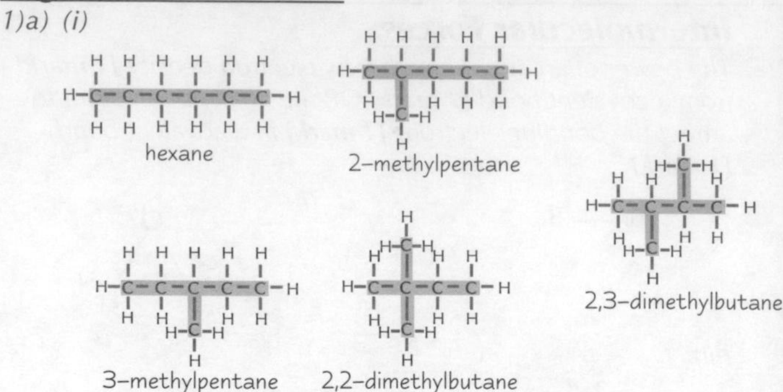

[1 mark for each correctly drawn isomer, 1 mark for each correct name]

(ii) *The same molecular formula **[1 mark]** but different arrangements of the carbon skeleton **[1 mark]**.*
Watch out — the atoms can rotate around the single C–C bonds, so these two aren't isomers — they're just the same molecule bent a bit.

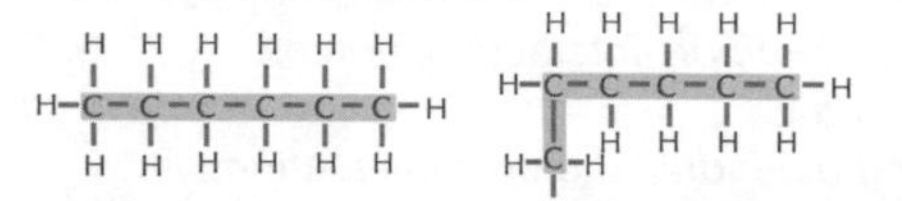

b) (i)

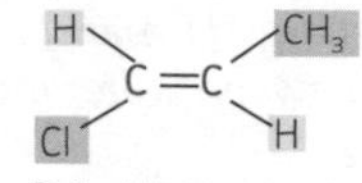

Z-1-chloropropene [or cis-1-chloropropene]

E-1-chloropropene [or trans-1-chloropropene]

[1 mark for each correctly drawn isomer, 1 mark for each correct name]

(ii)

2-chloropropene

3-chloropropene

[1 mark for each correctly drawn isomer, 1 mark for each correct name]

c) *A group of compounds **[1 mark]** that can be represented by the same general formula OR having the same functional group OR with similar chemical properties **[1 mark]**.*

Page 53 — Atom Economy and Percentage Yield

1)a) *2 is an addition reaction **[1 mark]***
b) *For reaction 1: % atom economy*
*$= M_r(C_2H_5Cl) \div [M_r(C_2H_5Cl) + M_r(POCl_3) + M_r(HCl)]$ **[1 mark]***
*$= [(2 \times 12) + (5 \times 1) + 35.5] \div [(2 \times 12) + (5 \times 1) + 35.5 + 31 + 16 + (3 \times 35.5) + 1 + 35.5] \times 100\%$ **[1 mark]***
*$= (64.5 \div 254.5) \times 100\% = 25.3\%$ **[1 mark]***
c) *The atom economy is 100% because there is only one product (there are no by-products) **[1 mark]***

2)a) *There is only one product, so the theoretical yield can be calculated by adding the masses of the reactants **[1 mark]**. So theoretical yield = 0.275 + 0.142 = 0.417 g **[1 mark]***
b) *percentage yield = (0.198 ÷ 0.417) × 100 = 47.5% **[1 mark]***
c) *Changing reaction conditions will have no effect on atom economy **[1 mark]**. Since the equation shows that there is only one product, the atom economy will always be 100% **[1 mark]**.*
Atom economy is related to the type of reaction — addition, substitution, etc. — not to the quantities of products and reactants.

AS Answers

Page 55 — Alkanes

1)a) *One with no double bonds OR the maximum number of hydrogens OR single bonds only* **[1 mark]**.
It contains only hydrogen and carbon atoms **[1 mark]**.

b) $C_2H_{6(g)} + 3½O_{2(g)} \rightarrow 2CO_{2(g)} + 3H_2O_{(g)}$
[1 mark for correct symbols, 1 mark for balancing]

2)a) *Nonane is an alkane* **[1 mark]**.

b) *Nonane will have a higher boiling point than 2,2,3,3-tetramethylpentane* **[1 mark]** *because the molecules of branched-chain alkanes like 2,2,3,3-tetramethylpentane are less closely packed together than their straight-chain isomers, so they have weaker Van der Waals forces holding them together* **[1 mark]**.

c) (i) $C_9H_{20} + 9½O_2 \rightarrow 9CO + 10H_2O$ **[1 mark]**

(ii) *Carbon monoxide binds to haemoglobin in the blood in preference to oxygen* **[1 mark]**, *so less oxygen can be carried around the body, leading to oxygen deprivation* **[1 mark]**.

d) *Nonane is a larger molecule than methane, so it contains more bonds* **[1 mark]**. *The energy released is due to the breaking and then reforming of bonds* **[1 mark]**.

Page 57 — Petroleum

1)a) (i) *There's greater demand for smaller fractions* **[1 mark]** *for motor fuels* **[1 mark]**. *Alternatively: There's greater demand for alkenes* **[1 mark]** *to make petrochemicals/ polymers* **[1 mark]**.

(ii) *E.g.* $C_{12}H_{26} \rightarrow C_2H_4 + C_{10}H_{22}$ **[1 mark]**.
There are loads of possible answers — just make sure the C's and H's balance and there's an alkane and an alkene.

b) (i) *Branched-chain alkanes, cycloalkanes, arenes* **[1 mark for each]**.

(ii) *They reduce knocking (autoignition)* **[1 mark]**.

(iii)

2-methylbutane

2, 2-dimethylpropane

[1 mark for each structure, 1 mark for each name]

Page 59 — Fossil Fuels

1)a) *Fermentation* **[1 mark]**

b) *It is carbon neutral because the carbon dioxide taken in as the plant grows* **[1 mark]** *is equal to the carbon dioxide released as the fuel burns* **[1 mark]**.

c) *Poorer countries may try to earn money by growing suitable crops, meaning less land will be available for them to grow food* **[1 mark]**. *Clearing forests to grow crops leads to loss of biodiversity and damage to the environment* **[1 mark]**.

2)a) *Burning them is very exothermic* **[1 mark]**.

b) *Burning fossil fuels releases carbon dioxide* **[1 mark]**, *which is a greenhouse gas, so it enhances the greenhouse effect* **[1 mark]**.

c) *Fossil fuels take millions of years to form* **[1 mark]**, *so once they have been used, they cannot be replaced* **[1 mark]**.

d) *They are the raw material for many organic chemicals* **[1 mark]**.

Page 61 — Alkanes — Substitution Reactions

1)a) *Free radical substitution.* **[1 mark]**

b) $CH_4 + Br_2 \xrightarrow{U.V\ light} CH_3Br + HBr$ **[1 mark[**

c) $Br\cdot + CH_4 \rightarrow HBr + CH_3\cdot$ **[1 mark]**
$CH_3\cdot + Br_2 \rightarrow CH_3Br + Br\cdot$ **[1 mark]**

d) (i) *Two methyl radicals bond together to form an ethane molecule.* **[1 mark]**

(ii) *Termination step* **[1 mark]**

(iii) $CH_3\cdot + CH_3\cdot \rightarrow CH_3CH_3$ **[1 mark]**

e) *Tetrabromomethane* **[1 mark]**

2) $CH_3CH_3 + Br_2 \xrightarrow{U.V\ light} CH_3CH_2Br + HBr$ **[1 mark]**

Initiation: $Br_2 \xrightarrow{U.V\ light} 2Br\cdot$ **[1 mark]**

Propagation: $CH_3CH_3 + Br\cdot \rightarrow CH_3CH_2\cdot + HBr$ **[1 mark]**
$CH_3CH_2\cdot + Br_2 \rightarrow CH_3CH_2Br + Br\cdot$ **[1 mark]**

Termination: $CH_3CH_2\cdot + Br\cdot \rightarrow CH_3CH_2Br$
Or: $CH_3CH_2\cdot + CH_3CH_2\cdot \rightarrow CH_3CH_2CH_2CH_3$
[1 mark]

[1 mark for mentioning U.V.]
Watch out — you're asked for the reaction with ethane here. It's just the same as the methane reaction though.

Page 63 — Alkenes and Polymers

1)a) *The carbon atoms in ethene are joined by a double bond* **[1 mark]**, *consisting of:*
A sigma (σ) bond between the s-orbitals of the carbon atoms **[1 mark]**

A pi (π) bond between the p-orbitals **[1 mark]**

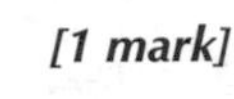

b) *Ethene is a planar molecule* **[1 mark]**. *The atoms cannot move around the C=C double bond* **[1 mark]** *because the π bond does not allow rotation* **[1 mark]**.

c) *Ethene is useful because it is very reactive* **[1 mark]** *and is the starting point for making many different polymers* **[1 mark]**.

d)

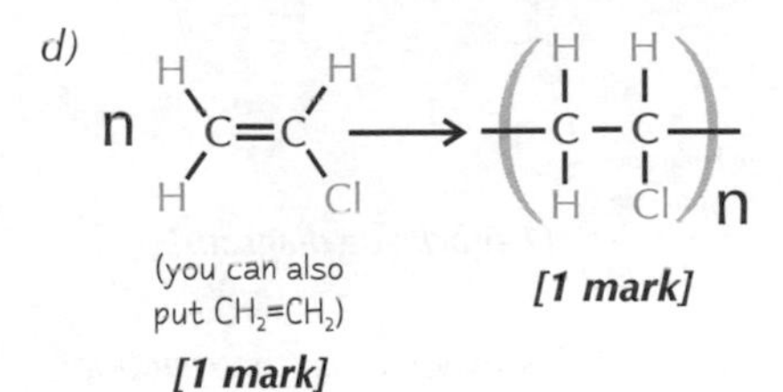

(you can also put CH_2=CH_2)
[1 mark]

[1 mark]

Page 65 — Polymers and the Environment

1)a) *Saves on landfill OR Energy can be used to generate electricity* **[1 mark for either]**

b) *Toxic gases produced* **[1 mark]**.
Scrubbers can be used **[1 mark]** *to remove these toxic gases.*

2) *Melted* **[1 mark]** *and remoulded* **[1 mark]** *OR Cracked* **[1 mark]** *and processed* **[1 mark]** *to make a new object.*

3) *Renewable raw material / Less energy used (in manufacture) / Less CO_2 produced over lifetime of polymer (if used to replace plastics that are usually burnt)*
[1 mark for each, up to a maximum of 2 marks]

AS Answers

Page 67 — Reactions of Alkenes

1)a) *Shake the alkene with bromine water* **[1 mark]**, *and the solution goes colourless if a double bond is present* **[1 mark]**.

b) *Electrophilic* **[1 mark]** *addition* **[1 mark]**.

c)

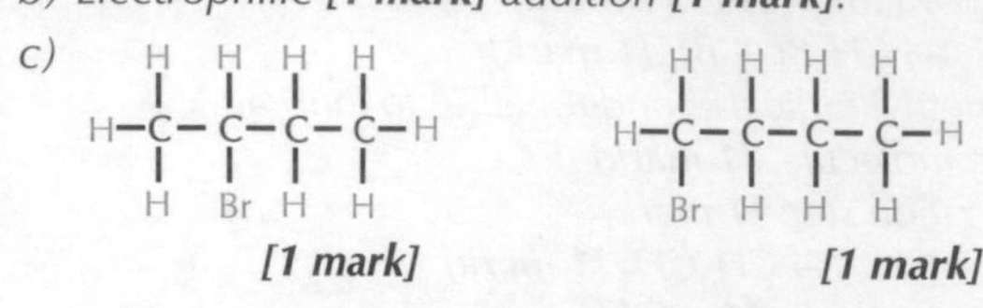

[1 mark] **[1 mark]**

Unit 2: Module 2 — Alcohols, Halogenoalkanes and Analysis

Page 69 — Alcohols

1)a) *Butan-1-ol* **[1 mark]**, *primary* **[1 mark]**

b) *2-methylpropan-2-ol* **[1 mark]**, *tertiary* **[1 mark]**

c) *Butan-2-ol* **[1 mark]**, *secondary* **[1 mark]**

d) *2-methylpropan-1-ol* **[1 mark]**, *primary* **[1 mark]**

2)a) *Primary* **[1 mark]**. *The -OH group is bonded to a carbon with one alkyl group/other carbon atom attached* **[1 mark]**.

b) (i) $C_6H_{12}O_{6(aq)} \rightarrow 2C_2H_5OH_{(aq)} + 2CO_{2(g)}$ **[1 mark]**

(ii) *Temperature between 30 and 40 °C* **[1 mark]**, *Anaerobic conditions OR air/oxygen excluded* **[1 mark]** *(Allow 'yeast' as an alternative to one of the above.)*

c) *Ethene is cheap and abundantly available / It's a low-cost process / it's a high-yield process / Very pure ethanol is produced / Fast reaction* **[1 mark each for up to two of these reasons]**. *This might change in the future as crude oil reserves run out or become more expensive* **[1 mark]**.

Page 71 — Oxidation of Alcohols

1)a)

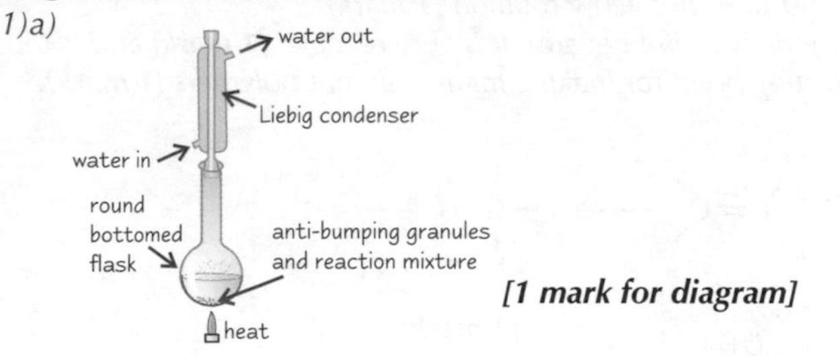

[1 mark for diagram]

You set up reflux apparatus in this way so that the reaction can be heated to boiling point **[1 mark]** *without losing any materials/reactants/products OR so vapour will condense and drip back into the flask* **[1 mark]**

b) (i) *Propanoic acid* **[1 mark]**

(ii) $CH_3CH_2CH_2OH + [O] \rightarrow CH_3CH_2CHO + H_2O$ **[1 mark]**
$CH_3CH_2CHO + [O] \rightarrow CH_3CH_2COOH$ **[1 mark]**

(iii) *Distillation* **[1 mark]**. *This is so aldehyde is removed immediately as it forms* **[1 mark]**.
If you don't get the aldehyde out quick-smart, it'll be a carboxylic acid before you know it.

c) (i)

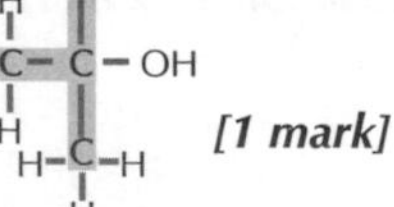

[1 mark]

(ii) *2-methylpropan-2-ol is a tertiary alcohol (which is more stable)* **[1 mark]**.

Page 74 — Halogenoalkanes

1)a) *Chlorofluorocarbons (CFCs)* **[1 mark]**

b) **[1 mark each for any three of the following]** *non-toxic, non-flammable, stable, volatile*

2 a) **[1 mark]**

b) **[1 mark]** **[1 mark]** **[1 mark]**

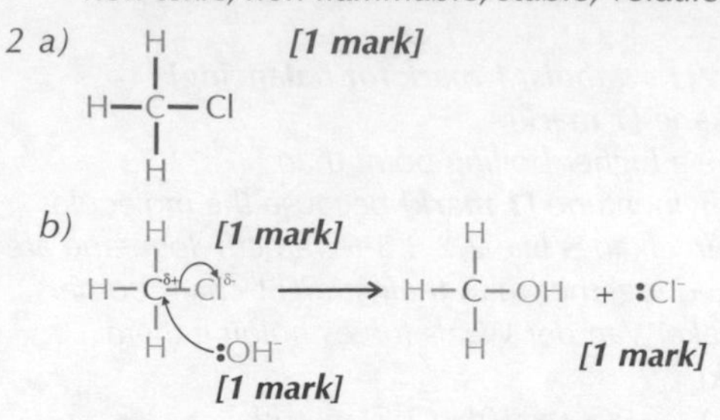

c) *A white* **[1 mark]** *precipitate* **[1 mark]**

d) *Iodomethane would be hydrolysed more quickly than chloromethane* **[1 mark]**.

Page 77 — Analytical Techniques

1 a) *A's due to an O–H group in a carboxylic acid* **[1 mark]**. *B's due to a C=O as in an aldehyde, ketone, acid or ester* **[1 mark]**.

b) *The spectrum suggests it's a carboxylic acid — it's got a COOH group* **[1 mark]**. *This group has a mass of 45, so the rest of the molecule has a mass of 29 (74 – 45), which is likely to be* C_2H_5 **[1 mark]**. *So the molecule could be* C_2H_5COOH *— propanoic acid* **[1 mark]**.

2)a) *44* **[1 mark]**

b) *X has a mass of 15. It is probably an methyl group/*CH_3. **[1 mark]**
Y has a mass of 29. It is probably an ethyl group/C_2H_5. **[1 mark]**

c) **[1 mark]**

```
    H   H   H
    |   |   |
H — C — C — C — H
    |   |   |
    H   H   H
```

d) *If the compound was an alcohol, you would expect a peak with m/z ratio of 17* **[1 mark]**, *caused by the OH fragment* **[1 mark]**.

Unit 2: Module 3 — Energy

Page 79 — Enthalpy Changes

1)

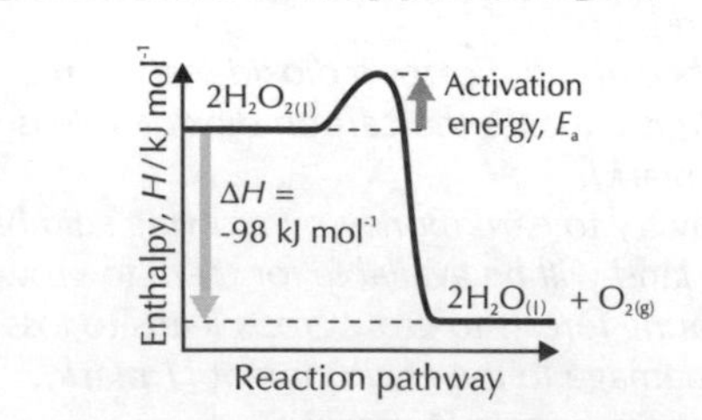

Reactants lower in energy than products **[1 mark]**. *Activation energy correctly labelled* **[1 mark]**. *ΔH correctly labelled with arrow pointing downwards* **[1 mark]**.
For an exothermic reaction, the ΔH arrow points downwards, but for an endothermic reaction it points upwards. The activation energy arrow always points upwards though.

AS Answers

2)a) $CH_3OH_{(l)} + 1\frac{1}{2}O_{2(g)} \rightarrow CO_{2(g)} + 2H_2O_{(l)}$
Correct balanced equation **[1 mark]**.
Correct state symbols **[1 mark]**.
It is perfectly OK to use halves to balance equations. Make sure that only 1 mole of CH_3OH is combusted, as it says in the definition for ΔH_c°.

b) $C_{(s)} + 2H_{2(g)} + \frac{1}{2}O_{2(g)} \rightarrow CH_3OH_{(l)}$
Correct balanced equation **[1 mark]**.
Correct state symbols **[1 mark]**.

c) H_2O should be formed under standard conditions (i.e. liquid, not gas) **[1 mark]**. Only 1 mole of C_3H_8 should be shown according to the definition of ΔH_c° **[1 mark]**.
You really need to know the definitions of the standard enthalpy changes off by heart. There are loads of nit-picky little details they could ask you questions about.

3)a) $C_{(s)} + O_{2\ (g)} \rightarrow CO_{2\ (g)}$
[1 mark for correct equation, 1 mark for correct state symbols]

b) It has the same value because it is the same reaction **[1 mark]**.

c) 1 tonne = 1 000 000 g
1 mole of carbon is 12 g
so 1 tonne is 1 000 000 ÷ 12 = 83 333 moles **[1 mark]**
1 mole releases 393.5 kJ
so 1 tonne will release 83333 × 393.5 = 32 791 666 kJ **[1 mark]**

Page 81 — More on Enthalpy Changes

1) No. of moles of $CuSO_4$ = $\frac{0.2 \times 50}{1000}$ = 0.01 moles **[1 mark]**
From the equation, 1 mole of $CuSO_4$ reacts with 1 mole of Zn.
So, 0.01 moles of $CuSO_4$ reacts with 0.01 moles of Zn **[1 mark]**.
Heat produced by reaction = $mc\Delta T$
= 50 × 4.18 × 2.6 = 543.4 J **[1 mark]**
0.01 moles of zinc produces 543.4 J of heat, therefore 1 mole of zinc produces $\frac{543.4}{0.01}$ **[1 mark]** = 54 340 J ≈ 54.3 kJ
So the enthalpy change is **−54.3 kJ mol⁻¹** (you need the minus sign because it's exothermic) **[1 mark for correct number, 1 mark for minus sign]**.
It'd be dead easy to work out the heat produced by the reaction, breathe a sigh of relief and sail on to the next question. But you need to find out the enthalpy change when 1 mole of zinc reacts. It's always a good idea to reread the question and check you've actually answered it.

2)a) A chemical reaction always involves bond breaking which needs energy / is endothermic **[1 mark]** and bond making which releases energy / is exothermic **[1 mark]**. Whether the reaction is exothermic or endothermic depends on whether more energy is used to break bonds or released by forming new bonds over the whole reaction **[1 mark]**.

b) Use the formula $Q = mc\Delta T$ **[1 mark]**
m = 1 kg = 1000 g
c = 4.18 J $g^{-1}K^{-1}$
6 g of carbon is 0.5 moles **[1 mark]**
So Q = 0.5 × 393.5 = 196.75 kJ = 196 750 J **[1 mark]**
So 196 750 = 1000 × 4.18 × ΔT
ΔT = 196 750 ÷ 4180 = 47.1 K **[1 mark]**

Page 83 — Enthalpy Calculations

1) ΔH_r° = sum of ΔH_f°(products) − sum of ΔH_f°(reactants)
= [0 + (3 × −602)] **[1 mark]** − [−1676 + (3 × 0)] **[1 mark]**
= −130 kJ mol⁻¹ **[1 mark]**
Don't forget the units. It's a daft way to lose marks.

2) ΔH_r° = ΔH_c°(glucose) − 2 × ΔH_c°(ethanol) **[1 mark]**
= [−2820] − [(2 × −1367)] **[1 mark]**
= −86 kJ mol⁻¹ **[1 mark]**

3) ΔH_f° = sum of ΔH_c°(reactants) − ΔH_c°(propane) **[1 mark]**
= [(3 × −394) + (4 × −286)] − [−2220] **[1 mark]**
= −106 kJ mol⁻¹ **[1 mark]**

4) Total energy required to break bonds = (4 × 435) + (2 × 498) = 2736 kJ **[1 mark]**
Energy released when bonds form = (2 × 805) + (4 × 464) = 3466 kJ **[1 mark]**
Net energy change = +2736 + (−3466) = −730 kJmol⁻¹
[1 mark for correct numerical value, 1 mark for correct unit]

Page 85 — Reaction Rates

1) Increasing the pressure will increase the rate of reaction **[1 mark]** because the molecules will be closer together, so they are more likely to collide, and therefore more likely to react **[1 mark]**.

2)a) X **[1 mark]**.
The X curve shows the same total number of molecules as the 25°C curve, but more of them have lower energy.

b) The smaller area to the right of the activation energy line shows fewer molecules **[1 mark]** will have enough energy to react **[1 mark]**. / The shape of the curve shows fewer molecules **[1 mark]** have the required activation energy **[1 mark]**.

Page 87 — Catalysts

1)a)

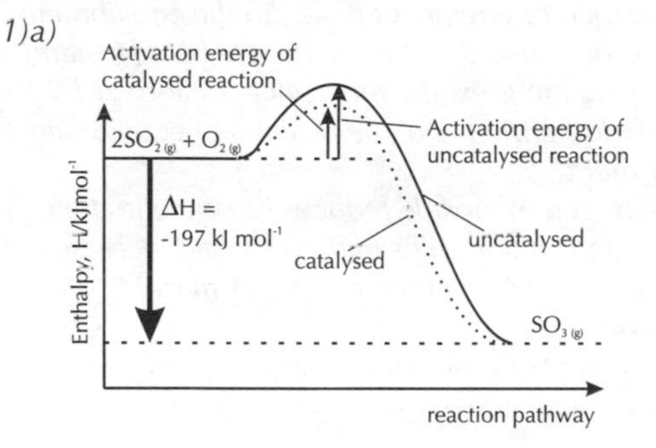

Curve showing activation energy **[1 mark]**. This must link reactants and products. Showing exothermic change (products **lower** in energy than reactants), with ΔH correctly labelled and a **downward** arrow **[1 mark]**. Correctly labelling activation energy (from reactants to highest energy peak) **[1 mark]**.
Label your axes correctly. (No, not the sharp tools for chopping wood or heads off — you know what I mean.)

b) See the diagram above. Reaction profile showing a **greater** activation energy than for the catalysed reaction **[1 mark]**.
Remember — catalysts lower the activation energy.
So uncatalysed reactions have greater activation energies.

c) Catalysts increase the rate of the reaction by providing an alternative reaction pathway **[1 mark]**, with a lower activation energy **[1 mark]**.

AS Answers

2)a) $2H_2O_{2(l)} \rightarrow 2H_2O_{(l)} + O_{2(g)}$
Correct symbols ***[1 mark]*** *and balancing equation* ***[1 mark]****.*
You get the marks even if you forgot the state symbols.

b)
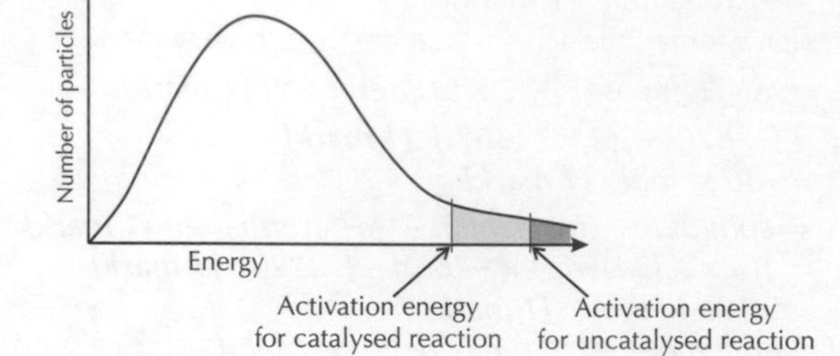

Correct general shape of the curve ***[1 mark]****. Correctly labelling the axes* ***[1 mark]****. Activation energies marked on the horizontal axis — the catalysed activation energy must be lower than the uncatalysed activation energy* ***[1 mark]****.*
You don't have to draw another curve for the catalysed reaction. Just mark the lower activation energy on the one you've already done.

c) *Manganese(IV) oxide lowers the activation energy by providing an alternative reaction pathway* ***[1 mark]****. So, more reactant molecules have the activation energy* ***[1 mark]****, meaning there are more successful collisions in a given period of time, and so the rate increases* ***[1 mark]****.*

Page 89 — Dynamic Equilibrium

1)a) *If a reaction at equilibrium is subjected to a change in concentration, pressure or temperature, the equilibrium will shift to try to oppose (counteract) the change.* ***[1 mark]****.*
Examiners are always asking for definitions so learn them — they're easy marks.

b) (i) *There's no change* ***[1 mark]****. There's the same number of molecules/moles on each side of the equation* ***[1 mark]****.*
(ii) *Reducing temperature removes heat. So the equilibrium shifts in the exothermic direction to release heat* ***[1 mark]****. The reverse reaction is exothermic (since the forward reaction is endothermic). So, the position of equilibrium shifts left* ***[1 mark]****.*
(iii) *Removing nitrogen monoxide reduces its concentration. The equilibrium position shifts right to try and increase the nitrogen monoxide concentration again* ***[1 mark]****.*

c) *No effect* ***[1 mark]****.*
Catalysts don't affect the equilibrium position. They just help the reaction to get there sooner.

Unit 2: Module 4 — Resources

Page 91 — Green Chemistry

1)a) *Ethanol is from a renewable resource (sugar cane) but petrol is from crude oil, which is non-renewable.* ***[1 mark]***
Ethanol from sugar cane can be a carbon neutral fuel. ***[1 mark]***

b) *The climate in some countries is unsuitable for growing sugar cane.* ***[1 mark]***

c) *Land used for food production may be taken up growing the sugar cane, increasing the cost of food.* ***[1 mark]***
Forests may be cleared to plant sugar cane. ***[1 mark]***

2)a) *If a catalyst is used:*
less energy is needed for the process ***[1 mark]***
more efficient processes are made possible ***[1 mark]***

b) *The process has 100% atom economy as only one product is made* ***[1 mark]*** *so no waste is produced* ***[1 mark]****.*

Page 93 — The Greenhouse Effect

1)a) *Water vapour* ***[1 mark]****, carbon dioxide* ***[1 mark]****, methane.* ***[1 mark]***

b) *The molecule/bond absorbs infrared radiation and the bond's vibrational energy increases.* ***[1 mark]***
Energy is transferred to other molecules by collision. ***[1 mark]***
The average kinetic energy of the molecules increases, so the temperature increases. ***[1 mark]***

c) *How much radiation one molecule of the gas absorbs* ***[1 mark]***
How much of the gas there is in the atmosphere ***[1 mark]***

2)a) *Increased use of fossil fuels* ***[1 mark]***
Increased deforestation ***[1 mark]***

b) *Scientists have produced evidence that the Earth's average temperature has dramatically increased in recent years* ***[1 mark]****.*

c) *Capturing CO_2 and storing it in underground rock formations/ storing it deep in the ocean/converting it to stable minerals / Developing alternative fuels / Increasing photosynthesis e.g. by increasing growth of phytoplankton.*
[1 mark each method, up to a maximum of 2 marks]

Page 96 — The Ozone Layer and Air Pollution

1)a) *Ozone is formed by the effect of UV radiation from the Sun on oxygen molecules.* ***[1 mark]*** *The oxygen molecules split to form oxygen free radicals* ***[1 mark]*** *which react with more oxygen molecules to form ozone.* ***[1 mark]***

b) *UV radiation can cause skin cancer.* ***[1 mark]*** *The ozone layer prevents most harmful UV radiation from the Sun from reaching the Earth's surface.* ***[1 mark]***

c) *The ozone molecules interact with UV radiation to form an oxygen molecule and a free oxygen radical* ***[1 mark]***
$O_3 + h\nu \rightarrow O_2 + O\bullet$ ***[1 mark]***
The radical produced then forms more ozone with an O_2 molecule.
$O_2 + O\bullet \rightarrow O_3$ ***[1 mark]***

2)a) $N_{2(g)} + O_{2(g)} \rightarrow 2NO_{(g)}$ ***[1 mark for correct reactants and products, 1 mark for correct balancing]***

b) $2NO_{(g)} + O_{2(g)} \rightarrow 2NO_{2(g)}$ ***[1 mark for correct reactants and products, 1 mark for correct balancing]***

c) *Acid rain OR smog* ***[1 mark]***

A2 Answers

Unit 4: Module 1 — Rings, Acids and Amines

Page 103 — Benzene

1 a) C_7H_8 or $C_6H_5CH_3$ **[1 mark]**
b) Aromatic OR arene **[1 mark]**
c) A: 1,3-dichlorobenzene **[1 mark]**
B: nitrobenzene **[1 mark]**
C: 2,4-dimethylphenol **[1 mark]**

2 a) The model suggests that there should be two different bond lengths in the molecule, corresponding to C=C and C–C **[1 mark]**
b) X-ray diffraction **[1 mark]**
c) X-ray diffraction shows that all the carbon–carbon bond lengths in benzene are actually the same, which doesn't fit the Kekulé model. **[1 mark]**

Page 105 — Reactions of Benzene

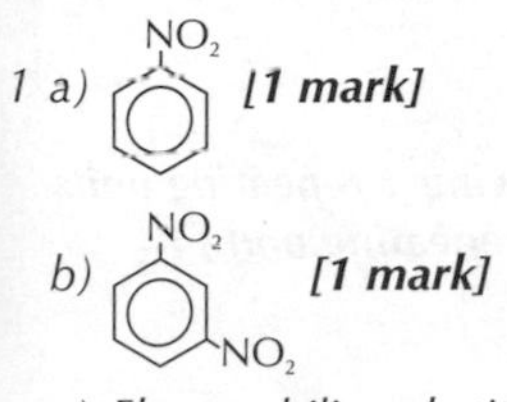

1 a) **[1 mark]**
b) **[1 mark]**
c) Electrophilic substitution **[1 mark]**
d)

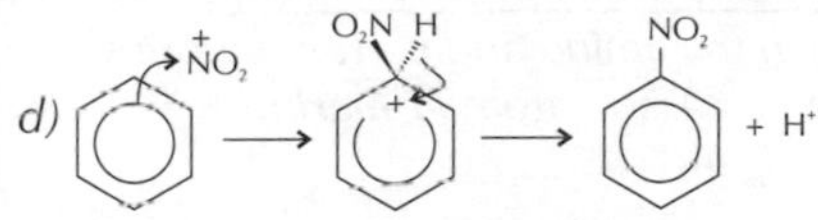

[3 marks available — 1 mark for each stage above given either as a diagram or a description.]

You need to get used to working out structures of compounds from their names. In 1,3-dinitrobenzene, dinitro tells you that the nitro group NO_2 appears twice and the 1 and 3 tell you where the two nitro groups appear.

2 a) Cyclohexene would decolorise bromine water, benzene would not. **[1 mark]**
b) Bromine reacts with the cyclohexene – the bromine atoms are added to the cyclohexene molecules, forming a dibromocycloalkane and leaving a clear solution **[1 mark]**. Benzene does not react. **[1 mark]**
c) i) It acts as a halogen carrier **[1 mark]**

ii)

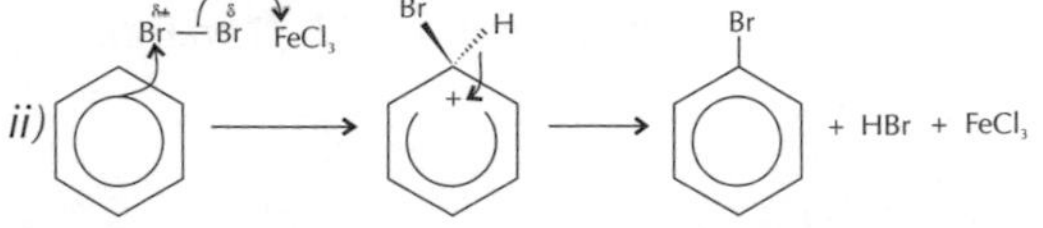

[3 marks available — 1 mark for each stage above given either as a diagram or a description.]

Page 107 — Phenols

1 a)

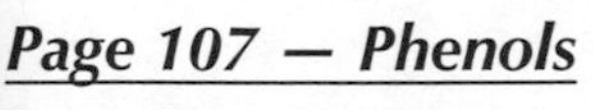

[1 mark]

b) $2K + 2C_7H_7OH \rightarrow 2C_7H_7OK + H_2$ **[1 mark]**
c) 4.8 dm^3 of H_2 = 4.8 ÷ 24 = 0.2 moles **[1 mark]**
From eqn:
2 moles of C_7H_7OH give 1 mole of H_2
So 0.4 moles of C_7H_7OH give 0.2 moles of H_2 **[1 mark]**
M_r of C_7H_7OH is $(7 \times 12) + (7 \times 1) + 16 + 1 = 108$
Mass of 0.4 moles of C_7H_7OH is 108 × 0.4 = 43.2g **[1 mark]**

2 a) With benzene, there will be no reaction **[1 mark]** but with phenol a reaction will occur which decolorises the bromine water / gives a precipitate / smells of antiseptic **[1 mark for any of these observations]**
b) 2,4,6-tribromophenol **[1 mark]**
c) Electrons from one of oxygen's p-orbitals overlap with the benzene ring's delocalised system, increasing its electron density **[1 mark]**. This makes the ring more likely to be attacked by electrophiles. **[1 mark]**
d) Electrophilic substitution **[1 mark]**

Page 110 — Aldehydes and Ketones

1 a) Y is a ketone **[1 mark]**.
It cannot be an acid as it has a neutral pH **[1 mark]** and it cannot be an aldehyde as there is no reaction with Tollens' reagent OR because an aldehyde heated under reflux would be oxidised to an acid **[1 mark]**.
b) It must be a secondary alcohol as they are the only ones oxidised to ketones. **[1 mark]**
c) E.g. add Brady's reagent **[1 mark]** and then find the melting point of the precipitate that is formed. This will identify the ketone **[1 mark]**.

2 a) $H_3C—C—C_2H_5$ (with =O on the central C)

[2 marks for correct answer or 1 mark for correct functional group.]

b) Butan-2-ol **[1 mark]** $H_3C—C—C_2H_5$ (with OH on the central C) **[1 mark]**

3 a) Reflux propan-1-ol with acidified potassium dichromate(VI) **[1 mark]**.
b) To make propanal you would gently heat the same mixture as in a), but with excess alcohol and distil the aldehyde off as it is produced. **[1 mark]**
c) Testing the pH would identify the acid **[1 mark]**.
Tollens' reagent **[1 mark]** will then distinguish between the alcohol and the aldehyde as only the aldehyde will give a silver mirror when heated with Tollens' reagent **[1 mark]**.

Page 113 — Carboxylic Acids and Esters

1 a) 2-methylpropyl ethanoate **[1 mark]**
b) Food flavouring/perfume **[1 mark]**
c) Ethanoic acid **[1 mark]** $H_3C–C(=O)OH$ **[1 mark]**

2-methylpropan-1-ol **[1 mark]** $HO–CH_2–CH(CH_3)–CH_3$ **[1 mark]**

This is acid hydrolysis **[1 mark]**
d) With sodium hydroxide, sodium ethanoate is produced, but in the reaction in part (c), ethanoic acid is produced **[1 mark]**.

2 a) Propanol **[1 mark]**
b) Ethanoic acid **[1 mark]**
Ethanoic anhydride **[1 mark]**
c)

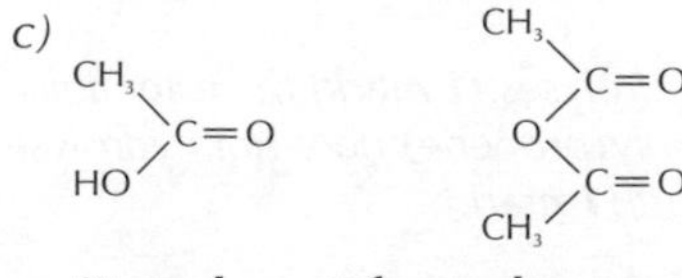

[2 marks — 1 for each correctly drawn molecule]

Esters are pretty complicated molecules, but remember that they're basically an alcohol (-OH group) and carboxylic acid (-COOH) with H_2O removed.

A2 Answers

Page 116 — Fatty Acids and Fats

1 a) $C_{17}H_{35}COOH$ or $C_{18}H_{36}O_2$ **[1 mark]**

b) i) Triglycerides **[1 mark]**

ii) The triester is saturated as there are no double bonds OR because stearic acid and glycerol are both saturated / contain no double bonds. **[1 mark]**

2 a) 'Cis' means that the hydrogen atoms attached to the carbon atoms either side of the double bond are on the same side **[1 mark]**

b) Trans fatty acids increase the levels of 'bad' cholesterol/ decrease the level of 'good' cholesterol in the body. **[1 mark]** 'Bad' cholesterol increases the risk of heart disease and strokes/'good' cholesterol decreases the risk of heart disease and strokes. **[1 mark]**

Page 119 — Amines

1 a) propylamine, dipropylamine and tripropylamine **[2 marks available — 1 each for any two.]**

b) Fractional distillation **[1 mark]**

c) React nitrobenzene with tin **[1 mark]** and hydrochloric acid **[1 mark]** under reflux **[1 mark]**, then react the product with sodium hydroxide **[1 mark]**

2 a) Azo dyes **[1 mark]**

b) The -N=N- group **[1 mark]**

c) NH_2–C_6H_4–SO_3Na or $N^+{\equiv}N$–C_6H_4–SO_3Na **[1 mark]**

OH–$C_{10}H_6$–SO_3Na **[1 mark]**

As a first step make sure you remember the basic overall reaction, e.g. phenylamine + phenol = azo dye, then learn the actual equations and reaction conditions.

Unit 4: Module 2 — Polymers and Synthesis

Page 121 — Polymers

1 a) Addition polymerisation **[1 mark]**

b)

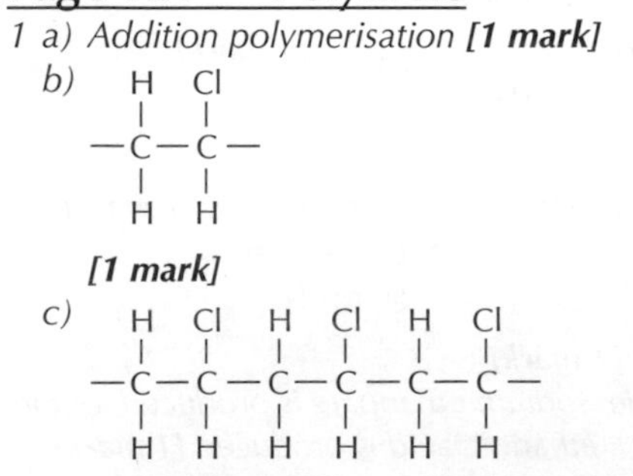

[1 mark]

c) **[1 mark]**

d) Polymers formed from alkenes are very stable **[1 mark]** — they cannot be broken down easily, e.g. by hydrolysis **[1 mark]**.

2 Water in the body slowly hydrolyses **[1 mark]** the ester links **[1 mark]** in the polyester. Poly(propene) does not hydrolyse/ is chemically unreactive/inert **[1 mark]**.

Page 123 — Polyesters and Polyamides

1 a) i) $H_2N-(CH_2)_6-NH_2$ and $HO-C(=O)-(CH_2)_4-C(=O)-OH$

[1 mark each]

ii) Amide/peptide link **[1 mark]**

b) i) $-[C(=O)-(CH_2)_4-C(=O)-O-(CH_2)_6-O]-$

or

$-[O-(CH_2)_6-O-C(=O)-(CH_2)_4-C(=O)]-$ **[1 mark]**

ii) For each link formed, one water molecule is eliminated **[1 mark]**.

2 a) $\cdots-[C(=O)-CH(H)-O]-C(=O)-CH(H)-O-C(=O)-CH(H)-O-\cdots$

repeating unit

[2 marks — 1 mark for correctly drawing 3 repeating units and 1 mark for correctly labelling 1 repeating unit.]

b) Polyesters **[1 mark]**

Page 125 — Amino Acids and Proteins

1 a) An amino acid in which the amino and carboxyl groups are attached to the same carbon atom **[1 mark]**.

b)

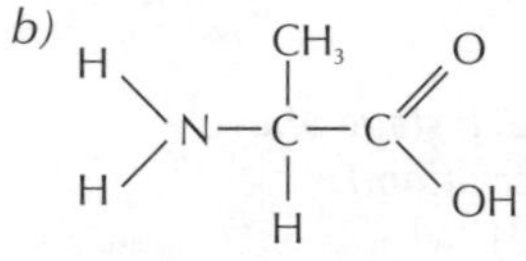

[1 mark — structure of NH_2 and COOH not required]

c) The CH_3 group **[1 mark]**.

d) Alanine has a central carbon atom with four different groups attached to it OR it has enantiomers/optical isomers. **[1 mark]**

e)

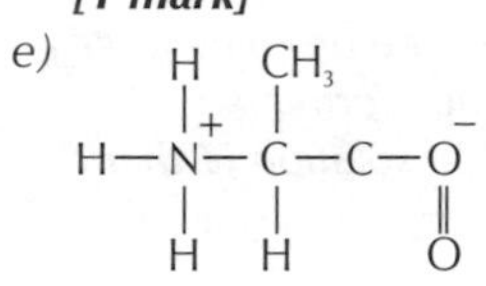

[1 mark]

2 a) A molecule made by joining 2 amino acids **[1 mark]**.

b)

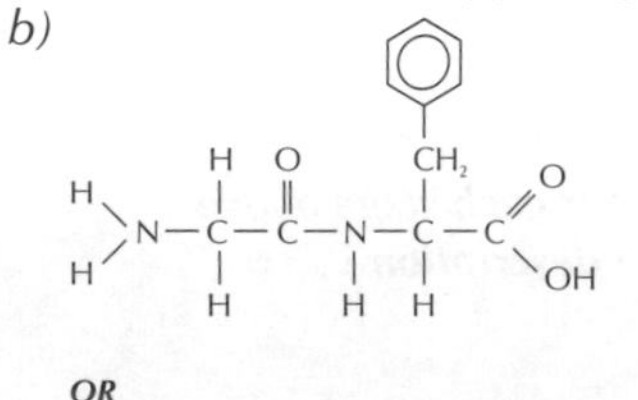

OR

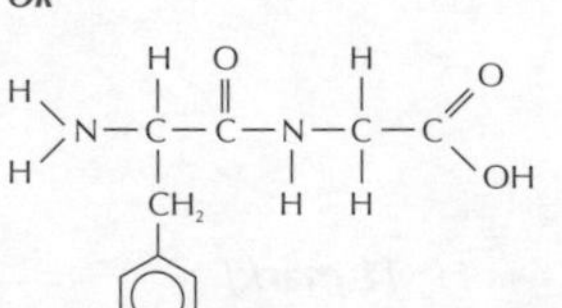

[1 mark — structure of NH_2 and COOH not required]

c) A condensation reaction involves a water molecule being removed as two other molecules are joined **[1 mark]**.

d) Hydrolysis **[1 mark]**.

e) Glycine is not chiral as the central carbon atom is not joined to four different groups **[1 mark]**.

A2 Answers

Page 127 — Organic Synthesis

1 Step 1: The methanol is refluxed **[1 mark]** with $K_2Cr_2O_7$ **[1 mark]** and acid **[1 mark]** to form methanoic acid **[1 mark]**.
Step 2: The methanoic acid is reacted under reflux **[1 mark]** with ethanol **[1 mark]** using an acid catalyst **[1 mark]** to make ethyl methanoate.

2 Step 1: React propane with bromine **[1 mark]** in the presence of UV light **[1 mark]**. Bromine is toxic and corrosive **[1 mark]** so great care should be taken. Bromopropane is formed **[1 mark]**.
Step 2: Bromopropane is then refluxed **[1 mark]** with sodium hydroxide solution **[1 mark]**, again a corrosive substance so take care **[1 mark]**, to form propanol.

Page 129 — Functional Groups

1 a) A — hydroxyl **[1 mark]**
B — hydroxyl and alkenyl **[1 mark]**
C — hydroxyl and phenyl **[1 mark]**
b) C **[1 mark]**
c) A **[1 mark]**
d) C **[1 mark]**

2 a) carbonyl / ketone OR amide ($-CONH_2$) **[1 mark]** and (primary) amine / amino **[1 mark]**
b) i) C=O **[1 mark]**
ii) It is a double-ended molecule **[1 mark]** with 2 amine / amino groups **[1 mark]**.

3 a) phenyl **[1 mark]**, hydroxyl **[1 mark]** and ester **[1 mark]**
b) ester **[1 mark]**
c) Expanded structure (with C and H atoms showing):

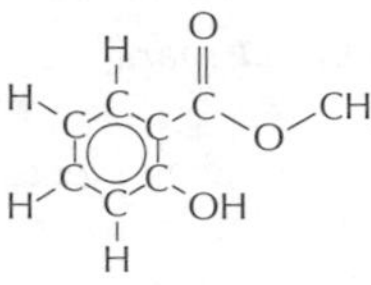

So molecular formula of methyl salicylate is $C_8H_8O_3$. **[1 mark]**

d)

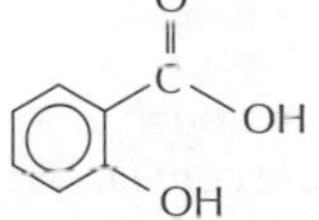

[1 mark]

Look back at esters on p111 if you had trouble with this one.

Page 131 — Stereoisomerism and Chirality

1 a)

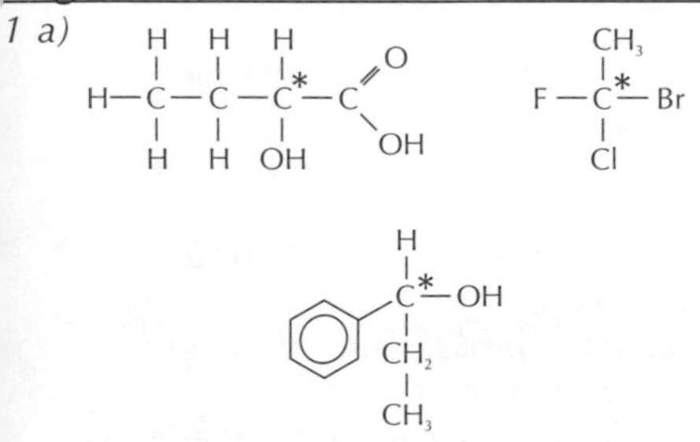

* is the chiral carbon atom, or chiral centre.
[3 marks available — 1 mark for each correct diagram.]

b)

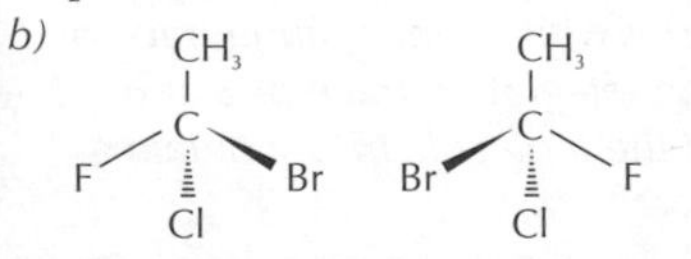

[2 marks available — 1 mark for each correct diagram.]

c) They rotate plane-polarised light **[1 mark]**.

2 a)

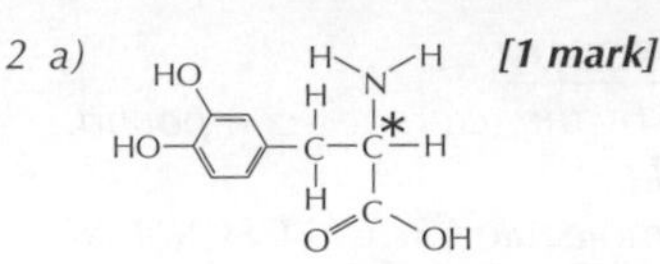

[1 mark]

The chiral carbon is the one with 4 different groups attached.

b) i) Smaller doses needed **[1 mark]**, fewer side-effects (because there's no D-DOPA enantiomer) **[1 mark]**.
ii) A mixture containing equal quantities of each enantiomer of an optically active compound **[1 mark]**.
iii) The other enantiomer may have unexpected and dangerous side effects or completely different effects **[1 mark]**.

Unit 4: Module 3 — Analysis

Page 133 — Chromatography

1 a) $R_f \text{ value} = \dfrac{\text{Distance travelled by spot}}{\text{Distance travelled by solvent}}$ **[1 mark]**

R_f value of spot A = $7 \div 8 = 0.875$ **[1 mark]**

The R_f value has no units, because it's a ratio.

b) Substance A has moved further up the plate because it's less strongly adsorbed **[1 mark]** onto the stationary phase **[1 mark]** than substance B **[1 mark]**.

2 a) The peak at 5 minutes **[1 mark]**.
b) The mixture may contain another chemical with a similar retention time, which would give a peak at 5 minutes **[1 mark]**.

3 a) The mixture is injected into a stream of carrier gas, which takes it through a tube over the stationary phase **[1 mark]**. The components of the mixture dissolve in the stationary phase **[1 mark]**, evaporate into the mobile phase **[1 mark]**, and redissolve, gradually travelling along the tube to the detector **[1 mark]**.
b) The substances separate because they have different solubilities in the stationary phase **[1 mark]**, so they take different amounts of time to move through the tube **[1 mark]**.
c) The areas under the peaks will be proportional to the relative amount of each substance in the mixture OR the area under the benzene peak will be three times greater than the area under the ethanol peak **[1 mark]**.

Page 136 — Mass Spectrometry and Chromatography

1 a) 88 **[1 mark]**
b) A has a mass of 43, so it's probably $CH_3CH_2CH_2$ **[1 mark]**.
B has a mass of 45, so it's probably COOH **[1 mark]**.
C has a mass of 73, so it's probably CH_2CH_2COOH **[1 mark]**.
c) Since the molecule is a carboxylic acid that contains the three fragments that you found in part (b), it must have this structure:

```
   H  H  H
   |  |  |     O
H—C—C—C—C//
   |  |  |    \
   H  H  H     OH
```

[1 mark]

This is butanoic acid **[1 mark]**.

2 a) Several alkanes/compounds have similar GC retention times **[1 mark]**, but different masses, so they would give different peaks on a mass spectrum **[1 mark]**.
b) Formula of molecular ion: $CH_3CH_3^+$ **[1 mark]**
Mass = $(12 \times 2) + (1 \times 6) = 24 + 6 = 30$ **[1 mark]**
c) i) $CH_3CH_2^+$ **[1 mark]**
ii) CH_3^+ **[1 mark]**

A2 Answers

Page 139 — NMR Spectroscopy

1 a) *The peak at $\delta = 0$ is produced by the reference compound, tetramethylsilane/TMS* ***[1 mark]****.*

b) *All three carbon atoms in the molecule $CH_3CH_2CH_2NH_2$ are in different environments* ***[1 mark]****. There are only two peaks on the carbon-13 NMR spectrum shown* ***[1 mark]****.*
The ^{13}C NMR spectrum of $CH_3CH_2CH_2NH_2$ would have three peaks because this molecule has three carbon environments.

c) *The peak at $\delta \approx 25$ represents carbons in C–C bonds* ***[1 mark]****. The peak at $\delta \approx 40$ represents a carbon in a C–N bond* ***[1 mark]****. The spectrum has two peaks, so the molecule must have two carbon environments* ***[1 mark]****. So the structure of the molecule must be:*

```
   H  NH2 H
   |   |   |
H—C—C—C—H   [1 mark].
   |   |   |
   H   H   H
```

The 2 carbon environments in the molecule are CH_3–$CH(NH_2)$–CH_3 and $CH(NH_2)$–$(CH_3)_2$.

Page 141 — More NMR Spectroscopy

1 a) *A CH_2 group adjacent to a halogen* ***[1 mark]****.*
You've got to read the question carefully — it tells you it's an alkyl halide. So the group at 3.6 p.p.m. can't have oxygen in it. It can't be halogen-CH_3 either, as this has 3 hydrogens in it.

b) *A CH_3 group* ***[1 mark]****.*

c) *CH_2 added to CH_3 gives a mass of 29, so the halogen must be chlorine with a mass of 35* ***[1 mark]****. So a likely structure is CH_3CH_2Cl* ***[1 mark]****.*

d) *The quartet at 3.6 p.p.m. is caused by 3 protons on the adjacent carbon* ***[1 mark]****. The n + 1 rule tells you that 3 protons give 3 + 1 = 4 peaks* ***[1 mark]****. Similarly the triplet at 1.3 p.p.m. is due to 2 adjacent protons* ***[1 mark]*** *giving 2 + 1 = 3 peaks* ***[1 mark]****.*

Page 143 — Infrared Spectroscopy

1 a) *A's due to an O–H group in a carboxylic acid* ***[1 mark]****.*
B's due to a C=O as in an aldehyde, ketone, acid or ester ***[1 mark]****.*
C's due to a C–O as in an alcohol, ester or acid ***[1 mark]****.*

b) *The spectrum suggests it's a carboxylic acid, so it's got a COOH group* ***[1 mark]****. This group has a mass of 45, so the rest of the molecule has a mass of 74 – 45 = 29, which is likely to be C_2H_5* ***[1 mark]****. So the molecule could be C_2H_5COOH – propanoic acid* ***[1 mark]****.*

2 a) *X is due to an O–H group in an alcohol or phenol* ***[1 mark]****.*
Y is due to C–H bonds ***[1 mark]****.*
Z is due to a C–O group in an alcohol, ester or acid ***[1 mark]****.*

b) *The spectrum suggests it's an alcohol, so it's got an OH group* ***[1 mark]****. This group has a mass of 17, so the rest of the molecule has a mass of 46 – 17 = 29, which is likely to be C_2H_5* ***[1 mark]****. So the molecule could be C_2H_5OH — ethanol* ***[1 mark]****.*

Page 145 — More on Spectra

1 a) *Mass of molecule = 73* ***[1 mark]****.*
You can tell this from the mass spectrum — the mass of the molecular ion is 73.

b) *Structure of the molecule:*

```
   H  H
   |  |     NH2
H–C–C–C           [1 mark]
   |  |     O
   H  H
```

Explanation: Award ***1 mark*** *each for the following pieces of reasoning, up to a total of* ***[5 marks]****:*
The infrared spectrum of the molecule shows a strong absorbance at about 3200 cm^{-1}, which suggests that the molecule contains an amine or amide group.
It also has a trough at about 1700 cm^{-1}, which suggests that the molecule contains a C=O group.
The ^{13}C NMR spectrum tells you that the molecule has three carbon environments.
One of the ^{13}C NMR peaks has a chemical shift of about 170, which corresponds to a carbonyl group in an amide.
The ^1H NMR spectrum has a quartet at $\delta \approx 2$, and a triplet at $\delta \approx 1$ — to give this splitting pattern the molecule must contain a CH_2CH_3 group.
The ^1H NMR spectrum has a singlet at $\delta \approx 6$, corresponding to H atoms in an amine or amide group.
The mass spectrum shows a peak at m/z = 15 which corresponds to a CH_3 group.
The mass spectrum shows a peak at m/z = 29 which corresponds to a CH_2CH_3 group.
The mass spectrum shows a peak at m/z = 44 which corresponds to a $CONH_2$ group.

2 a) *Mass of molecule = 60* ***[1 mark]****.*

b) *Structure of the molecule:*

```
   H  H  H
   |  |  |
H–C–C–C–OH   [1 mark]
   |  |  |
   H  H  H
```

Explanation: Award ***1 mark*** *each for the following pieces of reasoning, up to a total of* ***[5 marks]****:*
The ^{13}C NMR spectrum tells you that the molecule has three carbon environments.
One of the ^{13}C NMR peaks has a chemical shift of 60 — which corresponds to a C–O group.
The infrared spectrum of the molecule has a trough at about 3300 cm^{-1}, which suggests that the molecule contains an alcoholic OH group.
It also has a trough at about 1200 cm^{-1}, which suggests that the molecule also contains a C–O group.
The mass spectrum shows a peak at m/z = 15 which corresponds to a CH_3 group.
The mass spectrum shows a peak at m/z = 17 which corresponds to an OH group.
The mass spectrum shows a peak at m/z = 29 which corresponds to a C_2H_5 group.
The mass spectrum shows a peak at m/z = 31 which corresponds to a CH_2OH group.
The mass spectrum shows a peak at m/z = 43 which corresponds to a C_3H_7 group.
The ^1H NMR spectrum has 4 peaks, showing that the molecule has 4 proton environments.
The ^1H NMR spectrum has a singlet at $\delta \approx 2$, corresponding to H atoms in an OH group.
The ^1H NMR spectrum has a sextuplet with an integration trace of 2 at $\delta \approx 1.5$, a quartet with an integration trace of 2 at $\delta \approx 3.5$, and a triplet with an integration trace of 3 at $\delta \approx 1$ — to give this splitting pattern the molecule must contain a $CH_3CH_2CH_2$ group.

A2 Answers

Unit 5: Module 1 — Rates, Equilibrium and pH

Page 147 — Rate Graphs and Orders

1 a) *1st order* **[1 mark]**

b)

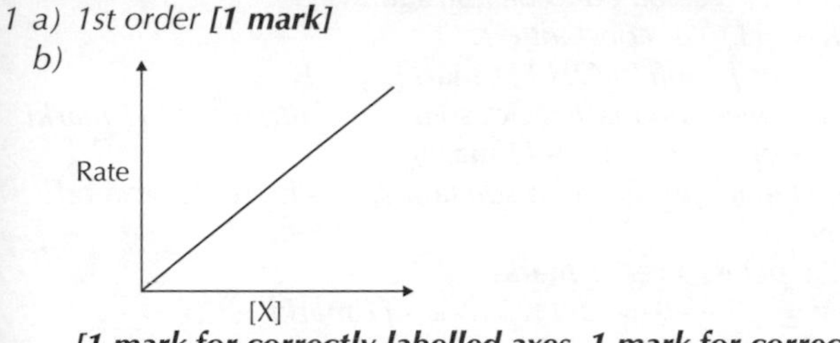

[1 mark for correctly labelled axes, 1 mark for correct line]

c) *The volume* **[1 mark]** *of hydrogen gas produced in a unit time* **[1 mark]**.

2 a) *E.g. Gas volume of* $O_{2(g)}$ **[1 mark]** *using, e.g. a gas syringe* **[1 mark]**.

b)

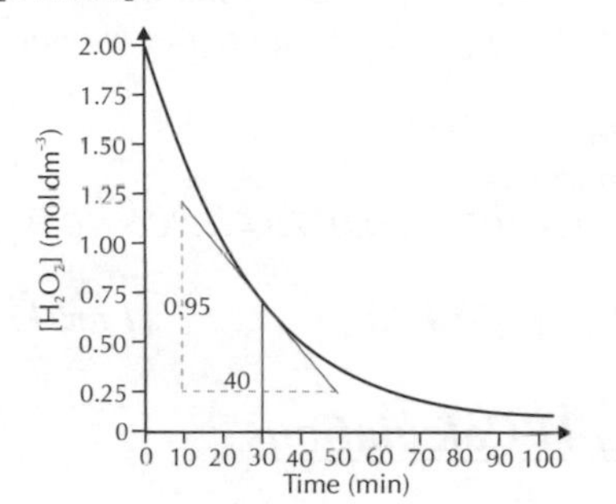

Rate after 30 minutes $= 0.95 \div 40$

≈ 0.024 **[1 mark]** *mol dm*$^{-3}$ *min*$^{-1}$ **[1 mark]**

Accept rate within range 0.024 ± 0.005.

[1 mark for $[H_2O_{2(aq)}]$ on y-axis and time on x-axis. 1 mark for points accurately plotted. 1 mark for best-fit smooth curve. 1 mark for tangent to curve at 30 minutes.]

Page 149 — Initial Rates and Half-Life

1 *Experiments 1 and 2:*

[D] doubles and the initial rate quadruples (with [E] remaining constant) **[1 mark]**. *So it's 2nd order with respect to [D]* **[1 mark]**.

Experiments 1 and 3:

[E] doubles and the initial rate doubles (with [D] remaining constant) **[1 mark]**. *So it's 1st order with respect to [E]* **[1 mark]**.

Always explain your reasoning carefully — state which concentrations are constant and which are changing.

2 a)

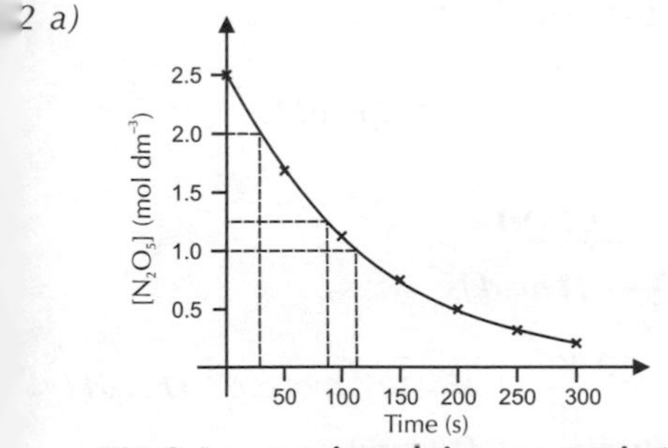

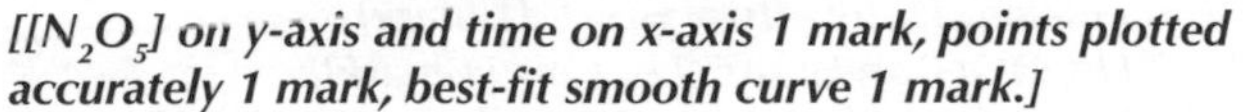

[$[N_2O_5]$ on y-axis and time on x-axis 1 mark, points plotted accurately 1 mark, best-fit smooth curve 1 mark.]

b) i) *Horizontal dotted line from 1.25 on y-axis to curve and vertical dotted line from curve to x-axis* **[1 mark]**.

Time value = 85s **[1 mark, allow 85 ± 2]**

ii) *Vertical dotted line from curve at 2.0 mol dm*$^{-3}$ *and same at 1.0 mol dm*$^{-3}$ **[1 mark]**. *Time value difference = 113 (± 2) – 28 (± 2) = 85* **[1 mark, allow 85 ± 4]**

c) *The order of reaction = 1* **[1 mark]** *because the half-life of ≈ 85 is independent of concentration* **[1 mark]**.

Page 151 — Rate Equations

1 a) *Rate* $= k[NO_{(g)}]^2[H_{2(g)}]$

[1 mark for correct orders, 1 mark for the rest]

b) i) $0.00267 = k \times (0.004)^2 \times 0.002$ **[1 mark]**

$k = 8.34 \times 10^4$ dm^6 mol^{-2} s^{-1}

[1 mark for answer, 1 mark for units].

Units: k = mol dm^{-3} s^{-1} $\div$ [(mol dm^{-3})^2 $\times$ (mol dm^{-3})]

= mol^{-2} dm^6 s^{-1}.

ii) *It would decrease* **[1 mark]**.

If the temperature decreases, the rate decreases too. A lower rate means a lower rate constant.

2 a) *Rate* $= k[CH_3COOC_2H_5][H^+]$ **[1 mark]**

b) $2.2 \times 10^{-3} = k \times 0.25 \times 2.0$ **[1 mark]**

$k = 2.2 \times 10^{-3} \div 0.5 = 4.4 \times 10^{-3}$ **[1 mark]**

The units are:

k = (mol dm^{-3} s^{-1}) $\div$ (mol dm^{-3})(mol dm^{-3})

= mol^{-1} dm^3 s^{-1} **[1 mark]**

c) *If the volume doubles, the concentration of each reactant halves to become 1 mol dm*$^{-3}$ *and 0.125 mol dm*$^{-3}$ *respectively* **[1 mark]**.

So the rate $= 4.4 \times 10^{-3} \times 1 \times 0.125 = 5.5 \times 10^{-4}$ mol dm^{-3} s^{-1} **[1 mark]**.

3 *The rate equation is: Rate* $= k[X][Y]$ **[1 mark]**

Rate is proportional to [X], so increasing [X] by 3.33% increases the rate by 3.33% also **[1 mark]**.

However, increasing the temperature by 10K doubles the rate. So temperature has a greater effect **[1 mark]**.

Page 153 — Rates and Reaction Mechanisms

1 a) *rate* $= k[H_2][ICl]$ **[1 mark]**

b) i) *One molecule of* H_2 *and one molecule of ICl* **[1 mark]**. *If the molecule is in the rate equation, it must be in the rate-determining step* **[1 mark]**.

ii) *Incorrect* **[1 mark]**. H_2 *and ICl are both in the rate equation, so they must both be in the rate-determining step OR the order of the reaction with respect to ICl is 1, so there must be only one molecule of ICl in the rate-determining step* **[1 mark]**.

2 a) *The rate equation is first order with respect to HBr and* O_2 **[1 mark]**. *So only 1 molecule of HBr (and* O_2*) is involved in the rate-determining step* **[1 mark]**. *There must be more steps as 4 molecules of HBr are in the equation* **[1 mark]**.

b) $HBr + O_2 \rightarrow HBrO_2$ *(rate-determining step)* **[1 mark]**

$HBr + HBrO_2 \rightarrow 2HBrO$ **[1 mark]**

$HBr + HBrO \rightarrow H_2O + Br_2$ **[1 mark]**

$HBr + HBrO \rightarrow H_2O + Br_2$ **[1 mark]**

Part b) is pretty tricky — you need to do a fair bit of detective work and some trial and error. Make sure you use all of the clues in the question...

A2 Answers

Page 155 — The Equilibrium Constant

1 a) $K_c = \frac{[NH_3]^2}{[N_2][H_2]^3}$
[1 mark for correct reactants on top and bottom, 1 mark for correct powers.]

b) $K_c = \frac{0.150^2}{1.06 \times 1.41^3}$ **[1 mark]**
$= 7.6 \times 10^{-3}$ **[1 mark]** $mol^{-2} dm^6$ **[1 mark]**

2 a) i) moles = mass ÷ M_r = 42.5 ÷ 46 = 0.92 **[1 mark]**
ii) moles of O_2 = mass ÷ M_r = 14.1 ÷ 32 = 0.44 **[1 mark]**
moles of NO = 2 × moles of O_2 = 0.88 **[1 mark]**
moles of NO_2 = 0.92 – 0.88 = 0.04 **[1 mark]**

b) $[O_2]$ = 0.44 ÷ 22.8 = 0.019 mol dm^{-3}
[NO] = 0.88 ÷ 22.8 = 0.039 mol dm^{-3}
$[NO_2]$ = 0.04 ÷ 22.8 = 1.75×10^{-3} mol dm^{-3} **[1 mark]**

$K_c = \frac{[NO]^2[O_2]}{[NO_2]^2}$ **[1 mark]**

$\Rightarrow K_c = \frac{(0.039)^2 \times (0.019)}{(1.75 \times 10^{-3})^2}$ **[1 mark]** = 9.4 **[1 mark]** mol dm^{-3} **[1 mark]**

(Units = $(mol\ dm^{-3})^2 \times (mol\ dm^{-3}) \div (mol\ dm^{-3})^2 = mol\ dm^{-3}$)

Page 157 — More on the Equilibrium Constant

1 a) T_2 is lower than T_1 **[1 mark]**.
A decrease in temperature shifts the position of equilibrium in the exothermic direction, producing more product **[1 mark]**.
More product means K_c increases **[1 mark]**.
A negative ΔH means the forward reaction is exothermic — it gives out heat.

b) The yield of SO_3 increases **[1 mark]**. (A decrease in volume means an increase in pressure. This shifts the equilibrium position to the right.) K_c is unchanged **[1 mark]**.

2 a) $K_c = \frac{[CO][H_2]^3}{[CH_4][H_2O]}$
[1 mark for correct reactants on top and bottom, 1 mark for correct powers.]

b) i) K_c will increase **[1 mark]** as the forward (endothermic) reaction is favoured.
ii) No effect **[1 mark]**.

c) Increasing the pressure will move the reaction to the left, so amounts of CH_4 and H_2O will increase **[1 mark]** and amounts of CO and H_2 will decrease **[1 mark]**.

Page 159 — Acids and Bases

1 a) H^+ or H_3O^+ and SO_4^{2-} **[1 mark]**
b) $2H^+_{(aq)} + Mg_{(s)} \rightarrow Mg^{2+}_{(aq)} + H_{2(g)}$ **[1 mark]**
c) SO_4^{2-} **[1 mark]**
d) The acid dissociates / ionises in water as follows:
$H_2SO_4 + 2H_2O \rightleftharpoons 2H_3O^+ + SO_4^{2-}$ **[1 mark]**
The equilibrium position lies almost completely to the right **[1 mark]**.

2 a) $HCN + H_2O \rightleftharpoons H_3O^+ + CN^-$ **[1 mark]**
b) The equilibrium position lies to the left **[1 mark]**.
c) The pairs are HCN and CN^- **[1 mark]** AND H_2O and H_3O^+ **[1 mark]**.
d) H^+ **[1 mark]**

3 a) $NH_3 + H_2O \rightleftharpoons NH_4^+ + OH^-$ **[1 mark]**
b) An acid **[1 mark]** as it donates a proton **[1 mark]**
c) OH^- **[1 mark]**

Page 161 — pH

1 a) K_c is the equilibrium constant for water
$K_c = [H^+][OH^-] \div [H_2O]$ **[1 mark]**
But $[H_2O]$ is assumed to be constant.
So $K_c \times [H_2O]$ is constant = K_w
This gives $K_w = [H^+][OH^-]$ **[1 mark]**.

b) It's a strong monobasic acid, so $[H^+]$ = [HBr] = 0.32 **[1 mark]**
pH = $-\log_{10} 0.32$ = 0.49 **[1 mark]**

c) More H^+ are produced in solution as it is more dissociated **[1 mark]**
So the pH is lower **[1 mark]**.

2 a) 2.5 g = 2.5 ÷ 40 = 0.0625 moles **[1 mark]**
1 mole of NaOH gives 1 mole of OH^-
So $[OH^-]$ = [NaOH] = 0.0625 mol dm^{-3} **[1 mark]**

b) $K_w = [H^+][OH^-]$ **[1 mark]**
$[H^+] = 1 \times 10^{-14} \div 0.0625 = 1.6 \times 10^{-13}$ **[1 mark]**
pH = $-\log_{10}(1.6 \times 10^{-13})$ = 12.8 (1 d.p.) **[1 mark]**

c) The value of K_c and hence K_w is temperature dependent **[1 mark]**.

3 $K_w = [H^+][OH^-]$ **[1 mark]**
$[OH^-]$ = 0.0370 **[1 mark]**
$[H^+] = K_w \div [OH^-] = (1 \times 10^{-14}) \div 0.0370 = 2.70 \times 10^{-13}$ **[1 mark]**
pH = $-\log_{10}[H^+] = -\log_{10}(2.70 \times 10^{-13})$ = 12.57 **[1 mark]**

Page 163 — More pH Calculations

1 a) $K_a = \frac{[H^+][A^-]}{[HA]}$ **[1 mark]**

b) $K_a = \frac{[H^+]^2}{[HA]}$ ⇒ [HA] is 0.280 because very few HA will dissociate **[1 mark]**.

$[H^+] = \sqrt{(5.60 \times 10^{-4}) \times 0.280} = 0.0125$ mol dm^{-3} **[1 mark]**
pH = $-\log_{10}[H^+] = -\log_{10}(0.0125)$ = 1.90 **[1 mark]**

2 a) $[H^+] = 10^{-2.65} = 2.24 \times 10^{-3}$ mol dm^{-3} **[1 mark]**
$K_a = \frac{[H^+]^2}{[HX]}$ **[1 mark]** $= \frac{[2.24 \times 10^{-3}]^2}{[0.150]}$
$= 3.34 \times 10^{-5}$ **[1 mark]** mol dm^{-3} **[1 mark]**

b) $pK_a = -\log_{10}K_a = -\log_{10}(3.34 \times 10^{-5})$ **[1 mark]** = 4.48 **[1 mark]**

3 $K_a = 10^{-pKa} = 10^{-4.2} = 6.3 \times 10^{-5}$ **[1 mark]**
$K_a = \frac{[H^+]^2}{[HA]}$ **[1 mark]**

So $[H^+] = \sqrt{K_a[HA]}$
$= \sqrt{(6.3 \times 10^{-5}) \times (1.6 \times 10^{-4})} = \sqrt{1.0 \times 10^{-8}}$
$= 1.0 \times 10^{-4}$ mol dm^{-3} **[1 mark]**
pH = $-\log_{10}[H^+] = -\log_{10} 1.0 \times 10^{-4}$ = 4 **[1 mark]**

Page 165 — Buffer Action

1 a) $K_a = \frac{[C_6H_5COO^-][H^+]}{[C_6H_5COOH]}$ **[1 mark]**

$\Rightarrow [H^+] = 6.4 \times 10^{-5} \times \frac{0.40}{0.20} = 1.28 \times 10^{-4}$ mol dm^{-3} **[1 mark]**
pH = $-\log_{10}[1.28 \times 10^{-4}]$ = 3.9 **[1 mark]**

b) $C_6H_5COOH \rightleftharpoons H^+ + C_6H_5COO^-$ **[1 mark]**
Adding H_2SO_4 increases the concentration of H^+ **[1 mark]**.
The equilibrium shifts left to reduce the concentration of H^+, so the pH will only change very slightly **[1 mark]**.

A2 Answers

2 a) $CH_3(CH_2)_2COOH \rightleftharpoons H^+ + CH_3(CH_2)_2COO^-$ **[1 mark]**
b) $[CH_3(CH_2)_2COOH] = [CH_3(CH_2)_2COO^-]$,
so $[CH_3(CH_2)_2COOH] \div [CH_3(CH_2)_2COO^-] = 1$ **[1 mark]**
and $K_a = [H^+]$. $pH = -log_{10}[1.5 \times 10^{-5}]$ **[1 mark]** = 4.8 **[1 mark]**
If the concentrations of the weak acid and the salt are equal, they cancel from the K_a expression and the buffer pH = pK_a.

Page 167 — pH Curves, Titrations and Indicators

1 The pH at equivalence for nitric acid is 7, whereas the pH at equivalence for ethanoic acid is greater than 7.
The pH at the start for nitric acid is lower than for ethanoic acid.
The near-vertical bit is bigger/slightly steeper for nitric acid than for ethanoic acid.
[1 mark for each difference, up to a maximum of 2]

2 a) moles of ethanoic acid = (25.0 × 0.350) ÷ 1000
= 0.00875 **[1 mark]**
Volume of KOH = (number of moles × 1000) ÷ molar concentration = (0.00875 × 1000) ÷ 0.285 **[1 mark]**
= 30.7 cm^3 **[1 mark]**
b) Thymol blue **[1 mark]**. It's a weak acid/strong base titration so the equivalence point is above pH 8 **[1 mark]**.

Unit 5: Module 2 — Energy

Page 169 — Neutralisation and Enthalpy

1 a) $H_2SO_4 + 2NaOH \rightarrow Na_2SO_4 + 2H_2O$ **[1 mark]**
b) 200 ml of 2.75 mol dm^{-3} acid will contain 2.75 × 0.2
= 0.55 moles **[1 mark]**
The volume of the solution is 400 ml, so the mass of the solution is 0.4 kg, so m = 0.4.
$\Delta H = -mc\Delta T = -0.4 \times 4.18 \times 38 = -63.54$ kJ for 0.55 moles of acid **[1 mark]**
1 mole of water is given by 0.5 moles of acid.
So 0.5 moles of acid will produce 0.5 ÷ 0.55 × −63.54
= −57.76 kJ **[1 mark]**

2 a) $HCl + KOH \rightarrow H_2O + KCl$ **[1 mark]**
$H_2SO_4 + 2KOH \rightarrow K_2SO_4 + 2H_2O$ **[1 mark]**
$H_3PO_4 + 3KOH \rightarrow K_3PO_4 + 3H_2O$ **[1 mark]**
b) The enthalpy change of neutralisation is the energy change when 1 mole of water is formed by the reaction of an acid and a base. **[1 mark]**
1 mole of H_3PO_4 will give 3 moles of water, so the energy released per mole of acid will be 3 × 57 kJ, whereas H_2SO_4 would give 2 × 57 and HCl just 57 kJ. So the neutralisation of H_3PO_4 is the most exothermic and would result in the largest temperature rise. **[1 mark]**

3 $HCl + NaOH \rightarrow H_2O + NaCl$ **[1 mark]**
From the equation 1 mole of acid gives 1 mole of water.
$\Delta H = -mc\Delta T = -0.2 \times 4.18 \times 3.4 = -2.84$ kJ **[1 mark]**
In 100 ml of X mol dm^{-3} acid there are 0.1X moles. **[1 mark]**
So if 0.1X mol gives 2.84 kJ and 1 mole of acid gives 57 kJ then X = 2.84 ÷ (57 × 0.1) = 0.5 mol dm^{-3} **[1 mark]**

Page 171 — Lattice Enthalpy and Born-Haber Cycles

1 a)

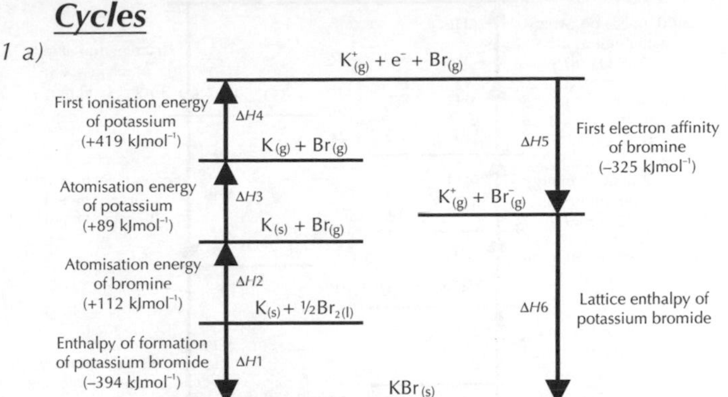

[1 mark for left of cycle. 1 mark for right of cycle. 1 mark for formulas/state symbols. 1 mark for correct directions of arrows.]

b) Lattice enthalpy, $\Delta H6 = -\Delta H5 - \Delta H4 - \Delta H3 - \Delta H2 + \Delta H1$
= −(−325) − (+419) − (+89) − (+112) + (−394) **[1 mark]**
= −689 **[1 mark]** kJ mol^{-1} **[1 mark]**
Award marks if calculation method matches cycle in part (a).

2 a)

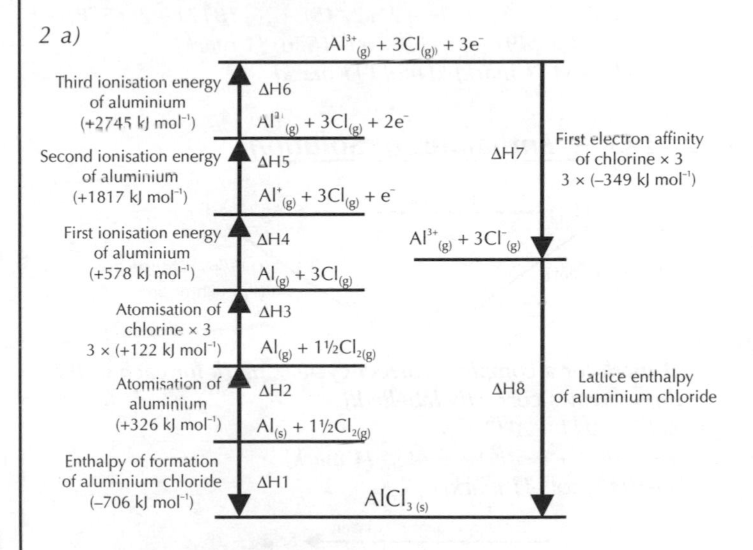

[1 mark for left of cycle. 1 mark for right of cycle. 1 mark for formulas/state symbols. 1 mark for correctly multiplying all the enthalpies. 1 mark for correct directions of arrows.]

b) Lattice enthalpy,
$\Delta H8 = -\Delta H7 - \Delta H6 - \Delta H5 - \Delta H4 - \Delta H3 - \Delta H2 + \Delta H1$
= −3(−349) − (+2745) − (+1817) − (+578) − 3(+122) − (+326) + (−706) **[1 mark]**
= −5491 **[1 mark]** kJ mol^{-1} **[1 mark]**

A2 Answers

3 a)

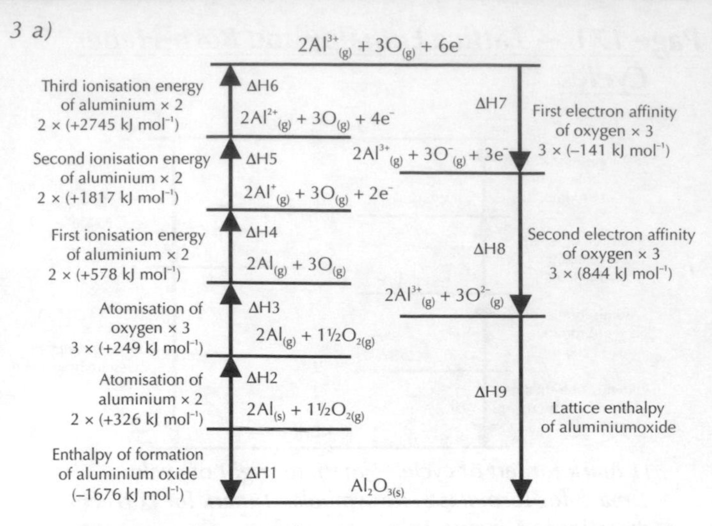

[1 mark for left of cycle. 1 mark for right of cycle. 1 mark for formulas/state symbols. 1 mark for correctly multiplying all the enthalpies. 1 mark for correct directions of arrows.]

b) Lattice enthalpy, *ΔH9 = –ΔH8 – ΔH7 – ΔH6 – ΔH5 – ΔH4 – ΔH3 – ΔH2 + ΔH1*
= –3(+844) – 3(–141) – 2(+2745) – 2(+1817) – 2(+578) – 3(+249) – 2(+326) + (–1676) **[1 mark]**
= –15 464 **[1 mark]** *kJ mol^{-1}* **[1 mark]**

Page 173 — Enthalpies of Solution

1 a)

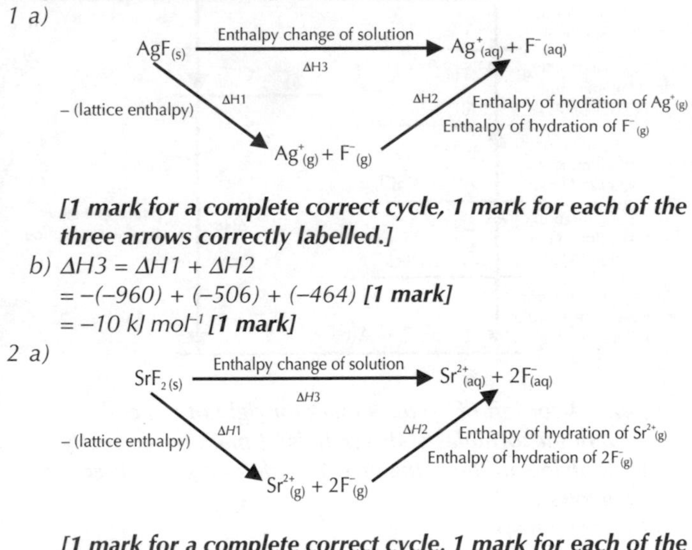

[1 mark for a complete correct cycle, 1 mark for each of the three arrows correctly labelled.]

b) *ΔH3 = ΔH1 + ΔH2*
= –(–960) + (–506) + (–464) **[1 mark]**
= –10 kJ mol^{-1} **[1 mark]**

2 a)

[1 mark for a complete correct cycle, 1 mark for each of the 3 arrows correctly labelled.]

Don't forget — you have to double the enthalpy of hydration for F^- because there are two in SrF_2.

b) *–(–2492) + (–1480) + (2 × –506)* **[1 mark]**
= 0 kJ mol^{-1} **[1 mark]**

3 By Hess's law:
Enthalpy change of solution ($MgCl_{2(s)}$)
= –lattice enthalpy ($MgCl_{2(s)}$)
+ enthalpy of hydration ($Mg^{2+}_{(g)}$)
+ [2 × enthalpy of hydration ($Cl^-_{(g)}$)] **[1 mark]**
= –(–2526) + (–1920) + [2 × (–364)] **[1 mark]**
= 2526 – 1920 – 728 = –122 kJ mol^{-1} **[1 mark]**

Page 175 — Free-Energy Change and Entropy Change

1 a) *The reaction is not likely to be spontaneous* **[1 mark]** *because there is a decrease in entropy* **[1 mark]**.
Remember — more particles means more entropy. There's 1½ moles of reactants and only 1 mole of products.

b) $\Delta S_{system} = 26.9 - (32.7 + 102.5)$ **[1 mark]**
$= -108.3$ **[1 mark]** *J K^{-1} mol^{-1}* **[1 mark]**

c) *Reaction is not likely to be spontaneous* **[1 mark]** *because ΔS_{system} is negative/there is a decrease in entropy* **[1 mark]**.

2 a) (i) $\Delta S_{system} = 48 - 70 = -22\ JK^{-1}mol^{-1}$ **[1 mark]**
$\Delta S_{surroundings} = -(-6000)/250 = +24\ JK^{-1}mol^{-1}$ **[1 mark]**
$\Delta S_{total} = \Delta S_{system} + \Delta S_{surroundings} = -22 + 24 = +2\ JK^{-1}mol^{-1}$ **[1 mark]**

(ii) $\Delta S_{surroundings} = -(-6000)/300 = +20\ JK^{-1}mol^{-1}$ **[1 mark]**
$\Delta S_{total} = \Delta S_{system} + \Delta S_{surroundings} = -22 + 20 = -2\ JK^{-1}mol^{-1}$ **[1 mark]**

b) *It will be spontaneous at 250 K, but not at 300 K* **[1 mark]**, *because ΔS_{total} is positive at 250 K but negative at 300 K* **[1 mark]**.

Page 177 — Redox Reactions

1 a) *Ti + (4 × –1) = 0, Ti = +4* **[1 mark]**
b) *(2 × V) + (5 × –2) = 0, V = +5* **[1 mark]**
c) *Cr + (4 × –2) = –2, Cr = +6* **[1 mark]**
d) *(2 × Cr) + (2 × –7) = –2, Cr = +6* **[1 mark]**

2 a) $2MnO^-_{4\,(aq)} + 16H^+_{(aq)} + 10I^-_{(aq)} \rightarrow 2Mn^{2+}_{(aq)} + 8H_2O_{(l)} + 5I_{2(aq)}$
[1 mark for correct reactants and products, 1 mark for correct balancing]
You have to balance the number of electrons before you can combine the half-equations. And always double-check that your equation definitely balances. It's easy to slip up and throw away marks.

b) *Mn has been reduced* **[1 mark]** *from +7 to +2* **[1 mark]**
I^- has been oxidised **[1 mark]** *from –1 to 0* **[1 mark]**

c) *Reactive metals have a tendency to lose electrons, so are good reducing agents* **[1 mark]**. *I^- is already in its reduced form* **[1 mark]**.

Page 179 — Electrode Potentials

1 a) *Iron* **[1 mark]** *as it has a more negative electrode potential/ it loses electrons more easily than lead* **[1 mark]**
b) *The iron half-cell* **[1 mark]** *as it loses electrons* **[1 mark]**
c) *Standard cell potential = –0.13 – (–0.44) = +0.31 V* **[1 mark]**

2 a) *+0.80 V – (–0.76 V) = 1.56 V* **[1 mark]**
b) *The concentration of Zn^{2+} ions or Ag^+ ions was not 1.00 mol dm^{-3}* **[1 mark]**. *The pressure wasn't 100 kPa* **[1 mark]**.
c) $Zn_{(s)} + 2Ag^+_{(aq)} \rightarrow Zn^{2+}_{(aq)} + 2Ag_{(s)}$ **[1 mark]**
d) *The zinc half-cell. It has a more negative standard electrode potential* **[1 mark]**.

Page 181 — The Electrochemical Series

1 a) $Zn_{(s)} + Ni^{2+}_{(aq)} \rightleftharpoons Zn^{2+}_{(aq)} + Ni_{(s)}$ **[1 mark]**
$E^\circ = (-0.25) - (-0.76) = +0.51\ V$ **[1 mark]**

b) $2MnO^-_{4\,(aq)} + 16H^+_{(aq)} + 5Sn^{2+}_{(aq)} \rightleftharpoons 2Mn^{2+}_{(aq)} + 8H_2O_{(l)} + 5Sn^{4+}_{(aq)}$ **[1 mark]**
$E^\circ = (+1.51) - (+0.15) = +1.36\ V$ **[1 mark]**

c) *No reaction* **[1 mark]**. *Both reactants are in their oxidised form* **[1 mark]**.

d) $Ag^+_{(aq)} + Fe^{2+}_{(aq)} \rightleftharpoons Ag_{(s)} + Fe^{3+}_{(aq)}$ **[1 mark]**
$E^\circ = (+0.80) - (+0.77) = +0.03\ V$ **[1 mark]**

A2 Answers

2 a) $KMnO_4$ **[1 mark]** because it has a more positive/less negative electrode potential **[1 mark]**

b) $MnO_4^- + 8H^+ + 5Fe^{2+} \rightarrow MnO_2 + 4H_2O + 5Fe^{3+}$ **[1 mark]**
$Cr_2O_7^{2-} + 14H^+ + 6Fe^{2+} \rightarrow 2Cr^{3+} + 7H_2O + 6Fe^{3+}$ **[1 mark]**

c) Cell potential for the first reaction is $+1.51 - 0.77 = 0.74$ V **[1 mark]**
Cell potential for the second reaction is $+1.33 - 0.77 = 0.56$ V **[1 mark]**

3 a) $Cu^{2+}_{(aq)} + 2e^- \rightleftharpoons Cu_{(s)}$ **[1 mark]**
$Ni^{2+}_{(aq)} + 2e^- \rightleftharpoons Ni_{(s)}$ **[1 mark]**

b) $0.34 - (-0.25) = 0.59$ V **[1 mark]**

c) If the copper solution was more dilute, the E of the copper half-cell would be lower, so the overall cell potential would be smaller. **[1 mark]**
If the nickel solution was more concentrated, the E of the nickel half-cell would be higher (more positive/ less negative), so the overall cell potential would be lower. **[1 mark]**

Page 184 — Storage and Fuel Cells

1 a)

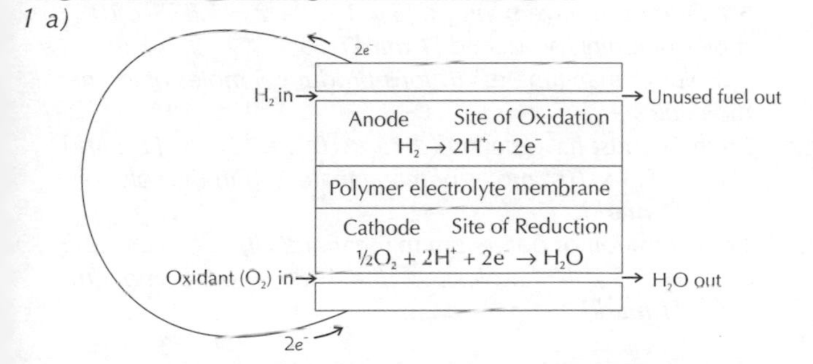

[1 mark for naming anode and cathode. 1 mark for anode half equation. 1 mark for cathode half equation. 1 mark for showing H_2/ fuel and O_2/ oxidant in and unused fuel and H_2O out. 1 mark for showing correct direction of flow of electrons in circuit.]

b) Correctly label anode as site of oxidation **[1 mark]** and cathode as site of reduction. **[1 mark]**

2 a) Possible advantages – more efficient/release less pollution **[1 mark]** Possible disadvantages – problems storing and transporting hydrogen/ manufacturing hydrogen currently requires energy from fossil fuels/ fuel cells are expensive **[1 mark]**

b) Hydrogen would be used as the basic fuel for everything (e.g. vehicles, buildings and electronic equipment) **[1 mark]**
Replacing fossil fuels (oil, gas) **[1 mark]**

c) Clean production of hydrogen from renewable power **[1 mark]**
Improved fuel cell design to reduce costs in manufacture and disposal **[1 mark]**
Building/ creating of hydrogen infrastructure – supply, delivery and storage in a safe way **[1 mark]**

3 a) $Pb + 2H_2SO_4 + PbO_2 \rightleftharpoons 2PbSO_4 + 2H_2O$
[2 marks, otherwise 1 mark for correct equation but not simplified by cancelling H^+]

b) Voltage = $1.68 - 0.35 = 1.33$ V **[1 mark]**

Unit 5: Module 3 — Transition Elements

Page 187 — Properties of Transition Elements

1 a) $1s^2\ 2s^2\ 2p^6\ 3s^2\ 3p^6\ 3d^{10}$ or $[Ar]\ 3d^{10}$ **[1 mark]**

b) No, it doesn't **[1 mark]**. A transition metal is an element that can form at least one stable ion with an incomplete d-subshell, but Cu^+ ions have a full 3d subshell **[1 mark]**.

c) copper(II) sulfate ($CuSO_{4(aq)}$) **[1 mark]**

2 a) $Fe^{3+}_{(aq)} + 3OH^-_{(aq)} \rightarrow Fe(OH)_{3(s)}$
[1 mark for balanced equation, 1 mark for state symbols].

b) i) +7 **[1 mark]**
ii) +4 **[1 mark]**
iii) +7 **[1 mark]**
Oxide ions, O^{2-}, have an oxidation state of –2. Potassium ions, K^+, have an oxidation state of +1. All three compounds are neutral, so the charge on the manganese ions must balance out the charges on the other ions in each compound.

c) The $FeCl_3$ solution is yellow **[1 mark]**. When you add the sodium hydroxide solution, an orange precipitate forms **[1 mark]**.
The orange precipitate is iron(III) hydroxide, $Fe(OH)_3$.

2 a) i) $1s^2\ 2s^2\ 2p^6\ 3s^2\ 3p^6\ 3d^5$ **[1 mark]**
ii) $1s^2\ 2s^2\ 2p^6$ **[1 mark]**

b) Iron can exist in two different oxidation states, Fe^{2+} and Fe^{3+} **[1 mark]**.
Aluminium can only exist in one oxidation state, Al^{3+} **[1 mark]**.
Iron can form coloured compounds/solutions **[1 mark]**.
Aluminium forms only colourless/clear/white compounds/solutions **[1 mark]**.

Page 189 — Complex Ions

1 a) A complex ion is a metal ion surrounded by coordinately bonded ligands **[1 mark]**.

b) A lone pair of electrons from the N atom is donated to/forms a coordinate bond with the central iron ion **[1 mark]**.

c)

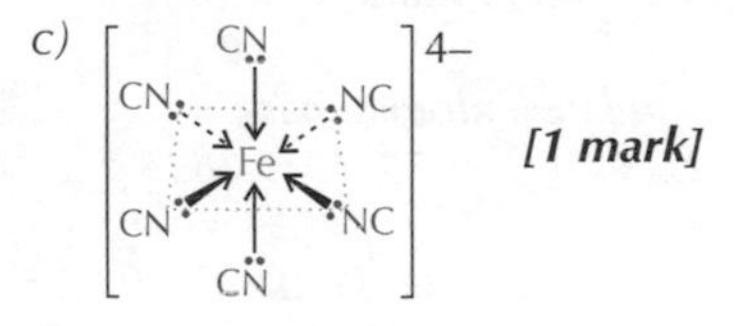

[1 mark]

2 a) A bidentate ligand has two lone pairs of electrons **[1 mark]**, so it can form two coordinate bonds with the central metal ion **[1 mark]**.

b) 6 **[1 mark]**
Each ethanedioate ligand forms two bonds with the Fe^{3+} ion — so that's 6 altogether.

c) Octahedral **[1 mark]**
Complex ions with a coordination number of 6 are usually octahedral.

3 a)

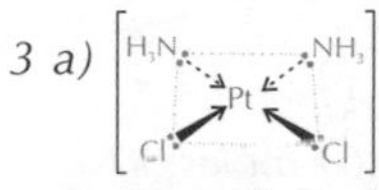

[1 mark for four correct ligands, 1 mark for square planar shape, 1 mark for chloride ions being adjacent]

b) The two chloride ligands are displaced **[1 mark]** and their places taken by two nitrogen atoms from the cancer cell's DNA **[1 mark]**. This stops the cell from reproducing **[1 mark]**. The cell is unable to repair the damage, and dies **[1 mark]**.

A2 Answers

Page 191 — Substitution Reactions

1 a) The equation for the formation of the ion is:

$Co^{2+}_{(aq)} + 6NH_{3(aq)} \rightleftharpoons [Co(NH_3)]^{2+}_{(aq)}$

So $K_{stab} = \frac{[(Co(NH_3)_6)^{2+}]}{[Co^{2+}][NH_3]^6}$ **[1 mark]**

b) $K_{stab} = \frac{[(Cu(CN)_6)^{4-}]}{[(Cu(H_2O)_6)^{2+}][CN^-]^6}$ **[1 mark]**

2 a) $[Cu(H_2O)_6]^{2+}$ **[1 mark]**.

b) i) $[Cu(H_2O)_6]^{2+}_{(aq)} + 4NH_{3(aq)} \rightleftharpoons [Cu(NH_3)_4(H_2O)_2]^{2+}_{(aq)} + 4H_2O_{(l)}$

[1 mark for correct formula of the new complex ion formed, 1 mark for the rest of the equation being correctly balanced]

ii) $K_{stab} = \frac{[(Cu(NH_3)_4(H_2O)_2)^{2+}]}{[(Cu(H_2O)_6)^{2+}][NH_3]^4}$

[1 mark for top of fraction correct, 1 mark for bottom of fraction correct]

3 a) i) The water **[1 mark]** ligand is replaced with an oxygen **[1 mark]** ligand.

ii) It is the basis of the oxygen transportation mechanism in the bloodstream **[1 mark]**.

b) Carbon monoxide will bind strongly to the haemoglobin complex **[1 mark]**. It is a strong ligand, and will not exchange with an oxygen (or water) ligand **[1 mark]**. The haemoglobin can't transport oxygen any more, so the cells of the body will get less oxygen **[1 mark]**.

Page 193 — Redox Reactions and Transition Elements

1 a) $MnO_4^- + 8H^+ + 5Fe^{2+} \rightarrow Mn^{2+} + 4H_2O + 5Fe^{3+}$

[1 mark for MnO_4^- and Mn^{2+} correct, 1 mark for $5Fe^{2+}$ and $5Fe^{3+}$ correct, 1 mark for $8H^+$ and $4H_2O$ correct].

b) Number of moles = (concentration × volume) ÷ 1000

Moles of MnO_4^- = (0.4 × 11.5) ÷ 1000 = 4.6×10^{-3} **[1 mark]**

Moles of Fe^{2+} = moles of MnO_4^- × 5 **[1 mark]** = 2.3×10^{-2} **[1 mark]**

c) Mass of substance = moles × relative atomic mass

Mass of iron in solution = $(2.3 \times 10^{-2}) \times 55.8 = 1.2834$ g **[1 mark]**

% iron in steel wool = (1.2834 ÷ 1.3) × 100 **[1 mark]** = 98.7 % **[1 mark]**

2 a) A redox reaction **[1 mark]**.

b) Number of moles = (concentration × volume) ÷ 1000

Number of moles = (0.5 × 10) ÷ 1000 **[1 mark]** = 0.005 **[1 mark]**

c) Number of moles = (concentration × volume) ÷ 1000

Number of moles = (0.1 × 20) ÷ 1000 **[1 mark]** = 0.002 **[1 mark]**

d) 1 mole of MnO_4^- ions needs 5 moles of electrons to be reduced.

So to reduce 0.002 moles of MnO_4^-, you need (0.002 × 5) = 0.01 moles of electrons **[1 mark]**.

The 0.005 moles of tin ions must have lost 0.01 moles of electrons as they were oxidised OR all of these electrons must have come from the tin ions **[1 mark]**.

Each tin ion changed its oxidation state by 0.01 ÷ 0.005 = 2 **[1 mark]**.

The oxidation state of the oxidised tin ions is (+2) + 2 = +4 **[1 mark]**.

Page 195 — Iodine-Sodium Thiosulfate Titrations

1 a) $IO_3^- + 5I^- + 6H^+ \rightarrow 3I_2 + 3H_2O$

[1 mark for correct reactants and products, 1 mark for balancing]

b) Number of moles = (concentration × volume) ÷ 1000

Number of moles of thiosulfate = (0.15 × 24) ÷ 1000 = 3.6×10^{-3} **[1 mark]**

c) 2 moles of thiosulfate react with 1 mole of iodine, so there were $(3.6 \times 10^{-3}) \div 2 = 1.8 \times 10^{-3}$ moles of iodine **[1 mark]**

d) 1/3 mole of iodate(V) ions produces 1 mole of iodine molecules **[1 mark]**

e) There must be $1.8 \times 10^{-3} \div 3 = 6 \times 10^{-4}$ moles of iodate(V) in the solution **[1 mark]**.

So concentration of potassium iodate(V) = 6×10^{-4} moles × 1000 ÷ 10 = 0.06 mol dm^{-3} **[1 mark]**.

2 Number of moles = (concentration × volume) ÷ 1000

Number of moles of thiosulfate = (0.3 × 12.5) ÷ 1000 = 3.75×10^{-3} **[1 mark]**

2 moles of thiosulfate react with 1 mole of iodine.

So there must have been $(3.75 \times 10^{-3}) \div 2 = 1.875 \times 10^{-3}$ moles of iodine produced **[1 mark]**.

2 moles of manganate(VII) ions produce 5 moles of iodine molecules

So there must have been $(1.875 \times 10^{-3}) \times (2 \div 5)$ **[1 mark]** = 7.5×10^{-4} moles of manganate(VII) in the solution **[1 mark]**.

Concentration of potassium manganate(VII) = (7.5×10^{-4} moles) ÷ (18 ÷ 1000) = 0.042 mol dm^{-3} **[1 mark]**

Index

Index

Index

Index

Some Useful Stuff

Useful Formulas...

Here's a handy reference guide to the formulas you'll need. You'll find them all inside the book too.

Mole equations: $\text{Number of moles} = \frac{\text{mass of substance}}{\text{molar mass}}$

$\text{Number of moles} = \frac{\text{concentration} \times \text{volume (cm}^3)}{1000}$

Enthalpy change: $q = mc\Delta T$ — q = heat lost or gained (joules), m = mass of water or solution (g), c = specific heat capacity of water (4.18 J $g^{-1}K^{-1}$), ΔT = temperature change of water/solution

Rate equation: $\text{Rate} = k[A]^m[B]^n$

Equilibrium constant: $K_c = \frac{[D]^d[E]^e}{[A]^a[B]^b}$

pH formula: $pH = -\log_{10}[H^+]$

Ionic product of water: $K_w = [H^+][OH^-]$

Acid dissociation constant: $K_a = \frac{[H^+][A^-]}{[HA]}$

pK_a formula: $pK_a = -\log_{10}K_a$

Enthalpy change (from calorimetry): $\Delta H = -mc\Delta T$

Entropy change of a system: $\Delta S_{system} = \Delta S_{products} - \Delta S_{reactants}$

Entropy change of the surroundings: $\Delta S_{surroundings} = \frac{-\Delta H}{T}$

Total entropy change: $\Delta S_{total} = \Delta S_{system} + \Delta S_{surroundings}$

Free energy change: $\Delta G = \Delta H - T\Delta S$

Organic Compounds...

These can be a real nightmare. You need to learn all the types in this table and practise naming them...

Homologous series	Prefix or Suffix	Examples	Functional Group
alkanes	-ane	Propane $CH_3CH_2CH_3$	n/a
branched alkanes	alkyl- (-yl)	methylpropane $CH_3CH(CH_3)CH_3$	n/a
halogenoalkanes (haloalkanes)	chloro- bromo- iodo-	chloroethane CH_3CH_2Cl	– X
alkenes	-ene	propene $CH_3CH{=}CH$	C = C
alcohols	-ol	ethanol CH_3CH_2OH	–OH
aldehydes	-al	ethanal CH_3CHO	–C(=O)H
ketones	-one	propanone CH_3COCH_3	>C = O
esters	alkyl -oate	propyl ethanoate $CH_3COOCH_2CH_2CH_3$	–C(=O)O–
carboxylic acids	-oic acid	ethanoic acid CH_3COOH	–C(=O)OH
cycloalkanes	cyclo- -ane	cyclohexane C_6H_{12}	n/a
arenes	phenyl- (-benzene)	ethylbenzene $C_6H_5C_2H_5$	(benzene ring)
amines	-amine (primary)	methylamine CH_3NH_2	$-NH_2$
	di- -amine (secondary)	dimethylamine $(CH_3)_2NH$	>NH
	tri- -amine (tertiary)	trimethylamine $N(CH_3)_3$	>N–

The Periodic Table

Relative Atomic Mass → (top number in each cell)
Atomic number → (bottom number in each cell)

Periods	Group 1	Group 2											Group 3	Group 4	Group 5	Group 6	Group 7	Group 0
1								1.0 H Hydrogen 1										4.0 He Helium 2
2	6.9 Li Lithium 3	9.0 Be Beryllium 4											10.8 B Boron 5	12.0 C Carbon 6	14.0 N Nitrogen 7	16.0 O Oxygen 8	19.0 F Fluorine 9	20.2 Ne Neon 10
3	23.0 Na Sodium 11	24.3 Mg Magnesium 12											27.0 Al Aluminium 13	28.1 Si Silicon 14	31.0 P Phosphorus 15	32.1 S Sulfur 16	35.5 Cl Chlorine 17	39.9 Ar Argon 18
4	39.1 K Potassium 19	40.1 Ca Calcium 20	45.0 Sc Scandium 21	47.9 Ti Titanium 22	50.9 V Vanadium 23	52.0 Cr Chromium 24	54.9 Mn Manganese 25	55.8 Fe Iron 26	58.9 Co Cobalt 27	58.7 Ni Nickel 28	63.5 Cu Copper 29	65.4 Zn Zinc 30	69.7 Ga Gallium 31	72.6 Ge Germanium 32	74.9 As Arsenic 33	79.0 Se Selenium 34	79.9 Br Bromine 35	83.8 Kr Krypton 36
5	85.5 Rb Rubidium 37	87.6 Sr Strontium 38	88.9 Y Yttrium 39	91.2 Zr Zirconium 40	92.9 Nb Niobium 41	95.9 Mo Molybdenum 42	98.9 Tc Technetium 43	101.1 Ru Ruthenium 44	102.9 Rh Rhodium 45	106.4 Pd Palladium 46	107.9 Ag Silver 47	112.4 Cd Cadmium 48	114.8 In Indium 49	118.7 Sn Tin 50	121.8 Sb Antimony 51	127.6 Te Tellurium 52	126.9 I Iodine 53	131.3 Xe Xenon 54
6	132.9 Cs Caesium 55	137.3 Ba Barium 56	138.9 La Lanthanum 57	178.5 Hf Hafnium 72	180.9 Ta Tantalum 73	183.9 W Tungsten 74	186.2 Re Rhenium 75	190.2 Os Osmium 76	192.2 Ir Iridium 77	195.1 Pt Platinum 78	197.0 Au Gold 79	200.6 Hg Mercury 80	204.4 Tl Thallium 81	207.2 Pb Lead 82	209.0 Bi Bismuth 83	210.0 Po Polonium 84	210.0 At Astatine 85	222.0 Rn Radon 86
7	223.0 Fr Francium 87	226.0 Ra Radium 88	227.0 Ac Actinium 89	261 Rf Rutherfordium 104	262 Db Dubnium 105	266 Sg Seaborgium 106	264 Bh Bohrium 107	277 Hs Hassium 108	268 Mt Meitnerium 109	271 Ds Darmstadium 110	272 Rg Roentgenium 111							

The Lanthanides	140.1 Ce Cerium 58	140.9 Pr Praseodymium 59	144.2 Nd Neodymium 60	144.9 Pm Promethium 61	150.4 Sm Samarium 62	152.0 Eu Europium 63	157.3 Gd Gadolinium 64	158.9 Tb Terbium 65	162.5 Dy Dysprosium 66	164.9 Ho Holmium 67	167.3 Er Erbium 68	168.9 Tm Thulium 69	173.0 Yb Ytterbium 70	175.0 Lu Lutetium 71
The Actinides	232.0 Th Thorium 90	231.0 Pa Protactinium 91	238.0 U Uranium 92	237.0 Np Neptunium 93	239.1 Pu Plutonium 94	243.1 Am Americium 95	247.1 Cm Curium 96	247.1 Bk Berkelium 97	252.1 Cf Californium 98	252 Es Einsteinium 99	257 Fm Fermium 100	258 Md Mendelevium 101	259 No Nobelium 102	260 Lr Lawrencium 103